AF183109

Coastal Informatics:
Web Atlas Design and Implementation

Dawn J. Wright
Oregon State University, USA

Ned Dwyer
University College Cork, Ireland

Valerie Cummins
University College Cork, Ireland

Information Science
REFERENCE

INFORMATION SCIENCE REFERENCE
Hershey · New York

Director of Editorial Content: Kristin Klinger
Director of Book Publications: Julia Mosemann
Acquisitions Editor: Lindsay Johnston
Development Editor: Joel Gamon
Typesetter: Travis Gundrum
Production Editor: Jamie Snavely
Cover Design: Lisa Tosheff
Printed at: Lightning Source

Published in the United States of America by
Information Science Reference (an imprint of IGI Global)
701 E. Chocolate Avenue
Hershey PA 17033
Tel: 717-533-8845
Fax: 717-533-8661
E-mail: cust@igi-global.com
Web site: http://www.igi-global.com

Copyright © 2011 by IGI Global. All rights reserved. No part of this publication may be reproduced, stored or distributed in any form or by any means, electronic or mechanical, including photocopying, without written permission from the publisher. Product or company names used in this set are for identification purposes only. Inclusion of the names of the products or companies does not indicate a claim of ownership by IGI Global of the trademark or registered trademark.

Library of Congress Cataloging-in-Publication Data

Coastal informatics : web atlas design and implementation / Dawn Wright, Ned Dwyer, and Valerie Cummins, editors.
p. cm.
Includes bibliographical references and index. Summary: "This book examines state-of-the-art developments in coastal informatics (e.g., data portals, data/ metadata vocabularies and ontologies, metadata creation/ extraction/ cross-walking tools, geographic and information management systems, grid computing) and coastal mapping (particularly via Internet map servers and web-based geographical information and analysis)"-- Provided by publisher. ISBN 978-1-61520-815-9 (hardcover) -- ISBN 978-1-61520-816-6 (ebook) 1. Coasts--Geographic information systems. 2. Coastal mapping. 3. Management information systems. I. Wright, Dawn J., 1961- II. Dwyer, Ned. III. Cummins, Valerie, 1974- GC10.4.R4C63 2010
526.0914'6--dc22

2009052431

British Cataloguing in Publication Data
A Cataloguing in Publication record for this book is available from the British Library.

All work contributed to this book is new, previously-unpublished material. The views expressed in this book are those of the authors, but not necessarily of the publisher.

Editorial Advisory Board

Greg Benoit, *California Coastal Commission, USA*
Jim Good, *Oregon State University and Oregon Ocean Policy Advisory Council, USA*
Torill Hamre, *Nansen Environmental and Remote Sensing Center (NERSC), Norway*
Alejandro Iglesias-Campos, *European Topic Centre on Land Use and Spatial Information (ETC-LUSI), European Environment Agency, Spain*
Eamonn Ó Tuama, *Global Biodiversity Information Facility (GBIF), Denmark*
Janine Salwasser, *Oregon State University Natural Resources Digital Library Program, USA*
Pauline Weatherall, *GEBCO Digital Atlas Team, British Oceanographic Data Centre, UK*

List of Reviewers

Darius Bartlett, *University College Cork, Ireland*
Greg Benoit, *California Coastal Commission, USA*
Mike Blakemore, *Ecotec Research and Consulting Ltd. and Info-Dynamics Research Associates Ltd., UK*
Helen Bradley, *University College Cork, Ireland*
Boyan Brodaric, *Geological Survey of Canada, Canada*
Margaret Carlisle, *University of Aberdeen, UK*
Andrew Cooper, *University of Ulster, UK*
Renee Davis-Born, *Oregon State University, USA*
Torril Hamre, *Nansen Environmental and Remote Sensing Center (NERSC), Norway*
Muki Haklay, *University College London, UK*
David Hart, *University of Wisconsin, USA*
Christina Hoffman, *NOAA Coastal Services Center, USA*
Jeroen B.J. Huisman, *Water Noorderzijlvest, The Netherlands*
Alejandro Iglesias-Campos, *European Topic Centre on Land Use and Spatial Information (ETC-LUSI), European Environment Agency, Spain*
Philip Konings, *Aquaterra nv, Belgium*
Kathrin Kopke, *University College Cork, Ireland*
Audra Luscher, *NOAA Coastal Services Center, USA*
Susanna McMaster, *University of Minnesota, USA*
Jan Mees, *Flanders Marine Institute, Belgium*

Amy Merten, *NOAA Office of Response and Restoration, Coastal Response Research Center, USA*
Theuri Mwangi, *Nairobi Institute of Marine Science and United Nations Environment Programme (UNEP), Kenya*
Fatima Navas Concha, *University of Sevilla, Spain*
Eoin O'Grady, *Marine Institute, Ireland*
Eamonn Ó Tuama, *Global Biodiversity Information Facility (GBIF), Denmark*
Peter Pissierssens, *United Nations Education, Scientific, and Cultural Organization (UNESCO), Intergovernmental Oceanographic Commission (IOC), International Oceanographic Data and Information Exchange (IODE), Belgium*
Greg Reed, *Australian Ocean Data Centre Joint Facility and UNESSO IOC IODE, Australia*
Janine Salwasser, *Oregon State University Natural Resources Digital Library Program, USA*
Kathy Taylor, *Washington Department of Ecology, USA*
Kuuipo Walsh, *Oregon State University, USA*
Stephanie Watson, *Independent Consultant Texas A&M University and Marine Metadata Interoperability Project, USA*
James Wilson, *James Madison University, USA*

Table of Contents

Detailed Table of Contents

Following on the previous chapter, which describes what the basic features of a coastal web atlas are, this chapter goes on to present considerations and recommendations for actually implementing an atlas (i.e., design, development, deployment). This chapter includes lists of the advantages/disadvantages and applicability/execution challenges for various technical resources. And finally, it includes helpful information on open source versus proprietary software, as well as various technology standards.

This chapter provides a general definition of interoperability is the ability of diverse systems and/or organizations to work together, especially in the use and exchange of information. This chapter is about interoperability between computer systems, especially those systems that underlie a coastal web atlas. It reviews the relevant standards for interoperability between coastal web atlases, and gives practical guidelines on how to make atlases interoperable through the use of standards, web services, vocabulary words and ontologies. It concludes with a description of the International Coastal Atlas Network's interoperability prototype under development.

This chapter provides a brief overview of various coastal web atlas projects around the world, providing a contextual bridge to the atlas case studies of Chapters 6-14. A summary of the policy context within which many European atlases operate is followed by a summary of other efforts emerging in Australia, the Western Pacific, Africa, and the Caribbean.

Case study for the U.S. state of Oregon, focusing on the Oregon Coastal Atlas in action. This atlas, along with the Marine Irish Digital Atlas, has been online and in constant development for a long period of time, and therefore one of the more mature coastal atlases on the Internet. Each case study chapter describes the situation in country or state regarding the accessibility of coastal information, the motivation for the producing the, atlas, the knowledge gap that it is trying to fill, the intended audience for the atlas, and where possible, how is it financed. Case study chapters also identify issues of data collection, system design, usage and associated statistics, strengths and weakness of approaches to date, and future plans, including its relationship to ICAN.

Case study for Ireland focusing on the Marine Irish Digital Atlas in action. This atlas, along with the Oregon Coastal Atlas, has been online and in constant development for a long period of time, and therefore one of the more mature coastal atlases on the Internet. The case study follows the "template" of topics as described for Chapter 6.

Case study for the Chesapeake Bay region of the U.S. states of Virginia and Maryland, focusing on the Virginia Coastal Geospatial and Educational Mapping System and the Maryland Shorelines Online in action. The case study follows the "template" of topics as described for Chapter 6.

Case study for Spain, focusing on the SIGLA (Sistema de Información Geografica del Litoral Andaluz or Coastal Information System of Andalusia). The case study follows the "template" of topics as described for Chapter 6.

Section 3
Coastal Web Atlas Management and Governance Issues

This chapter transitions from coastal web atlas (CWA) case studies to atlas management and governance issues, by way of a summary of the International Coastal Atlas Network (ICAN). ICAN is a newly-founded informal group of over 30 organizations from over a dozen nations who have been meeting over the past two years to scope and implement data interoperability approaches to CWAs. Most of the atlases profiled in Section 2, Case Studies, are members of ICAN.

This chapter summarizes key projects and initiatives that are being implemented on very large scales (national/international) by national governments and commissions to build coastal spatial data infrastructures (SDIs). These include SDI efforts in the U.S. and Europe that are closely related to ICAN, and as such are of great value to its mission of developing interoperable atlases, providing along the way solutions for the integration of not only technologies, but people, institutions, and institutional objectives.

Foreword

Coasts, seas and oceans are being threatened by an unprecedented range of pressures including land-based coastal and marine pollution due to poorly managed sewage and industrial waste and agricultural run-off, fragmentation and habitat loss through unsustainable extraction practices and industrial zoning, over-exploitation of marine resources, invasive species infestations and climate change.

The impacts of these pressures have been widespread and generally adverse: over the past 50 years we have observed declines in the abundance of many high-profile, commercially important marine species, loss of genetic diversity, detected alterations in ecosystem functioning and reductions in critical habitats such as coral reefs, coastal wetlands and mangroves. Some 30% of coral reefs – which often have higher levels of biodiversity than tropical forests – have been seriously damaged through fishing, pollution, disease and coral bleaching. Some 35% of mangroves have disappeared over the past two decades and in some cases up to 80% have been lost nationally through conversion to aquaculture and storms. More than 50% of wild marine fisheries are fully exploited, with a further 25% over-exploited.

And yet, coasts are the home of 50% of the world's population whilst more than a billion people rely on fisheries as their main source of protein. In addition, the newly established evidence of a rapid acidification of our oceans starts bringing large-scale disruptions to key components of the food sources.

The damaged resilience and adaptive capacities of our coasts is informed by observations, science and continuity in information systems, such as coastal atlases. The whole point here though is that the "patient science" of relatively slow ecological and biological cycles of such complex systems can easily escape the attentions of fast public news gatherers. Let's face it: the proliferation of the www. and 24-hour news outlets, of scores of television and radio channels, and of personalized modes of receiving and delivering data and opinions has revolutionized the media through which complex science must pass to reach a multiplicity of publics. Dumbing down of much of the media and the seemingly shortening attention span of audiences is reducing the capacity to communicate complex science like that relating people to their dependency on coastal assets resources?

However, the problems arising from the new media technologies and configurations are accompanied by opportunities. For example, people with local and practical knowledge of coastal resources often experience and know about the reality of hazards well before the experts recognize them. They and citizen journalists can report on what's happening in their communities using new means of communication to get their knowledge out to wider publics in ways that were not possible just a decade ago. At the EEA we believe that if we are to tackle our environmental problems we need to move beyond conventional systems of data collection and management and adopt approaches such as the coastal web atlases described in this book. If we want to stimulate a change to the way we all live and confront natural processes it is no longer sufficient to develop passive lists or reports to "inform" citizens. Information is still too often made available as lists of figures or spreadsheets that only experts can interpret. Imagine if all the statistics that inform our evening weather forecasts were presented in this way, or all the data that drives popular software like Google or Facebook – do you think they would continue to be as popular – and be able to draw the benefits from participation?

We believe that the current achievements and future activities in relation to coastal web atlases will provide useful operational services to a large community of practitioners and users across the world. To encourage participation we need to present our information in a way everyone can understand; the environmental monitoring and reporting systems designed in the 20th century will not be able to cope with this increasing demand for higher quality, faster access, cost efficient systems to respond to today's emerging complex issues, e.g., climate change at our coastal door-step. Producers and providers of environmental data will have to move from centralized information management towards distributed data and information systems, both at a geographical scale, from local to global, as well as thematic integration. This book shows practical examples of how that is being achieved.

The EEA therefore looks forward to ongoing cooperation in these developments towards interoperable coastal information systems. This book will be tremendously useful in this regard, especially in view of services such as the methodological outcomes and content-based information, so as to help actions in the field of coastal zone integrated assessments, including coastal zone use potentials, vulnerabilities and adaptation needs to environmental changes.

Prof. Jacqueline McGlade
Executive Director
European Environment Agency

***Professor Jacqueline McGlade** became Executive Director of the European Environment Agency on June 1 2003. Prior to this she was Natural Environment Research Council Professorial Fellow in Environmental Informatics in the Mathematics Department of University College London where her main areas of research included spatial data analysis and informatics, expert systems, environmental technologies and the international politics of the environment and natural resources. Previous appointments have included Director of the UK's Centre for Coastal & Marine Sciences, Director of Theoretical Ecology at the Forschungszentrum Jülich Germany, Associate Professor at the Honda funded International Ecotchnology Research Centre, Senior Scientist in the Federal Government of Canada and in the USA, Adrian Fellow at Darwin College, Cambridge and Professorships at Warwick University and Aachen. Professor McGlade has won various prizes including the Minerva Prize, the Swedish Jubileum Award and the Brno University Gold Medal. She also has Honorary degrees from Wales (Bangor) Kent and is a Fellow of the Linnean Society and the Royal Society for the Encouragement of Arts, Manufacture & Commerce. Professor McGlade has worked extensively in North America, south-east Asia and West Africa; she has published more than 100 research papers, written popular articles, presented and appeared in many radio and television programmes, including her own BBC series The Ocean Planet and Learning from Nature and more recently Our Arctic Challenge, a film about sport and tourism in Greenland. She has given public lectures worldwide on sustainable development, conflicts over environmental impacts of industrial and natural activities, environmental technologies and the use of multimedia in developing countries. Professor McGlade was Chairman of The Earth Centre and a Board Member of the Environment Agency. She is currently a Trustee of the Natural History Museum, and a member of the Environment Advisory Committee of the European Bank for Reconstruction and Development, UK-China Forum and UK-Japan 21st Century Group. She is also Director of the software company, View the World Ltd. Recent books: Advanced Ecological Theory (Blackwell 1999); The Gulf of Guinea Large Marine Ecosystem (Elsevier 2002).*

Preface

This book is based on the results of two international workshops jointly funded by the US National Science Foundation and the National Development Program of Ireland. They brought together key experts from Europe, the United States, Canada Africa, and Australia to examine state-of-the-art developments in coastal informatics (e.g., data portals, data/ metadata vocabularies and ontologies, metadata creation/ extraction/ cross-walking tools, geographic and information management systems, grid computing) and coastal mapping (particularly via Internet map servers and web-based geographical information and analysis). The first workshop, held in Cork, Ireland in July 2006, enabled participants to examine state-of-the-art developments in coastal web atlases (CWAs), and to assess the potential and the limitations of selected CWAs from the United States and Europe. Participants also shared several case studies and lessons learned, and established key issues and recommendations related to the design, data requirements, technology and institutional capacity needed for these atlases. This necessitated an examination of best practices for achieving interoperability between CWAs, which led international participants to a second workshop entitled "Coastal Atlas Interoperability," and held on the campus of Oregon State University in July 2007. At this second workshop, expert participants learned how to use controlled vocabularies and ontologies in order to build a common approach to managing and disseminating coastal data, maps and information, and concluded with the aim of designing and developing a demonstration interoperability prototype using the metadata catalogs of two mature atlases (the Oregon Coastal Atlas and the Marine Irish Digital Atlas).

The technical experts, scientists, decision makers and practitioners of the workshops in Ireland and Oregon decided to informally organize under the International Coastal Atlas Network (ICAN) and sought to continue the momentum with a third workshop. Based on the success of the group to this point, the European Environment Agency (EEA) sponsored and hosted this third event in 2008, at its headquarters in Copenhagen, Denmark, under the theme: "Federated Atlases: Building on the Interoperable Approach." Workshop participants discussed the progress-to-date on the ICAN interoperability prototype and agreed upon future technical activities. The relevant policy context within which ICAN must operate was presented, along with an overview of a number of related coastal and marine information management projects that could inform ICAN developments. In addition, the workshop took place around a two-day conference on Coastal Atlas Development, organized by the EEA itself, whose objective was to inform EEA partners about the development of coastal atlases and the emergence of ICAN in light of relevant European policy developments in the maritime sphere.

By this time, CWAs in general and ICAN in particular had captured the interest of scores of local, state and national governments, non-governmental organizations, research institutes, and universities, as well NOAA, certainly the EEA, and the UNESCO Intergovernmental Oceanographic Commission.

Workshop participants therefore discussed ways of disseminating some of the wealth of knowledge and expertise that had been growing within the ICAN group (which now stands at over 35 organizations from over 10 countries). It was decided that one effective way to do this would be through the publication of a book to review and present the latest developments in the new emerging field of coastal web atlases, to share best practices and lessons learned through a series of case studies, to give practical guidance on geographic data management and documentation through standards-based metadata, as well as guidance on how to make underlying geographic databases interoperable. This current publication is the result. We hope that readers will find this book of practical use in web atlas design, development and implementation, and will thus improve their spatial thinking in the coastal context. Hence, rather than a lengthy theoretical treatise on basic and futuristic research questions and problems, the book has been prepared more as a concise, ready reference, with collections of subject-specific instructions where appropriate.

The prime audience for the book is coastal resource managers and consultants, coastal scientists, coastal technologists (e.g., information technologists, GIS specialists, software developers), government researchers, and graduate students. The book should be especially valuable to coastal resource managers who need to tackle such topic-based issues (explaining environmental concepts to the public and reaching them with current information has always been a difficult task).

The book may also be suitable for intermediate, advanced courses in coastal/marine GIS or coastal zone management (i.e., courses toward a related BS/BSc, MS/MSc or PhD degree, in the classroom, but also potentially for distance education as well). The material in the book and the dedicated website should allow students to familiarize themselves with what CWA (web GIS) technology is, what are the basics of related disciplines, and how to use physical environmental and biological data available in the atlases in order to develop specific GIS applications and models. Course instructors may use the contents of the dedicated website either to present ready-to-use applications or to use the variety of included data for building new GIS applications.

Further expected contributions of the book include:

- Wide data dissemination to enhance scientific and technological understanding. The book should be great interest not only to the coastal/marine research and management community, but also to libraries, high schools, and outreach sites. Linkages in the book are made to parallel research in geographic information science, digital library development, and computer science. The presentation of lessons learned should help guide the development of new national and regional atlases, and improve decision-support systems.
- Advancing discovery and understanding; promoting teaching, training and learning through integration of research and education. The book may be useful as additional content to faculty course materials and to graduate research. We anticipate a number of student research topics and projects at both the M.S./M.Sc. and Ph.D. levels that may be aided by this book.
- Benefiting society. With the release of the Pew and U.S. Ocean Commission reports as well as the European Union Integrated Maritime Policy there is growing public awareness of the critical state of our coastal zones and fisheries. The book poses informatics solutions that seek to improve management practices and decision-making. Mapping plays a critical role in issues of national sovereignty, resource management, maritime safety, and hazard assessment.

This book is also accompanied by a dedicated website (International Coastal Atlas Network, http://ican.science.oregonstate.edu) which includes links to mature CWAs, and is building templates for CWA design, snippets of scripts and programming routines to achieve interoperability with partner atlases, and several other resources mainly for online GIS developments and online data providers. We hope that you find it useful!

Dawn J. Wright
Oregon State University, USA

Valerie Cummins
University College Cork, Ireland

Edward Dwyer
University College Cork, Ireland

Acknowledgment

We are very grateful to the US National Science Foundation (Award #0527216), the Marine Institute of Ireland's Marine RTDI Networking and Technology Transfer Initiative under the National Development Plan, the Coastal and Marine Resources Centre of University College Cork (CMRC), Ireland, Oregon State University, which provided initial funding to bring together key experts from Europe, the United States, Canada and Africa to begin discussions, and collaborations that led to the idea for this handbook.

We also thank the European Environment Agency, with additional support from the European Topic Centre on Land Use and Spatial Information (ETC-LUSI), the European Environmental Information and Observation Network (EIONET), SeaZone Solutions Limited, the European Platform for Coastal Research (ENCORA), and the Marine Institute, for additional funding to bring key international experts together for further discussions and collaborations that ultimately led to the production of this handbook. The additional support of the NOAA Coastal Services Center, and US National Science Foundation Award #0921950 is also gratefully acknowledged.

We thank the many external reviewers whose careful comments and suggestions greatly improved the chapters, as well as our External Advisory Board for their support throughout this project. And finally, we extend a special thanks to special thanks to Kathrin Kopke of the CMRC and Joel Gamon of IGI Global for editorial assistance, as well as to Dawn's pet dog Lydia for moral support during the final stages of the project.

Dawn J. Wright
Oregon State University, USA

Valerie Cummins
University College Cork, Ireland

Edward Dwyer
University College Cork, Ireland

Section 1
Principles

Chapter 1
Introduction

Dawn J. Wright
Oregon State University, USA

Valerie Cummins
University College Cork, Ireland

Edward Dwyer
University College Cork, Ireland

ABSTRACT

Coastal web atlas (CWA) development is introduced in this chapter as a relatively new field of technology, driven by a wide range of coastal policy issues such as population pressure and climate change. International interest in CWAs is demonstrated by the large number of CWA initiatives worldwide. However, there is a need to take stock of technological developments as well as other lessons learned. This chapter sets the scene in relation to these issues which in turn provides the context for describing the aims of the book. The aims of the book are articulated as presenting the latest developments in CWAs and helping readers to determine future needs in mapping and informatics for coastal management.

COASTAL WEB ATLASES DEFINED

In recent years significant momentum has occurred in the development of Internet resources for decision makers, scientists and the general public who are interested in the coast. Governments, industry sectors, academic institutions and non-governmental organizations (NGOs) have a tremendous stake in the development and management of geospatial data resources. Coastal mapping plays an important role in informing decision makers on issues such as national sovereignty, resource management, maritime safety and hazard assessment. A key aspect of this trend has been the development of coastal web atlases (CWAs), based on web-enabled geographic information systems (GISs, Figures 1 and 2). A CWA is defined in O'Dea *et al.*, (2007) as: *a collection of digital maps and datasets with supplementary tables, illustrations and information that systematically illustrate the coast, oftentimes with cartographic and decision support tools, all of which are accessible via the Internet.* These atlases organize and coordinate all of the above through a single portal or entry point, with a common design theme that is followed through all of the pages of a CWA site.

DOI: 10.4018/978-1-61520-815-9.ch001

Copyright © 2011, IGI Global. Copying or distributing in print or electronic forms without written permission of IGI Global is prohibited.

Figure 1. Example of a coastal web atlas, the Oregon Coastal Atlas, http://www.coastalatlas.net, showing opening page with map, tools, learn, and search sections

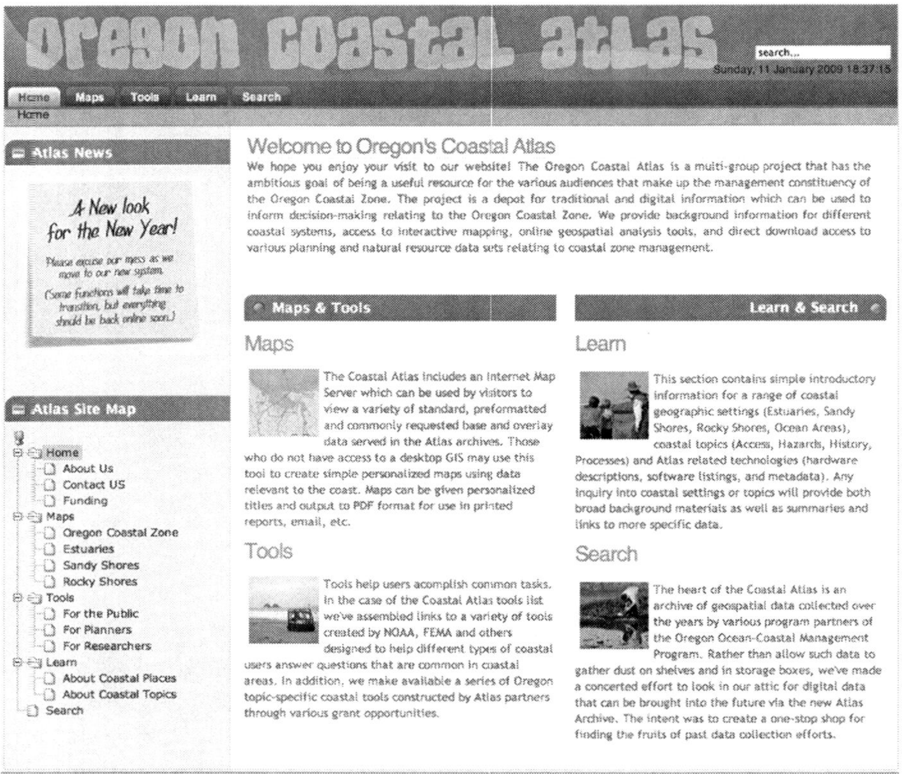

CWAs are also defined in Tikunov *et al.* (2008) (termed there as "atlas information systems") and with regard to their increasingly important role in national spatial data infrastructures (SDIs).

CWAs deal with a variety of thematic priorities (e.g., oil spills or recreational uses) and can be tailored to address the needs of a particular user group (e.g , coastal managers or educators). There are many benefits that CWAs can provide, including:

- **A portal** to coastal data and information from **diverse sources**;
- **Up to date** geospatial data which is frequently changing;
- A widely **accessible** coastal resource to a broad audience;
- A comprehensive and searchable **data catalogue**;
- Improved **efficiency** in finding data and information;
- An instrument for **spatial planning**;
- **Interactive tools and resources** which empower users to find their own answers;
- An **educational resource** which raises people's consciousness about coastal topics.

Driving factors for CWA development include the need for:

- Better planning to cater for increased **population pressures** in the coastal zone (e.g., the UN estimate that by 2020 75% of the world's population will be living within

Figure 2. Example of a coastal web atlas, the Oregon Coastal Atlas, http://www.coastalatlas.net,showing polygons of rapidly moving landslide regions resulting from a coastal hazard query.

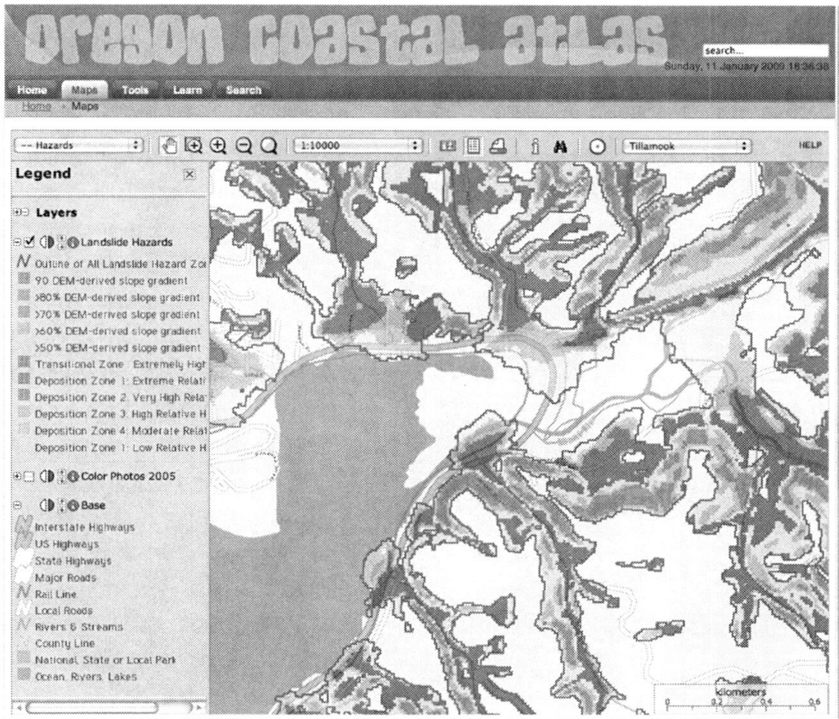

60 km of the coastal zone; United Nations, 1992; also World Resources Institute, 2001; Shi and Singh, 2003).

- Decision support systems in relation to **climate change** scenarios in vulnerable coastal regions.
- Information to facilitate assessments of **risk to natural hazards** (including tsunamis and floods).
- Access to data and maps to support **marine spatial planning** (MSP) as a tool for better coastal and marine area management (e.g., European Commission, 2007).
- Access to relevant data and information in coastal and ocean **capacity-building** initiatives worldwide, especially for effective **governance**, sustained economic support, and education and training opportunities (e.g., National Research Council, 2008).
- Maps of jurisdictional boundaries for maritime territories in support of claims related to the **United Nations Convention on the Law of the Sea** (UNCLOS), which has a deadline for submission of 2013.
- Information on **resource availability and exploitation** including habitat and species information, and ecological and community resilience as part of ecosystem-based management.
- Spatial information to underpin the development of **emerging offshore sectors** such as ocean energy, offshore aquaculture and marine bio-discovery.

SCIENTIFIC AND RESOURCE MANAGEMENT CONTEXT

The driving factors mentioned above have already resulted in the proliferation of CWA projects (such as those introduced in Section 2 of the book) that have been designed to address thematic (e.g., fisheries management, recreational use) or spatial areas of interest (e.g., country to local level). Important questions asked and potentially answered by coastal web atlases (given sufficient data and personnel), include these (in the realm of coastal erosion as an example):

From a coastal scientist...

1) What is the geomorphic evolution of the coast?
    a. What historic photography, geomorphology profiles, LIDAR surveys, shoreline surveys are available for study?
2) How many major erosion events due to severe storms have occurred within a defined section of shoreline in the past 50 years?
3) How have anthropogenic activities impacted natural coastal erosion processes?
4) Can a predictive model of hot spots be developed with the data available?

From a coastal resource manager...

1) What are the erosion rates along a geographically defined stretch of shoreline?
    a. Where are erosion hot spots based on geology and wave action?
    b. Where are erosion hot spots conflicting with human uses of the coast?
    c. For a defined planning window (e.g., 25 years) what is the anticipated extent and magnitude of coastal erosion risk along a designated stretch of shoreline?
    d. What is the potential for new development in the above designated risk zones? What actions can be taken to avoid, minimize or mitigate the placement of new development in predicted high-risk zones?
2) Where is the potential for habitat loss due to coastal erosion a significant risk?
    a. Where are ecologically vulnerable areas such as essential fish habitats, wetlands, beaches, wildlife refuges and conservation areas?

From a coastal private property owner...

1) What is the erosion rate along my stretch of shoreline?
    a. How close is my home to "the edge"?
    b. Will my home survive to the end of my mortgage?

From an emergency responder...

1) How big is an incoming storm / erosion-causing event?
    a. How do I alert affected areas?
2) What public infrastructure is threatened by chronic or severe erosion events (e.g. transportation networks, pubic utilities (waste water treatment facilities, power plants, etc.)?
3) Where are the best evacuation routes during major coastal storm events?

From a coastal geographic information scientist/data analyst...

1) What data and information can I make accessible regarding coastal erosion?
    a. What feature categories (historic photography, geomorphology profiles, LIDAR surveys, shorelines, plant/animal species, surveys etc.) exist in a designated area?
    b. Where do existing data reside? Can my system access the data?

   c. Can the data be shared (data ownership, permissions, licensing)?
   d. How well are existing data documented (reports, metadata), and does the documentation support the potential future uses of the data by my intended audience?

2) What analysis or visualization tools can I provide that can make use of available data to answer common questions from my audience(s) regarding coastal erosion?
3) Can I extract information from the atlas network to bolster data available to support coastal issues within my own program? Are current inventories sufficient?
4) Does my area of Interest extend beyond the geographic boundaries of the available Atlas system?
   a. If so, are there neighboring or regional Atlases that might have supplemental information that might be of use?
   b. If so, are the contents of neighboring or regional Atlases accessible to users of my Atlas, and the analysis or visualization tools it contains?

Significant capacity has been built in the field of coastal mapping and informatics in the last decade as a result of a steady advancement of GIS applications for coastal practitioners. For example, the European Commission is implementing the MARATLAS project as an educational tool and as a means of highlighting common maritime heritage. The web based MARATLAS will be multi-lingual and will link to other European mapping initiatives such as the Water Information System for Europe (WISE). In the U.S. NOAA's Coastal Services Center (CSC) has launched the "Digital Coast" initiative (http://www.csc.noaa.gov/digitalcoast), which seeks new ways to build the U.S. coastal and marine National Spatial Data Infrastructure or NSDI (Mapping Science Committee, 2001), so that U.S. coastal communities have improved access to organized and relevant data and tools needed to make more informed decisions. In addition, NOAA CSC has been working on an inventory and registry of planned, current and completed Federal and non-Federal mapping activities, including coastal web atlases.

Developments in resource management policy can also have implications for the roll out of CWAs. For example, the recent reports of the Pew Oceans Commission and the U.S. Commission on Ocean Policy (Pew Oceans Commission, 2003; Juda, 2005) have clearly shown that coastal communities are critical to the economy of the U.S., and to its overall health and well-being as a nation, and further that geographic technologies will be a fundamental, critical tool to address the threats of climate change, coastal hazards, overpopulation, and more. As such, the State Governors of Washington, Oregon and California signed the West Coast Governors' Agreement on Ocean Health (http://westcoastoceans.gov), which includes in its action plan, the need for harmonized ocean and coastal maps and information that also crosses administrative boundaries. In Europe, the European Commission has now published its vision for an integrated maritime policy for the European Union (European Commission, 2007), which calls for the development of an Atlas of the European Seas for use in regional ocean governance and management. Opportunities exist to facilitate such a development by providing for data interoperability among existing CWAs, and within the context of the Inspire Directive which establishes an infrastructure for spatial information in the European Community. Similar trends in resource management with implications for CWAs can be seen in other parts of the world, such as Australia, the Western Pacific, Africa and the Caribbean, as introduced in Chapter 5.

KEY ISSUES TO BE ADDRESSED

Despite the importance of key drivers described above, little has been done to take stock of the implications of these developments or to identify best practice in terms of creating an approach that takes lessons learned into consideration. Furthermore, the research community is still working toward providing widespread solutions to deal with common issues such as achieving full semantic interoperability of metadata and databases (where concepts, terminology, even abbreviations that are shared between two or more individuals, systems, or organizations are understood by all to mean the same thing; Gruber 1993; Egenhofer, 2002), reversing the lack of tool integration for coastal analysis and decision-making, and removing most impediments to effective use of online atlases for decision-support.

While multiple benefits are derived from tailor-made atlases (e.g., speedy access to multiple sources of coastal data and information; economic use of time by avoiding individual contact with different data holders), the potential exists to derive added value from the integration of disparate CWAs, to optimize decision making at a variety of levels and across themes (Chapter 4). And while digital data sets have continued to grow exponentially, our ability to derive meaning, knowledge and management decisions from all of these data in an analytical context remains poor (e.g., Aditya & Kraak, 2006, Deliiska, 2007; Athanasis *et al.*, 2008; Sahoo *et al.*, 2008). Hence, a fundamental research question that applies to CWAs is:

(1) *How best to achieve semantic interoperability so as to mitigate vague data queries, vague concepts or natural language semantics when retrieving and integrating data and information?*

For example, the terminology used to describe similar data in CWAs can vary widely between specialties or regions, which can complicate data searches and integration. Use of the word "*seabed*" in Europe versus use of the word "*seafloor*" to describe the same feature in North America is a good example of this scenario, as is the interchangeable use of "*coastline*" versus "*shoreline*" in both regions. Agreements on content/semantic interoperability must be developed to eliminate such problems, making searches between disparate, but mutually beneficial, projects feasible. The International Coastal Atlas Network (ICAN) aims to address such issues by providing a forum to share knowledge and experience among atlas developers (See Chapter 17). A technical group within the ICAN community have been developing a demonstration prototype as a proof-of-concept to inter-relate metadata and other information between two initial CWAs (the Marine Irish Digital Atlas or MIDA, http://mida.ucc.ie, and the Oregon Coastal Atlas or OCA, http://www.coastalatlas.net). The prototype is in the form of an Open Geospatial Consortium (OGC) catalogue services for the Web (CSW), where web map services (WMS) will be registered. But much more collaborative discussion about refining and extending this prototype is needed, which can best be achieved through the type of international collaboration accommodated through ICAN.

A second major question driving research is:

(2) *What are the best ways to share data and information across multiple distributed organizational and social contexts? And related to this, under what conditions do virtual organizations (such as ICAN) best foster and support transformative scientific research while also providing an effective spatial data infrastructure? Again, very little work has been done in any of these realms with respect to CWAs (e.g., there is a need for an assessment of CWAs used by different communities, and, where possible, to quantify the impact of CWA end user groups such as agencies, regions and states).*

AIM OF THE BOOK

This book is a book, as opposed to a research monograph. *Hence, rather than a lengthy theoretical treatise on basic and futuristic research questions and problems, it has been prepared more as a concise, ready reference, with collections of subject-specific instructions where appropriate.* The purpose of the book is to present the latest developments in the new field of coastal web atlases while also sharing best practices and lessons learned. This will, in turn, help readers to determine future needs in mapping and informatics for coastal management and improve spatial thinking in the coastal context. As such, this book provides a complete guide to CWA development and implementation including established principles and recommendations for atlas design, data requirements, software technology and institutional capacity, as well as best practice for achieving interoperability between CWAs (where concepts, terminology, and even abbreviations that are shared between two or more atlases are understood by all to mean the same thing).

The approach to the book is to present information according to three parts. Section 1 covers the principles of CWAs. Within Section 1, Chapter 2 covers the major features that make up an atlas, Chapter 3 covers how to implement those features, and Chapter 4 describes how to interoperate the services of one atlas with other atlases as part of a network, especially by way of international standards. The chapters of Section 2 provide examples of CWAs from around the world, including North America, Europe, Africa and the Caribbean. The chapters of Section 3 covers CWA management and governance issues, including how to best meet the needs of one's user community, how to make improvements in an atlas based on those user needs and make it grow, how to support an atlas with partnerships, funding, and the like so that it will mature, and how atlases play a key role in SDIs. Table 1 provides more detail by way of summarizing the content and importance of each chapter. This book is also accompanied by a dedicated website (International Coastal Atlas Network, http://ican.science.oregonstate.edu) which includes links to mature CWAs, and is building templates for CWA design, snippets

Table 1. Summary of the content of each chapter in this book

Section 1: Principles	
1: Introduction	General introduction to the entire book, including definition of a coastal web atlas and key issues surrounding it uses. The chapter also identifies the intended audience and gives a brief overview of topics and importance for all remaining chapters.
2: Coastal Web Atlas Features	This chapter features an overall summary with more detailed descriptions of common coastal web atlas features and the forms/functions they may take. This includes an extensive discussion of the various types of tools that one might find in a web atlas.
3: Coastal Web Atlas Implementation	Following on the previous chapter, which describes what the basic features of a coastal web atlas are, this chapter goes on to present considerations and recommendations for actually implementing an atlas (i.e., design, development, deployment). This chapter includes lists of the advantages/disadvantages and applicability/execution challenges for various technical resources. And finally, it includes helpful information on open source versus proprietary software, as well as various technology standards.
4: Coastal Atlas Interoperability	This chapter provides A general definition of interoperability is the ability of diverse systems and/or organizations to work together, especially in the use and exchange of information. This chapter is about interoperability between computer systems, especially those systems that underlie a coastal web atlas. It reviews the relevant standards for interoperability between coastal web atlases, and gives practical guidelines on how to make atlases interoperable through the use of standards, web services, vocabulary words and ontologies. It concludes with a description of the International Coastal Atlas Network's interoperability prototype under development.

continued on following page

Table 1. continued

Section 2: Coastal Web Atlas Case Studies Around the World	
5: Overview of Coastal Atlases	This chapter provides a brief overview of various coastal web atlas projects around the world, providing a contextual bridge to the atlas case studies of Chapters 6-14. A summary of the policy context within which many European atlases operate is followed by a summary of other efforts emerging in Australia, the Western Pacific, Africa, and the Caribbean.
6: Oregon, USA	Case study for the U.S. state of Oregon, focusing on the Oregon Coastal Atlas in action. This atlas, along with the Marine Irish Digital Atlas, has been online and in constant development for a long period of time, and therefore one of the more mature coastal atlases on the Internet. Each case study chapter describes the situation in country or state regarding the accessibility of coastal information, the motivation for the producing the, atlas, the knowledge gap that it is trying to fill, the intended audience for the atlas, and where possible, how is it financed. Case study chapters also identify issues of data collection, system design, usage and associated statistics, strengths and weakness of approaches to date, and future plans, including its relationship to ICAN.
7: Ireland	Case study for Ireland focusing on the Marine Irish Digital Atlas in action. This atlas, along with the Oregon Coastal Atlas, has been online and in constant development for a long period of time, and is therefore one of the more mature coastal atlases on the Internet. The case study follows the "template" of topics as described for Chapter 6.
8: Virginia and Maryland, USA	Case study for the Chesapeake Bay region of the U.S. states of Virginia and Maryland, focusing on the Virginia Coastal Geospatial and Educational Mapping System and the Maryland Shorelines Online in action. The case study follows the "template" of topics as described for Chapter 6.
9: Wisconsin, USA	Case study for the U.S. state of Wisconsin, focusing on the ongoing development of the Wisconsin Coastal Atlas, with a future eye toward a regional Great Lakes Coastal Atlas. The case study follows the "template" of topics as described for Chapter 6.
10: Belgium	Case study for Belgium, focusing on the Belgian Coastal Atlas, which was first published as a hardcopy book but then transitioned to the web. The case study follows the "template" of topics as described for Chapter 6.
11: Africa	Case study for the continent of Africa, focusing on the African Marine Atlas. The case study follows the "template" of topics as described for Chapter 6.
12: Caribbean	Case study for the Caribbean region, focusing on the Caribbean Marine Atlas. The case study follows the "template" of topics as described for Chapter 6.
13: UK	Case study for the United Kingdom, providing a brief overview of the origins and evolution of coastal web atlases throughout the country. The case study follows the "template" of topics as described for Chapter 6.
14: Spain	Case study for Spain, focusing on the SIGLA (Sistema de Información Geografica del Litoral Andaluz or Coastal Information System of Andalusia). The case study follows the "template" of topics as described for Chapter 6.
Section 3: Coastal Web Atlas Management and Governance Issues	
15: The International Coastal Atlas Network	This chapter transitions from coastal web atlas (CWA) case studies to atlas management and governance issues, by way of a summary of the International Coastal Atlas Network (ICAN). ICAN is a newly founded informal group of over 30 organizations from over a dozen nations who have been meeting over the past two years to scope and implement data interoperability approaches to CWAs. Most of the atlases profiled in Section 2, Case Studies, are members of ICAN.
16: Coastal Atlases in the Context of Spatial Data Infrastructures	This chapter summarizes key projects and initiatives that are being implemented on very large scales (national/international) by national governments and commissions to build coastal spatial data infrastructures (SDIs). These include SDI efforts in the U.S. and Europe that are closely related to ICAN, and as such are of great value to its mission of developing interoperable atlases, providing along the way solutions for the integration of not only technologies, but people, institutions, and institutional objectives.
17: Creating a Usable Atlas	Having covered some overarching management and governance issues for coastal web atlases, the book returns to the user level with 3 concluding chapters that guide the reader on how to create an atlas that is the most usable for its audience, how to make that seed effort grow, and how to maintain it. This chapter provides guidelines on how to better understand coastal web atlas users, how to undertake user-centered design and development for improved web site usability, and how to avoid major pitfalls with web interfaces.
18: Improving a Growing Atlas	This chapter covers aspects of atlas monitoring via web server statistics, user surveys, and other sorts of feedback mechanisms, and how to obtain improvement over time. Also covered are issues of scalability (how to accommodate increasing datasets and users), and the latest in reviewing/updating technology.

continued on following page

Table 1. continued

19: Supporting a Successful Atlas	This concluding chapter of the book is about to maintain a successful coastal web atlas. It discusses issues relating to securing long-term support for an atlas and provides guidance based on existing practice and experience with atlas developments at national and international levels. Specific topics include institutional capacity, institutional support, partnerships, funding, governance, and continued promotion. Also included is a discussion of data and metadata ownership issues, intellectual property rights, and the legal protection of atlas content.

of scripts and programming routines to achieve interoperability with partner atlases, and several other resources mainly for online GIS developments and online data providers.

REFERENCES

Aditya, T., & Kraak, M. J. (2006). Geospatial Data Infrastructure Portals: Using National Atlases as a Metaphor. *Cartographica*, *41*(2), 15–133.

Athanasis, N., Kalabokidis, K., Vaitis, M., & Soulakellis, N. (2008). Towards a semantics-based approach in the development of geographic portals. *Computers & Geosciences*. doi:.doi:10.1016/j.cageo.2008.01.014

Deliiska, B. (2007). Thesaurus and domain ontology of geoinformatics. *Transactions in GIS*, *11*(4), 637–651. doi:10.1111/j.1467-9671.2007.01064.x

Digital Coast. (2009). Retrieved October 31, 2009, from NOAA Coastal Services Center website http://www.csc.noaa.gov/digitalcoast

Egenhofer, M. (2002). *Toward the Semantic Geospatial Web, Tenth ACM International Symposium on Advances in Geographic Information Systems*. New York: ACM Press.

European Commission. (2007). *Communication from the Commission to the European Parliament, the Council, the European Economic and Social Committee and the Committee of the Regions: An Integrated Maritime Policy for the European Union*. Commission of the European Communities. Brussels, 10 October 2007, COM (2007) 575 final. Retrieved July 29, 2009 from the European Commission web site: http://eur-lex.europa.eu/LexUriServ/site/en/com/2007/com2007_0575en01.pdf

Gruber, T. (1993). A translation approach to portable ontology specifications. *Knowledge Acquisition*, *5*(2), 199–220. doi:10.1006/knac.1993.1008

International Coastal Atlas Network. (2009). Retrieved October 15, 2009, from the International Coastal Atlas Network web site: http://ican.science.oregonstate.edu.

Juda, L. (2005). The report of the U.S. Commission on Ocean Policy: State perspectives. *Coastal Management*, *34*(1), 1–16. .doi:10.1080/08920750500364930

Mapping Science Committee. (2001). *National Spatial Data Infrastructure Partnership Programs: Rethinking the Focus*. Washington, DC: National Academy Press.

Marine Irish Digital Atlas. (2009). Retrieved October 31, 2009, from Coastal and Marine Resources Centre, University College Cork web site: http://mida.ucc.ie.

National Research Council. (2008). *Increasing Capacity for Stewardship of Oceans and Coasts: A Priority for the 21st Century, Committee on International Capacity-Building for the Protection and Sustainable Use of Oceans and Coasts* (Feeley, M. H., & Pantoja, S. C., Eds.). Washington, D.C.: The National Academies Press.

O'Dea, L., Cummins, V., Wright, D., Dwyer, N., & Ameztoy, I. (2007). *Report on Coastal Mapping and Informatics Trans-Atlantic Workshop 1: Potentials and Limitations of Coastal Web Atlases.* University College Cork, Coastal & Marine Resources Centre: Cork, Ireland. Retrieved May 7, 2009, from the ICAN web site: http://ican.science.oregonstate.edu/node/47

Oregon Coastal Atlas. (2009). Retrieved October 31, 2009, from the Oregon Coastal Atlas web site: http://www.coastalatlas.net.

Pew Oceans Commission. (2003). *America's Living Oceans: Charting a Course for Sea Change. A Report to the Nation.* Arlington, VA: Pew Oceans Commission.

Sahoo, S. S., Bodenreider, O., Rutter, J. L., Skinner, K. J., & Sheth, A. P. (2008). An ontology-driven semantic mashup of gene and biological pathway information: Application to the domain of nicotine dependence. *Journal of Biomedical Informatics, 41*(5), 752–765. doi:10.1016/j.jbi.2008.02.006

Shi, H. & Singh, A. (2003). Status and interconnections of selected environmental issues in the global coastal zones. *AMBIO: A Journal of the Human Environment, 32*(2), 145–152.

Tikunov, V. S., Ormeling, F., & Konecny, M. (2008). Atlas information systems and geographical names information systems as contributants to spatial data infrastructure. *International Journal of Digital Earth, 1*(3), 279–290. doi:10.1080/17538940802291817

United Nations. (1992). *Agenda 21: The United Nations Programme of Action from Rio.* New York: United Nations.

West Coast Governors' Agreement on Ocean Health. (2009). Retrieved October 31, 2009, from West Coast Governors' Agreement on Ocean Health web site: http://westcoastoceans.gov.

World Resources Institute. (2001). *World Resources 2000–2001.* Washington, DC: World Resources Institute.

KEY TERMS AND DEFINITIONS

Coastal Web Atlas: A collection of digital maps and datasets with supplementary tables, illustrations and information that systematically illustrate the coast, oftentimes with cartographic and decision support tools, all of which are accessible via the Internet. Also known as web atlas, digital atlas, digital coastal atlas.

Capacity Building: With regard to coastal atlases, the actions involved in building local GIS infrastructure, including maintenance and update of GIS data layers, software, computer equipment and labs/offices, salaried personnel, and action items for project initiatives.

Coastal Informatics: A broad academic field encompassing the management and analysis of data collected from and representing the coast. Informatics can include spatial data infrastructure, hardware and software infrastructure, computational networking, modeling, and experimentation, the design and deployment of data portals and Internet mapping sites, as well as the creation, analysis, and understanding of data/metadata vocabularies and ontologies, metadata creation/extraction/cross-walking tools, geographic and information management systems, and grid computing.

Information Management: The means by which an organization, agency, or individual collects, documents, shares, and uses information. It often involves creating or identifying the appropriate resources to find quality information in order to fill gaps in knowledge. Coastal web atlases are therefore a key component of information management as applied to coast region of the world.

Marine Spatial Planning: A critical part of regional governance that involves the integrated, forward-looking planning and consistent decision-making regarding various uses of the coast and nearshore. Marine spatial planning must be guided by specific policies and regulations governing usage, the conditions that apply, and an eye toward what possible conflicts in use may arise. Marine spatial planning is often aided by a coastal web atlases which can provide data on habitats, species migrations, land and territorial sea use by humans, navigation, managed areas (parks, reserves, disposal sites, etc.), commercial and recreational fishing, and the like.

Regional Governance: Policies and initiatives that allow state and local governments to pursue concrete, practical steps toward more coordinated and holistic management of ocean and coastal resources. This often includes establishing lines of communication among regions to facilitate implementation of best practices, forming partnerships among regions, and deploying proven management principles and approaches (e.g., ecosystem-based management). Coastal web atlases often provide the datasets, mapping tools, and contextual information needed for effective regional governance

Research Collaboration Network: A collaboration of research scientists, resource managers, technical staffers, and other interested parties to develop a coordinated research network focused on a particular topic. The group fosters communication among those with common goals and interests, along with collaboration on common projects across disciplinary, organizational, institutional and geographical boundaries. The International Coastal Atlas Network is an example of a research collaboration network.

Spatial Data Infrastructure or SDI: A framework via an organization of people or government agencies, via the Internet, or via a series of guiding policies or standards to assist people with acquiring, processing, using, and preserving spatial data. The spatial data are often in geographic information system (GIS) format, are not, but not limited to this.

Chapter 2
Coastal Web Atlas Features

Elizabeth O'Dea
Washington State, USA

Tanya C. Haddad
Oregon Coastal Management Program, USA

Declan Dunne
University College Cork, Ireland

Kuuipo Walsh
Oregon State University, USA

ABSTRACT

A growing number of coastal web atlases (CWAs) for different regions exist around the world. These atlases are developed to meet the needs of a particular organization or audience. Each atlas developer faces the challenge of how best to design a web site that clearly communicates their content in an intuitive way. While most of these CWAs are developed independently of each other, many of them share common features. Interactive maps enable users to visualize data along the coast. A variety of geographic data are presented to inform users about the coastal environment and show professionals what data is available. Atlases include tools and supplemental text for users to learn more about the coast. This chapter provides an overview of common features which are found in existing coastal web atlases.

INTRODUCTION

The Internet is a valuable tool to communicate and share information on coastal regions, ranging in scale from estuaries to nations. Some resources consist of textual web pages that provide overviews of coastal issues, such as NOAA's Office of Ocean & Coastal Resource Management's web site (2009). Others serve as a data repository where people can find and download coastal geospatial data of interest, such as Ireland's Marine Data Online web site and Canada's Coastal and Ocean Information Network Atlantic (Irish Marine Institute, 2009; Atlantic Coastal Zone Information Steering Committee & GeoConnections, 2009). In some cases, coastal web resources provide an interactive web map where users can visualize and query features in the map, like the Oregon Coastal Atlas and the Marine Irish Digital Atlas (OCMP *et al.*, 2009; CMRC, 2009).

DOI: 10.4018/978-1-61520-815-9.ch002

Copyright © 2011, IGI Global. Copying or distributing in print or electronic forms without written permission of IGI Global is prohibited.

Among these various online resources, a growing number of coastal web atlases are appearing. But what differentiates a coastal web atlas from other online coastal resources? A coastal web atlas (CWA) is defined as *"a collection of maps with supplementary tables, illustrations and information which systematically illustrate the coast"* (O'Dea *et al.*, 2007, p. 1). It is more than just a data repository or an interactive map. A CWA can serve as a comprehensive web resource that pulls together textual information, related web links of interest, interactive maps, searchable geospatial data, and tools for targeted users such as coastal managers or recreationists. Each CWA is designed differently, incorporating any combination of these features provided in diverse ways.

MAP AREA

A key component of a coastal atlas is the map area of the main map web page. This is the area where the geographic data are displayed. Many coastal atlases have a central map that serves as the focus of the atlas, such as the Marine Irish Digital Atlas (MIDA) and the UK Coastal and Marine Resource Atlas (O'Dea *et al.*, 2007). Others, such as the Coastal Atlas Flanders-Belgium (Belpaeme & Konings, 2004), contain maps on multiple pages which focus on specific topics. No matter the design, the map area is the most important feature of the atlas. It allows users to visualize the geographic data and see how the features in those layers relate to each other spatially.

Size

The size of the map area can vary from occupying a small section on the page, such as the Gulf of Maine Mapping Portal (GoMMaP) where the map takes up about a quarter of the web page, to covering the entire web page, such as in the Oregon Coastal Atlas or the MarineMap Decision Support Tool (DM Solutions & GoMOOS, 2009; OCMP *et al.*, 2009; MarineMap Consortium, 2009). It is also possible to let the user choose the size of the map area they wish to interact with. In Oregon's North Coast Explorer, for example, the user has the option of viewing a small, medium or large map (Oregon State University, 2009). Similarly, the African Marine Atlas lets the user select the pixel size of the map: 400 x 300, 600 x 450, or 800 x 600 (ODINAFRICA, 2009). This flexibility accommodates different screen resolutions and allows the user some control over how the page appears on their monitor.

Quantity

An atlas may choose to display maps on multiple thematic web pages. Each map can then be customized to display specific data that illustrate a topic and accompanying text. A good example of this can be seen in the Coastal Atlas Flanders-Belgium, where the maps are provided on thematic pages such as Coastal Defense, Fisheries and Agriculture, and Multifunctional Use of the Sea (Belpaeme & Konings, 2004). The maps contain only layers of relevance to the specific theme, displayed on top of base data.

Maps are used on multiple pages for other reasons as well. Static maps highlighting selected features are sometimes displayed in a pop-up window when a user identifies a point, line or polygon on a map, such as in the MIDA (CMRC, 2009). This is particularly useful when maps contain many features in a small area, making it easier to see the selected feature associated with the tabular Identify results. The Oregon Coastal Atlas Beach Public Access Tool, for example, displays multiple static maps in a pop-up window. When a user selects a beach of interest, a page opens with a map of the larger surrounding area and nearby popular beaches, along with a second map that displays an orthophoto of the beach selected. Both of these maps help the user to orient themselves (OCMP *et al.*, 2009).

Interactivity

The ways that atlas users can interact with a map varies between coastal web atlases. Maps can be displayed as static images where the user has no control, such as the New York Coastal Atlas or the Coastal Atlas Flanders-Belgium (New York Department of State, 2009b; Belpaeme & Konings, 2004). Alternately, they can be interactive maps where the user is able to zoom in or out to an area of interest and interact with the data itself, as with the UK Coastal and Marine Resource Atlas and the MIDA (DEFRA, 2009; CMRC, 2009). Users can sometimes control which data layers they view in the map and even how those data are displayed (as described later in the chapter).

An atlas can use a combination of interactive maps and static maps. For example, the Oregon Coastal Atlas has an interactive map on its main mapping page. When the user queries some map features, such as public beach access locations or shoreline protective structures, the query results are accompanied by a static map of the feature selected.

Cartographic Features

There are a number of features that the map area can contain, such as:

- **An Overview Map:** Also known as a reference map, this shows the extent of the atlas area (state/country) and indicates the smaller area where the user is currently zoomed in (see MIDA in Figure 1). It can be a permanent feature or the user can have the possibility to turn it on or off. Sometimes it is located in a corner of the main map or it can be elsewhere in the web page, such as above the legend.
- **A North Arrow:** A north arrow may or may not appear on a map. If an arrow is not displayed then North can be presumed to be at the top of the page.
- **A Scale Bar:** This represents the scale at which the map is currently being viewed (see lower right corner of map in each Figure 1 example). The scale changes as a user zooms in or out. It can be displayed in a number of styles and distance units (miles, kilometers, etc.). In some cases a

Figure 1. These images show two different examples of the main map page in the Oregon Coastal Atlas (left) and the MIDA (right). (© 2009, Oregon Coastal Management Program & University College Cork. Used with permission.)

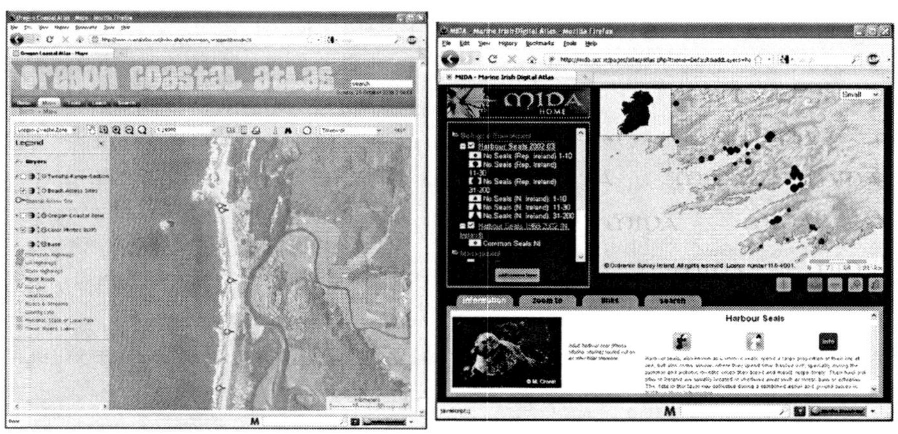

scale is written directly above or below a scale bar to denote the scale being viewed, for example 1:25,000. In some coastal atlases, the user can choose from a list of predetermined scales to zoom to, set by the coastal atlas developer.

- **Coordinates:** Coordinates can be displayed to indicate the geographic location of the pointer as it moves around the map area. These coordinates are useful for users who need locational coordinates for specific points. These coordinates may appear in a corner of the map, in a popup message or in the status bar at the bottom of the web page.
- **A Watermark:** A watermark is sometimes added for copyright purposes. A watermark appears as a semitransparent logo repeatedly over the map image (see MIDA example in Figure 1). Some data owners require that an atlas which displays their data have a watermark over their data to prevent misuse of the map image.
- **Copyright Statement:** As with a watermark, a copyright statement is sometimes required by data owners. It is commonly displayed in a corner of the map. This statement can be set to change, depending on which data are displayed in the map. There can also be multiple statements displayed at one time.

GEOGRAPHIC DATA

The maps displayed in the map area are composed of geographic datasets. Because the coastline straddles terrestrial and marine environments, coastal web atlases can include a very broad range of data themes from many different owners. The quantity of data layers contained in existing atlases ranges from a small number of layers (e.g., under 40 in the Coastal Atlas Flanders-Belgium) to a very large number (e.g., over 3,000 in the Oregon Coastal Atlas; O'Dea *et al.*, 2007). Atlases can make these data available to the user in a variety of ways cartographically and as downloadable data.

Data Types

The geographic data come in many types, including vector points, lines and polygons, as well as raster grids and imagery. Points represent locations such as archaeological sites or marine mammal sightings. Lines illustrate features such as armored shorelines or coastal walking trails. Polygons symbolize areas such as Marine Protected Areas or eelgrass habitat. Remotely sensed imagery and scanned base maps, such as satellite images and nautical charts, provide an invaluable backdrop to vector data that enable users to relate better to the area that they are viewing.

Vector data types such as point, line, or polygon datasets are most commonly displayed in ESRI Shapefile format, and increasingly in Google's KML format. Raster data types, on the other hand, tend to have a wider range of file formats, depending on the nature of the raster data in question. Orthophotography or scanned historic maps, for example, are commonly exchanged in some form of GeoTiff or Geo JPEG format, with or without various forms of file compression. Very large images may be compressed using proprietary formats such as MrSID (multi-resolution seamless image database) or ECW (Enhanced Compression Wavelet) formats. Other raster formats exist that conform to the processing requirements of specific geospatial software, such as ESRI GRID format or ERDAS IMAGINE .img files.

Data Themes

Coastal atlases can contain a number of data themes that are relevant to the coast and of interest to coastal atlas users. Base data such as infrastructure, topography/bathymetry, and satellite imagery are common features in atlases. Environmental management is also a common theme, containing data related to natural resources, fisheries, aquaculture, conservation, environmental monitoring, and human impacts on the coast. Scientific themes include biology, coastal habitats and the physical environment. Cultural themes such as recreation, culture and heritage, tourism, and industry are also sometimes included.

Within atlases, data can be grouped by theme to make them easier for users to find. For example, the MIDA lets the user choose a theme name from a pull-down list, such as protected areas or tourism and recreation. The map is then customized to that theme, displaying the data associated with it (CMRC, 2009). The Coastal Atlas Flanders-Belgium asks the user to select an atlas theme they wish to see from the atlas table of contents. They are then taken to a web page with information and maps on that theme (Belpaeme & Konings, 2004).

Data Sources

Coastal web atlases commonly display data from a wide variety of data owners. Government agencies often provide base data such as roads, rivers, coastline, topography, hydrography, and political boundaries. Federal agencies such as the United States' National Aeronautics and Space Administration (NASA) and National Oceanographic and Atmospheric Administration (NOAA), and international organizations such as the United Nations Educational, Scientific and Cultural Organization (UNESCO), also provide data that can be shared in atlases around the world. These include satellite imagery, climate and land cover data. Other data come from universities, non-governmental organizations (NGOs), and private consultancies, among others.

While it is common to display data from multiple owners, the number of data owners which contribute to CWAs can vary. In the case of the Washington Coastal Atlas, the data primarily comes from state and federal government agencies, such as the Departments of Ecology, Fish and Wildlife, Natural Resources, and Health, the U.S. Geological Survey (USGS), and NOAA. It also contains some data from private data suppliers for its roads (Washington State Department of Ecology, 2009). On the other end of the spectrum, some atlases provide data from a broad spectrum of owners. For example, over half of the data contained in the MIDA comes from government agencies, while the rest come from (in order of quantity) educational institutions, NGOs, and private organizations (See chapter 7).

Data are generally stored locally by the atlas host, creating a sort of repository. Atlas data can also come from remote locations where they are shared across a network and displayed seamlessly over local data in an atlas. In some cases, data is pulled from existing map services which share certain geographic data. For example, NOAA's nowCOAST site makes its data, such as oceanographic observations and precipitation accumulation, available to other web GIS sites through their ArcIMS Image Services (NOAA Coastal Services Center, 2009).

Variations in Data Quality

Data quality is difficult to define, as discussed by Bartlett and Bruce (2001). For GIS data, quality can concern "*scale, accuracy, data precision, logical consistency, dataset completeness* and *data lineage*" (Dwyer *et al.*, 2003). It is inevitable that the data will be of varying quality when displaying diverse data from many sources, as is done in coastal web atlases.

Data owners use different standards and methods when creating geographic data. The scale at which the data was produced varies, depending on the purpose for which it was created or what resources were available. To address this, some atlases, such as the UK Coastal and Marine Resource Atlas and the MIDA, set scale factor limits which restrict viewing certain data when a user zooms to an area above or below a specific scale. This is discussed further in the Data Availability section.

Also, available data may not cover the full geographic extent of the coastal atlas. For example, high resolution data on a topic may be available only for an estuary rather than the entire country. Another data quality issue is the data's attributes. Data may offer very few attribute fields, and its contents may be encoded and difficult to interpret. Quality of the accompanying metadata – or entire lack of metadata – is also often an issue (as discussed later in this chapter).

How Data are Displayed

Atlas developers have the added challenge of making a web map cartographically appealing and easy to interpret. With more layers comes more challenge. While designing the MIDA, developers took into careful consideration which layers would most likely be viewed together and attempted to accommodate overlapping map layers. For example, three important datasets representing protected areas are commonly viewed together yet sometimes all three overlap. The developer designed the polygon fill to be different colored hatchmarks, each at a different angle, with a transparent background in order to see all of the layers at once (O'Dea *et al.*, 2004, March). In some cases the user has more control of how the layers appear in the map. In the Oregon Coastal Atlas, for example, the user can change the transparency of each data layer to enhance their viewing of the data (OCMP *et al.*, 2009).

To avoid cluttering the map, atlases can limit the number of layers that can be viewed at one time. For example, the Oregon Coastal Atlas limits the user to viewing to five or six data layers at a time, depending on the theme selected (OCMP *et al.*, 2009). The African Marine Atlas asks the user to choose one background layer to view at one time and then select one or multiple additional layers from a list (ODINAFRICA, 2009). On the other extreme, atlases can set no limit on the number of layers that can be viewed at one time in an atlas. This is the case in the MIDA, where the user has the ability to customize the list of layers to meet their needs by adding and removing layers (discussed later in this chapter). Another alternative, previously discussed, is the use of grouped data by theme and letting the user select the theme of interest.

When a user opens an atlas, typically they can control the layers that they see in the map by turning them on and off in the layer list next to the map. Data layers can be scale dependant, only displaying when a user zooms to a particular scale. This occurs when the atlas developer sets a minimum or maximum scale factor limit for a layer. It is set by the atlas developer when data are not appropriate for users to view at small or large scales due to data quality issues or lack of full geographic extent over the complete atlas area. For example, the MIDA has scale factor limits set for its base maps, which change as the user zooms in closer to an area on the map (CMRC, 2009). Some atlases indicate to the user the layers which have scale factor limits. Some, such as the British Columbia Coastal Resource Information Management System and the UK Coastal and Marine Resource Atlas (Figure 2), grey out the tick box or the layer name in the layer list to denote that it exists but is not available at the current scale (GeoBC Spatial Analysis Branch, 2009; Maritime & Coastguard Agency & DEFRA, 2009). This is meant to inform the user that the layer is available but not visible at that scale. The New York Ocean and Great Lakes Atlas provides a "layers unavailable at zoom level" subset in its layer list (NYOGLECC, 2009). Some atlases, such as the MIDA, do not denote which layers are hidden despite scale factors being set.

Data Availability to User

Sometimes atlases provide a way for users to search for specific data of interest. The African Coastal Atlas implements Google to search its web content (UNESCO *et al.*, 2009). MIDA uses custom PHP scripts to search its PostGreSQL

Figure 2. The figure shows a few examples of the many ways in which layer lists and legends can be presented in a coastal web atlas. Belgium's example is simply a legend that the user can pull down or hide. The UK's atlas displays a permanent but dynamic legend and gives the user the option to add layers of interest from a pop-up window. Washington's atlas combines legend and layer list into one, with the ability to collapse lists by data theme using the folder icon to the left. (© 2009, Coordination Center for Integrated Coastal Zone Management; MAGIC (http://www.magic.co.uk); & Washington State Department of Ecology. Used with permission.)

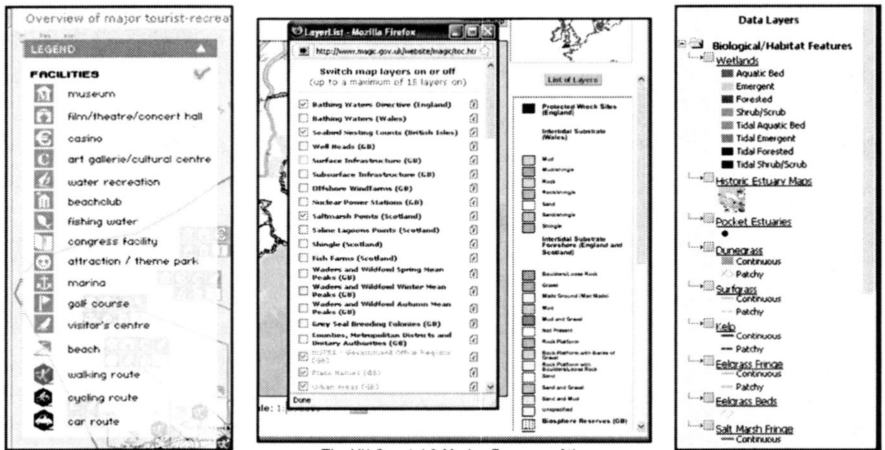

The Coastal Atlas Flanders-Belgium — The UK Coastal & Marine Resource Atlas — The Washington Coastal Atlas

database containing geographic metadata information (O'Dea *et al.*, 2004, May). NYOGLA uses ESRI's GeoPortal Extension to enable data searches by users (NYOGLECC, 2009). In addition to metadata, some atlas search results return downloadable geographic data, such as MIDA and the Oregon Coastal Atlas (CMRC, 2009; OCMP *et al.*, 2009). It is common to provide the data in one format, such as ESRI Shapefiles. The NYOGLA makes available for download ESRI Shapefiles, MapInfo Tab files, and Google Earth KML files (NYOGLECC, 2009). The Coastal Atlas Flanders-Belgium provides the option of downloadable PDF maps or geographic data (Belpaeme & Konings, 2004). Other atlases even provide access to their data through Web Services, which enable data stored on a server to be displayed in any web mapping application that uses the same Web Service standards. For example, NOAA's nowCOAST shares some of its data using ArcIMS Image Services, which enable users to view data through desktop ArcGIS and ArcExplorer applications. In the near future nowCOAST developers intend to implement an Open Geospatial Consortium (OGC) Web Mapping Service (WMS) and Web Feature Service (WFS), which is discussed further in this chapter (NOAA Coastal Services Center, 2009).

MAP LEGENDS AND LAYER LISTS

A map legend defines the symbols and colors used to display map features so that users can understand better the geographic data displayed in the map. A layer list gives the user control of the layers which are viewed in the map, typically with boxes that can be turned on or off. CWAs have one or both of these. Map legends and layer lists are typically displayed beside the map frame on the main map page, as can be seen in the examples given in Figure 1. Some atlases, such as those using ArcIMS, display layer and legend lists as tabbed frames next to the map so that users can easily display one or the other. In some cases, such as with the Coastal Atlas Flanders-Belgium, the list is displayed over the map itself (Figure 2) (Belpaeme & Konings, 2004). Sometimes they appear in pop-up windows instead, in order to reduce clutter on the main map page, as is seen in the UK Coastal & Marine Resource Atlas in Figure 2. Alternatively the map legend and layer list can be combined in order to save space and avoid the user having to flip back and forth between layer list and legend tabs, such as with the MIDA as well as the Washington Coastal Atlas (Figure 2).

The flexibility of the layer list and legend in a CWA depends on the atlas design and the level of interactivity allowed for atlas users. Map legends tend to be static images within the map page, and change if the layer list is modified. Layer lists tend to be more interactive by enabling users to turn layers on and off using the tick boxes next to the layer names. Users can also sometimes access more information about the layers themselves from this layer list by selecting links to metadata, downloadable files or further textual information. In some atlases layers can be grouped by data themes of interest to users. In the case of the MIDA, users have the added capability of searching for and adding data layers of interest to the layer list from the data archive, as well as removing data layers that are not of interest (CMRC, 2009). Another way in which the layer list can provide flexibility to the user is by allowing the user to rearrange the draw order of the layers as well as adjust their transparency. A good example of these functions can be seen in the Oregon Coastal Atlas (OCA, 2009).

ATLAS TOOLS

The term "Tools" in a CWA setting is typically used in one of two ways:

- To describe map tools, or map interface features that allow users to do things with maps, or
- To describe decision tools, or standalone tools, often but not always built around maps, which allow users to examine and work on a particular topic, dataset or decision.

The first category is relatively constrained to a short list of commonly used map interface widgets (i.e., pan, measure, identify feature, etc.), though there are variations across different map interface software. The second category can target specific topics that might be of concern in the coastal zone. A few examples of the most common types will be covered here.

Map Tools

Map interface tools often fall into a few broad groupings. These include tools for navigating the map, modifying the composition of the map, and interacting with map information. Figure 3 illustrates the toolbar from the Oregon Coastal Atlas, as a typical example, and numbers the various elements.

Map Navigation

Tools that allow a user to navigate the map both by panning ("horizontal zooming") and zooming ("vertical zooming") come in many flavors. In the map interface toolbar above, the two panning tools are sequential panning (#2) and non-sequential panning (#14). The terms sequential and non-sequential refer to the method with which a user can get to a new place in the map. A non-sequential tool like a *Zoom to Location* list allows the user to quickly shortcut to a specific bookmarked location on the map, which can save the user time depending on their needs. A sequential pan tool such as the *Hand/Pan* tool requires the user to refine a search by dragging the map progressively

Figure 3. The tools in the Oregon Coastal Atlas map toolbar allow a variety of methods for navigating the map, querying its contents and saving results. (© 2009, Oregon Coastal Management Program. Used with permission.)

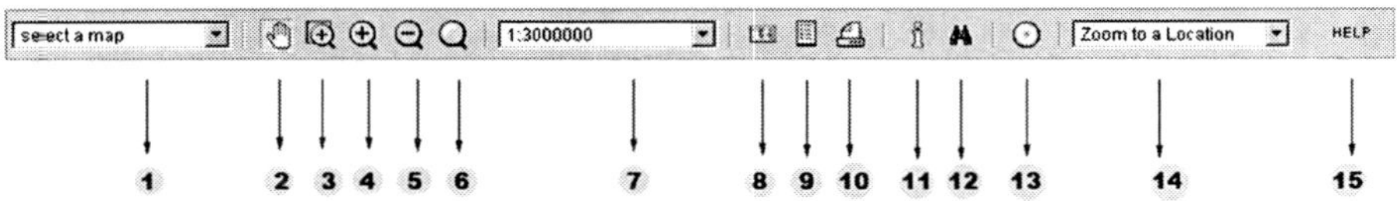

until the desired location is found (Harrower and Sheesley, 2005). Alternatively a user may also be able to sequentially zoom by selecting arrows located around the map perimeter to move in the respective direction. The vertical zooming options are many. In the example shown above, the *Define Extent* (#3) and *Zoom in to Point* (#4) tools are non-sequential and sequential zoom-in tools respectively, while *Zoom out from Point* tool (#5) is a sequential zoom out tool. The *Zoom to Extent* tool (#6) allows the user to go back to the default starting zoom level of the mapping application, while the *Zoom to Scale* tool (#7) allows the user to jump between specific pre-set zoom levels/scales.

Map Composition

CWAs that allow users to change the contents of the map include tools that enable the changing of map layers displayed or allow the changing of the entire map theme. In Figure 3, tool #1 allows the user to pick the map theme that will be viewed. Each map theme includes a specific set of layers that may be viewed. Tool #9 is the *Layer List* tool that allows users to see the layers available in the map and toggle the layers on and off. Additional functionality of layer lists are discussed in the map legends and layer lists section.

Map Information

The last group of map interface tools includes any tool relating to getting information from a map, and saving or passing on that information. Such tools include *Map Legend* (#9), *Overview/Reference Map* (#8), map *Identify* tools (#11), *Map Query/Search* tools (#12), *Map Sharing* tools (#13), *Map Printing* tools (#10) and finally *Map Help* (#15). *Map Legend*, *Overview/Reference Maps*, and *Identify* tools have been covered elsewhere in this chapter. A *Map Query/Search* tool allows the user to search for a known feature within a map. A *Map Sharing* tool enables the user to export the current map view as a URL embedded in an email. A *Map Printing* tool allows the user to export the current map to a variety of image types for use in other documents, or as a formatted PDF that includes a custom title plus legend, for use in meetings, presentations or as an attachment to documents.

Decision Tools

Decision Tools are tools designed for specific user needs and focus on a particular purpose. They may be integrated into the main map page or be a standalone tool located elsewhere in the atlas web site. These tools can provide a targeted resource for information that can be of use to a specific group, such as coastal planners.

Report Tools

One of the most common types of tools provided in coastal atlases is a report tool. These tools may be map based, text based, or a combination of both. They work by allowing the user to select

Figure 4. The Washington Coastal Atlas's land cover change tool generates reports on the percent change in Forest cover and impervious surface over time. (© 2009, Washington State Department of Ecology. Used with permission.)

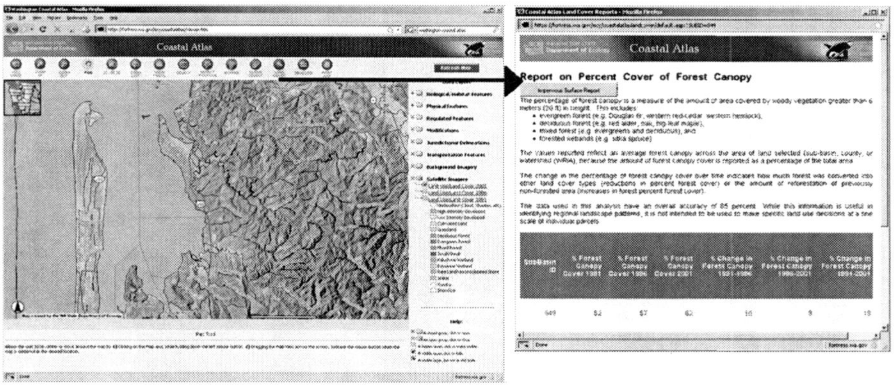

an area of interest and then use that area to query a dataset or series of datasets and databases to report basic information about that location. The Washington Coastal Atlas has a land cover change tool, seen in Figure 4, which is part of the main map interface and allows the user to generate reports on the percent change in forest cover and impervious surface over time (Washington State Department of Ecology, 2009). Report tools can be extremely valuable for CWA users who are seeking to summarize quickly known information about a particular topic in a particular place.

Analysis/Decision Support Tools

Map analysis and decision support tools are an area of considerable literature (for example Figure 5, Wright & Scholtz, 2005), the scope of which can only be touched upon here. To some extent the limitations of the web have cut down on the sophistication of tool features and functionality that could be delivered to CWA users. However, as broadband Internet speeds proliferate, more and more advanced tools are being constructed and used to engage users in a wide variety of coastal analyses and decision processes. As an example the California MarineMap Decision Support Tool project was developed for the Marine Life Protection Act (MLPA) initiative. It is a tool that enables stakeholders to participate in the ongoing process of designating marine reserves off of the California coast. Certain functions are available to all users while other more advanced participatory functions are only available to registered stakeholders. The tool allows users to browse information about California state coastal waters, draw suggestions (polygons) for MPA areas, and evaluate networks of suggested MPAs (MarineMap Consortium, 2009).

The Oregon Coastal Atlas provides a good variety of examples of additional tools that are of use to their audience. The Beach Water Quality Explorer enables users to see the latest water quality testing results for any beach of interest. In the Oregon Coastal Access Information System, a user can select a county along with the facilities they require and the activities they are interested in. The resulting page shows them all public beach access sites which meet their criteria. Also among the many tools available is the Coastal Inundation Visualization Tool, which provides access to and analysis of near real-time data from tide gages, ocean buoys and weather forecasts along

Figure 5. MarineMap Decision Support Tool (DST). (© 2009, MarineMap Consortium. Used with permission.)

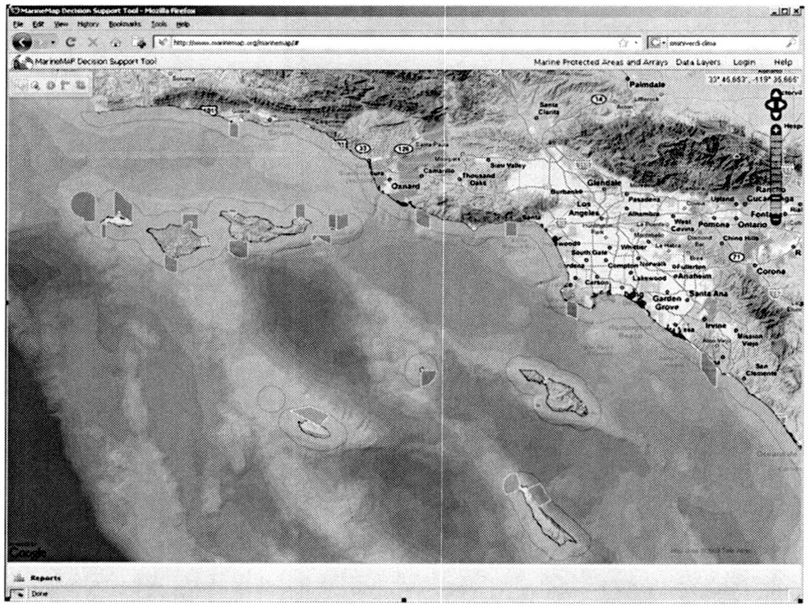

with other coastal geographic data for Tillamook County, Oregon (Haddad *et al.*, 2006).

Volunteered Geographic Information Tools

It has become increasingly common in recent years for map tools to allow users to contribute their own geographic data for sharing and display to others. While not yet available in coastal atlases per se, Volunteered Geographic Information (VGI) tools have a range of applications and are an interesting emerging field of study within geography (Goodchild, 2007). The European Environment Agency's web site "Eye on Earth" (Figure 6) has a tool that allows users to browse the latest bathing water quality status. In addition, the tool allows users to contribute their own opinion of bathing water quality. Both datasets (actual water measurements and user opinions) are viewable side by side (European Environment Agency, 2009). The Community Mapping Network in British Columbia, Canada provides similar input opportunities from local community members by allowing registered users to add points, lines or polygons and associated tabular data using Web Map Services. This comes with risks of data reliability when map features come from multiple data sources and people with varying levels of expertise (Canessa *et al.*, 2007).

ATTRIBUTE TABLES

Atlas users often have access to attribute tables which contain the data behind map features. Table fields can contain information such as names, coordinates, quantities and other details. For example, a layer representing marine protected areas (MPAs) might have fields for MPA name, year it was established and the size of area included.

Geographic layers can contain very few or many fields. Atlas developers sometimes select a subset of fields that are most informative, in

Figure 6. The EEA web site "Eye on Earth" has a tool that allows users to contribute their own opinion of bathing water quality. (© 2009, European Environment Agency. Used with permission.)

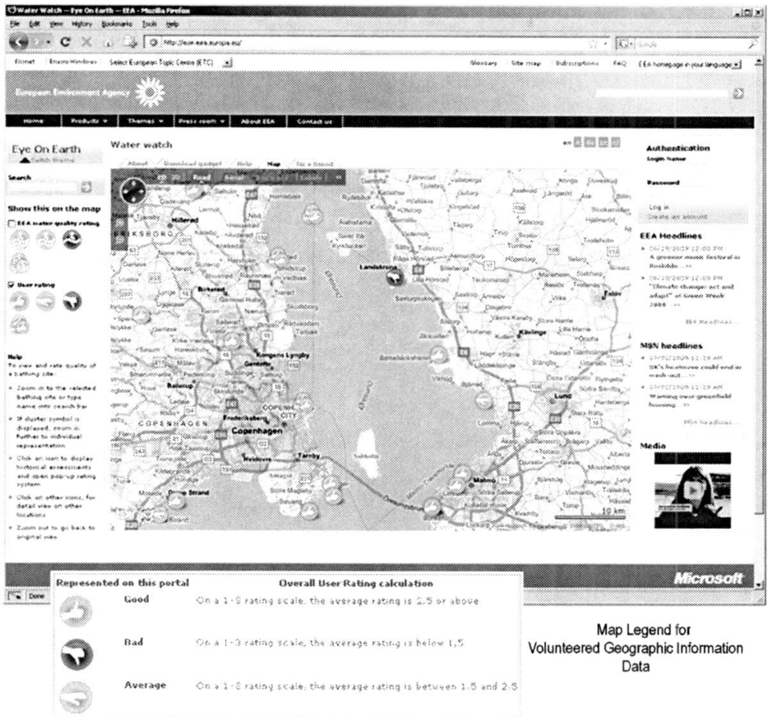

order not to overwhelm the user with too much information. In some cases, attributes are codes that mean nothing to the atlas user. In such cases, atlas developers can work around this by modifying the layer's attribute table to include a field that defines the code. Other atlases display the attributes as the owner provides the data. In such cases, accompanying metadata, if of good quality, can provide the definitions.

How do atlases provide access to attribute tables? Typically a user opens an attribute table by using an *Identify* tool and then clicking on a point, line, or polygon feature in the map. With more advanced CWAs, users can use a *Query* tool to search for specific attributes in tables, such as in the Oregon Coastal Atlas and nowCOAST (OCMP *et al.*, 2009; NOAA Coastal Services Center, 2009). When the table opens it appears in a popup window or in a frame of the map page. In other cases, such as the UK Coastal and Marine Resource Atlas, the selected map feature is highlighted either in the main map or in a new map accompanying the table results. Some atlases, such as the MIDA, use a drill-down tool which selects features from all of the layers under a point that the user selects on the map. This results in multiple tables being displayed in a list (CMRC, 2009).

METADATA

Metadata, or documentation about geographic data, is a crucial component of a coastal web atlas. Metadata provide the source description for each of the layers in an atlas, such as the data owner, the date it was produced, and the scale at which it was created. Metadata inform atlas users about the quality of geographic data. They can be provided to the user in one or multiple levels of data content:

- **Abstract Metadata:** Contain brief overviews which give the user a quick snapshot of data contents. This level is good for any skill of user. They are typically a subset of metadata standard elements.
- **Discovery Metadata:** Contain moderate levels of information on data content, for the more curious user. These are also typically a subset of metadata standard elements.
- **Full Metadata:** Contain highly detailed information which is generated by the data owner and considered complete. They assist advanced users in finding and working with data that are relevant to their specific needs (O'Dea *et al.*, 2007).

Metadata Standards and Quality

There are different geographic metadata standards that data owners can use to document their data. When different data owners use the same standards, it is easier to share data and metadata between applications and organizations. Existing atlases use standards that are common for their region. For example, American atlases such as the Washington Coastal Atlas and the Oregon Coastal Atlas use the Federal Geographic Data Consortium (FGDC) standard, which is the recommended standard for the United States (Washington State Department of Ecology, 2009; OCMP *et al.*, 2009). The UK Coastal and Marine Resource Atlas and the MIDA use the International Standardization Organization (ISO) 19115 metadata standard and the associated ISO 19139 formatting standard, which are global geographic metadata standards (Maritime & Coastguard Agency & DEFRA, 2009; CMRC, 2009). These are discussed later in this chapter as well as in Chapter 4.

The level of metadata quality varies between existing atlases, and also within a single atlas. This is primarily due to diverse data owners rather than the atlas developers themselves. Data owners document their data differently, with some providing minimal information, if any, and others providing fully detailed metadata. In some cases the lack of metadata requires developers to contact data owners directly to collect enough information, otherwise unavailable, to make a layer useful. As an example, MIDA developers had to do this for some layers in order to be able to provide complete Discovery Metadata, which was their set of minimum metadata requirements (O'Dea *et al.*, 2004, May).

Access to Metadata

Atlases work around varying quality of metadata in different ways. Some atlases provide no metadata at all, like British Columbia's COINPacific Web Mapping Application (BCMSRM, 2009). Others may only provide a brief but consistent summary of the metadata for all of the data contained. The Oregon Coastal Atlas provides only a detailed level of metadata for the data available (OCMP *et al.*, 2009). Alternately atlases may provide tiered levels of metadata information. The Washington Coastal Atlas provides two levels of metadata: a metadata summary that contains the layer description, source, scale at which it was created, date, and attribute information where possible; and it provides a link to "Detailed Layer Information" that takes the owner to full metadata on the Department of Ecology or data owner's web site, if it exists in full form (Washington State Department of Ecology, 2009). The MIDA uses the three-tiered system previously mentioned that shows progressive levels of information, starting with Abstract Metadata, then Discovery Metadata, and finally Full Metadata (O'Dea *et al.*, 2004, May).

Metadata can be displayed as web pages, either stored within the CWA or referenced as links to the relevant data owner's web page displaying the metadata. Examples of this can be seen in the African Marine Atlas as well as the full metadata in the Washington Coastal Atlas (UNESCO *et al.*, 2009; Washington State Department of Ecology, 2009). Alternatively, metadata can be stored in a database and then displayed dynamically in a web page template. This is the case in the MIDA, which uses Abstract and Discovery Metadata templates to extract selected metadata fields from a PostGreSQL database based on XML tags (O'Dea *et al.*, 2004, May).

Some CWAs make their metadata exportable. Metadata can either accompany downloadable datasets, such as within zipped Shapefiles, or be downloaded separately as HTML or exportable XML files. Atlases can design their exportable metadata to comply with the OGC Catalog Services for the Web (CSW) standard, which enables sharing of metadata with other databases and search engines. The International Coastal Atlas Network is developing a prototype among various partners to demonstrate the sharing of metadata using OGC standards (Lassoued *et al.*, 2008).

INFORMATION AND EXTRAS

Additional relevant information adds value to the map display by helping to highlight specific coastal topics. Just as traditional paper atlases often include illustrations, informative tables, and explanatory text to compliment maps, CWAs may provide additional quantitative and qualitative information to enhance understanding. When a CWA begins as a bound collection of maps and then transitions to be accessible on the Internet, the tables, illustrations and text are often web-enabled along with the maps.

Existing CWAs do not always contain supplemental information. There are some that call themselves atlases, however do not contain supplemental information and extras that differentiate an atlas from solely a web mapping application. In CWAs that *do* contain additional information and extras, things commonly included are textual information, tables and charts, multi-media, and relevant links and resources.

Textual Information

Atlases sometimes contain a large amount of text that provides meaningful and helpful information. Textual material may highlight specific coastal topics, give general and detailed descriptions of topics and issues relevant to the atlas purpose, or tell stories around the maps to engage users. For example, the Coastal Atlas Flanders-Belgium focuses on thematic topics such as fisheries and agriculture, tourism and recreation, and coastal defense (see example in Figure 7) (Belpaeme & Konings, 2004). MIDA users can access additional information on topics such as marine safety and protected areas (see MIDA example in Figure 7) (CMRC, 2009). The Oregon North Coast Explorer, which focuses on watersheds and fisheries, provides stories about interesting research and activities that shed light on what makes the Oregon North Coast Basin and its ecosystems unique (Institute for Natural Resources, 2005). The Oregon Coastal Atlas has a "Learn" tab on its web site that links to detailed information about coastal places and topics (OCMP *et al.*, 2009).

The ways that atlas users can access textual information varies between coastal web atlases. MIDA highlights specific coastal topics in a frame on their main map page, which changes when a user selects different layers in the map legend (Figure 7). This frame links out to more detailed textual pages contained in the MIDA InfoPort. The Coastal Atlas Flanders-Belgium highlights specific coastal topics on web pages under the General Information and Themes and Activities sections (Figure 7). In the North Sea Atlas web site, explanatory text is accessed interactively between the map and information by clicking on the "Read more about this map" link (North Sea Interdepartmental Consultation Directorate, 2009).

Figure 7. (l) The Coastal Atlas Flanders-Belgium focuses on thematic topics which use maps to illustrate information, while making related geospatial data, charts and illustrations available; (r) The MIDA centers on an interactive map where users can view data by theme and access additional information on the topics (users can also access maps from the text pages via links). (© 2009, Coordination Center for Integrated Coastal Zone Management & University College Cork. Used with permission.)

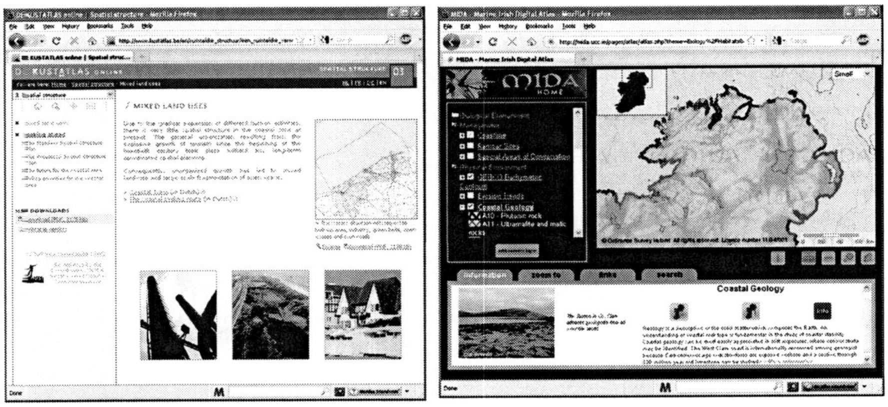

Tables and Charts

Large amounts of information can be presented in atlases in a way that is compact and easy to understand through charts and tables. For example, the Coastal Atlas Flanders-Belgium provides static charts of evolution of fish supplies and different species which make up fisheries. In the Oregon North Coast Explorer, geographic data is integrated with up-to-date statistics to support understanding and decision making. Whereas the Coastal Atlas Flanders-Belgium provides static charts, the Oregon North Coast Explorer dynamically generates tables or bar charts based on multiple statistics and locations (Belpaeme & Konings, 2004; Oregon State University, 2009).

Multimedia

Some web atlases incorporate multimedia features in a wide variety of formats such as illustrations, photos, video, and audio. Animation provides a unique way to illustrate coastal processes and data sources. A small handful of MIDA InfoPort pages (comprehensive informational pages) have animated images to illustrate points in the text. For example, under the Physical Environment section, the satellite imagery page has a NASA animation that shows how satellites travel around the earth (CMRC, 2009). In the Oregon Coastal Atlas's Tsunami Hazards section, an USGS animation shows a hypothetical Tsunami along the Pacific Northwest Coast (OCMP *et al.*, 2009). In the Oregon North Coast Explorer, an interactive flash tool gives a sense of how investments in fish passage restoration have benefited salmon in a particular watershed in the North Coast basin (Oregon State University, 2009). New York State's Division of Coastal Resources Online is part of a multi-media package which provides videos on waterfront revitalization, re-use of abandoned buildings, and watershed planning (New York Department of State, 2009a).

Relevant Links/Resources

There are a number of relevant links and additional resources, both internally and externally related to a CWA web site, that an atlas may contain. These can include:

- **A glossary:** An alphabetical list of coastal related terms, such as the one contained in Oregon's North Coast Explorer (Oregon State University, 2009).
- **Help pages and tutorials:** These explain how to use the coastal atlas and navigate the maps. The help page and glossary can be linked from the main map page, as is done in the New York Ocean and Great Lakes Atlas and the British Columbia Coastal Resource Information Management System (NYOGLECC, 2009; GeoBC Spatial Analysis Branch, 2009).
- **Links:** Atlases may provide links to relevant web sites, organizations, experts, reports, and documents. For example, the Coastal Atlas Flanders-Belgium provides links to information and maps on coastal sustainability indicators on the European Union funded SAIL (Schéma d'Aménagement Intégré du Littoral) project web site (Belpaeme & Konings, 2004).
- **Resources for specific user groups:** Atlases can provide resources by interest, such as coastal management, education, and tourism. For example, the Center for Coastal Resources Management in Virginia developed tools for coastal management, including the Shoreline Manager's Assessment Kit (SMAK) and the Wetlands Mitigation Targeting Tool (Center for Coastal Resources Management, 2009).

These links to additional information and resources take advantage of the benefits of the Internet, including providing access to the most current information from outside organizations that regularly update their web content.

TECHNOLOGY

In order to adequately manage all of the web content, tools and data, a web atlas requires a robust server and effective software in order to operate well. For CWA developers, the choice of these technologies takes into account criteria such as available financial and technical resources, audience needs and limitations. Existing CWAs use a variety of technology combinations based on their organizational needs.

Hardware

Disk space, physical memory, central processing units (CPUs), and Internet bandwidth are key hardware components. A web site with large traffic, large datasets, large dynamic processing requirements, etc. requires a powerful hardware infrastructure to support these loads. Hardware development is constantly progressing. For example, the server which first hosted the MIDA was purchased in 2003. It consisted of a single Xeon CPU, 1GB of physical memory and 350GB of SCSI disk space. In 2008, a new server was purchased. For a similar price, the new server consists of two Quad Core Xeon CPUs (i.e., eight processors), 16GB of physical memory, and 3TB of SCSI disk space. The MIDA web site was migrated to this new server, which has improved the performance of the MIDA. It is particularly noticeable when there is concurrent access and multiple tasks running on the server.

Web Servers and Operating Systems

Every hardware platform requires an operating system. The most common operating systems are Microsoft Windows or a version of UNIX (e.g., Linux, Solaris, Mac OS X, etc.). A web server is required in order to host a web site on an operating system. Apache HTTP server and Microsoft Internet Information Services (IIS) are the most widely used web servers (Netcraft, 2009). Apache operates on most operating systems while IIS only operates on Windows. On their own, these web servers only serve static content. Dynamic frameworks including JavaServer Pages (JSP),

Java Servlets, PHP, ASP, ASP.NET, and CGI are required in order to generate dynamic content. The MIDA web site is based on Linux, Apache HTTP server, and a PHP software environment (as described in Chapter 7), while the Washington Coastal Atlas uses Windows, Microsoft IIS server and ASP.NET for some of its tools (O'Dea *et al.*, 2007; Purce *et al.*, 2009).

Map Software

A web mapping software package is required in order to generate dynamic maps for a coastal atlas. This software can be either proprietary or open source. Advantages of proprietary software include an 'out of box' web mapping solution, user manuals, and consultancy support. Disadvantages include the cost, proprietary technology, and the requirement of programming skills to customize the interface. Advantages of open source are that it is free from commercial licensing fees, and is non-proprietary technology (meaning the code is available and customizable). Disadvantages include that it does not generally provide an 'out of box' web mapping solution, documentation can be inadequate, and programming skills are required to both build and customize the interface. ESRI ArcIMS and University of Minnesota (UMN) MapServer are good examples of proprietary and open source solutions, respectively. UMN MapServer is used by both the MIDA and the Oregon Coastal Atlas, while ESRI ArcIMS is used by the Washington Coastal Atlas (O'Dea *et al.*, 2004, March; Haddad *et al.*, 2006; Purce *et al.*, 2009).

Database/Content Management Systems

A coastal web atlas is comprised of a variety of content including HTML pages, imagery, and dynamic maps. In order to generate dynamic maps, underlying geographic datasets are required by the web mapping software. Geographic datasets displayed in the maps can either be stored simply as files or within a geodatabase or spatial database. For example, MIDA utilizes a file-based approach using Shapefiles to store vector data and GeoTiffs for raster data and imagery, while Ireland's Marine Data Online uses an ArcSDE/SQL Server Spatial database (Wright *et al.*, 2007).

Non-geographic web site content such as HTML pages and images can either be managed manually or else be managed in a more formalized system such as a content management system (CMS). Open source CMS systems include WordPress (http://wordpress.org) as a blogging system, phpBB (http://www.phpbb.com) as a bulletin board system, and Drupal (http://drupal.org) and Joomla (http://www.joomla.org) as full web portal systems. For example, the Oregon Coastal Atlas uses the Joomla content management system to manage that web site's content (Purce *et al.*, 2009).

Design and Testing

An atlas should be designed in line with latest international standards. For metadata, international standards include ISO 19115 and ISO 19139, as mentioned previously. ISO 19115 defines the general purpose metadata standard for geographic information, while 19139 defines how it is implemented in XML. For example, the MIDA metadata system uses the ISO 19115 standard (as described in Chapter 7). The metadata XML tags used by MIDA to contain the metadata were taken from a draft of the ISO 19139 standard.

In order for an atlas to remotely access maps, datasets and metadata hosted by third parties on the fly, OGC standards are important. OGC standards include Web Map Services (WMS), Web Feature Services (WFS), Web Coverage Services (WCS), and Catalog Services for Web (CSW). WMS enables remote access to maps as images. Alternatively, both WFS and WCS enable remote access to the actual data which allow the user more interaction with the data themselves. A

WFS enables access to feature data which includes the vector geometry (i.e., points, lines, polygons) plus their attributes, while a WCS enables access to gridded data (e.g., GeoTIFF containing data). In addition, the CSW standard enables discovery and access to remote metadata catalogs. Metadata standards and OGC web services are described in more detail in Chapter 4.

Using OGC services, coastal web atlases can display data from external data sources, such as NASA and NOAA services. CWAs can also share the data they contain with their users through OGC services. This has been tested by some atlases but has not yet been implemented by most. Problems encountered between distributed map services and CWAs with different projections or datums have limited implementation, such as in the experience of both the MIDA and the Washington Coastal Atlas. While not yet providing OGC services, NOAA nowCOAST shares some of its data using an ArcIMS Image Service, which can be viewed in ESRI ArcMap and ArcExplorer applications (NOAA Coastal Services Center, 2009). The International Coastal Atlas Network is developing a coastal atlas mediator prototype to demonstrate the use of OGC CSW, WMS and WFS to search and view data from any coastal web atlas that is connected, which is discussed later in this book (Lassoued *et al.*, 2008).

CONCLUSION

With the growth of web mapping software, a variety of coastal web atlases and web mapping applications have been developed to address the needs of coastal communities and managers. Most of these CWAs have been developed independently and have unique ways of delivering data and information. Comparing the similarities and differences in approaches to sharing content and tools provides useful information for those aiming to develop new CWAs, as well as for existing CWAs to consider new approaches in their design.

REFERENCES

Atlantic Coastal Zone Information Steering Committee & GeoConnections. (2009). *Coastal and Ocean Information Network Atlantic*. Retrieved May 7, 2009, from Coinatlantic web site: http://coinatlantic.ca/

Bartlett, D., & Bruce, E. (2001, July). *An Evaluation of Progress in Coastal Policies at the National Level. A Transatlantic and Euro-Mediterranean Perspective*. Paper presented at Quality Control in Coastal GIS, NATO Advanced Research Workshop, Ljubljana, Slovenia.

Belpaeme, K., & Konings, P. (Eds.). (2004). *The Coastal Atlas Flanders-Belgium*. The Coordination Center for Coastal Zone Management. Retrieved May 7, 2009, from Kustatlas website http://www.kustatlas.be

British Columbia Ministry of Sustainable Resource Management (BCMSRM). (2009). *COIN-Pacific Web Mapping Application*. Retrieved June 29, 2009, from Land Information B.C. website http://aardvark.gov.bc.ca/apps/coin/

Canessa, R., Butler, M., Leblanc, C., Stewart, C., & Howes, D. (2007). Spatial Information Infrastructure for Integrated Coastal and Ocean Management in Canada. *Coastal Management*, *35*(1), 105–142. doi:10.1080/08920750600970537

Center for Coastal Resources Management. (2009). *Shoreline Manager's Assessment Kit*. Virginia Institute of Marine Science. Retrieved on July 3, 2009, from VIMS website: http://ccrm.vims.edu/gis_data_maps/interactive_maps/disclaimer_smak.html

Coastal & Marine Resources Centre (CMRC). (2009). *The Marine Irish Digital Atlas*. University College Cork. Retrieved May 7, 2009, from MIDA web site: http://mida.ucc.ie

Dwyer, N., O'Dea, L., & Cummins, V. (2003, October). *The Marine Irish Digital Atlas: A Web Portal to Coastal and Marine Data in Ireland.* Paper presented at CoastGIS, Genoa, Italy.

European Environment Agency. (2009). *Water Watch: Eye On Earth.* Retrieved July 3, 2009, from EEA web site: http://eoe.eea.europa.eu/

GeoBC Spatial Analysis Branch. (2009). *British Columbia Coastal Resource Information Management System.* Integrated Land Management Bureau, Ministry of Agriculture and Lands. Retrieved June 29, 2009, from Spatial Analysis Branch B.C. website http://www.ilmb.gov.bc.ca/cis/coastal/others/crimsindex.htm

Goodchild, M. (2007). Citizens as sensors: The world of volunteered geography. *GeoJournal, 69*(4), 211–221. doi:10.1007/s10708-007-9111-y

Haddad, T. C., Wright, D. J., Dailey, M., Klarin, P., Marra, J., Dana, R., & Revell, D. (2006). The tools of the Oregon Coastal Atlas. In Wright, D. J., & Scholz, A. J. (Eds.), *Place Matters: Geospatial Tools for Marine Science* (pp. 133–149). Corvallis, OR: Oregon State University Press.

Institute for Natural Resources. (2005). *North Coast Basin Prototype Web site Project Completion Report.* Oregon State University. Retrieved May 7, 2009, from OSU website http://inr.oregonstate.edu/download/OWEB_project_completion_report.pdf

Irish Marine Institute. (2009). *Marine Data Online.* Retrieved May 7, 2009, from Marine Data Online website http://www.marinedataonline.ie/

Lassoued, Y., Wright, D. J., Bermudez, L., & Boucelma, O. (2008). Ontology-based mediation of OGC Catalogue Service for the Web requests: A virtual solution for integrating coastal web atlases. In *Proceedings of ICSOFT 2008, the 3rd International Conference on Data and Software Engineering*, Berlin: Springer.

MarineMap Consortium. (2009). *MarineMap: Decision Support for Marine Spatial Planning.* Retrieved July 3, 2009, from marinemap web site: http://marinemap.org/

Maritime & Coastguard Agency & DEFRA. (2006). *Coastal and Marine Resource Atlas - Final Report.* Retrieved May 7, 2009, from MAGIC web site: http://www.magic.gov.uk/Me1302_4030_FRP.pdf

Maritime & Coastguard Agency & DEFRA. (2009). *UK Coastal and Marine Resource Atlas.* Retrieved May 7, 2009, from MAGIC web site: http://www.magic.gov.uk/camra.html

Netcraft. (2009). Retrieved July 3, 2009, from Netcraft website http://news.netcraft.com/

New York Department of State. (2009a). *Coastal Resources Online.* Division of Coastal Resources. Retrieved June 29, 2009, from NYS DOS web site: http://nyswaterfronts.com/index.asp.

New York Department of State. (2009b). *New York Coastal Atlas.* Division of Coastal Resources. Retrieved May 7, 2009, from NYS DOS website http://nyswaterfronts.com/maps_relief.asp

New York Ocean and Great Lakes Ecosystem Conservation Council (NYOGLECC). (2009). *New York Ocean and Great Lakes Atlas.* Retrieved July 3, 2009, from NYOGATLAS website http://nyoglatlas.org

NOAA Coastal Services Center. (2009). *nowCOAST: Web Mapping Portal to Real-Time Coastal Observations and NOAA Forecasts.* Retrieved May 29, 2009, from NOAA website http://nowcoast.noaa.gov/

NOAA Office of Ocean & Coastal Resource Management. (2009). *The Many Faces of Coastal Issues.* Retrieved May 7, 2009, from NOAA website http://coastalmanagement.noaa.gov/issues/welcome.html

North Sea Interdepartmental Consultation Directorate. (2009). *North Sea Atlas*. Retrieved June 29, 2009, from NordZeeLoket website http://www.noordzeeatlas.nl/en/index.html

O'Dea, L., Cummins, V., & Dwyer, N. (2004, March). *Developing an Informational Web Portal for Coastal Data in Ireland: Data Issues in the Marine Irish Digital Atlas*. Paper presented at Oceanology International London 2004, 18 March, London, England. Retrieved June 24, 2009, from MIDA web site: http://mida.ucc.ie/assets/documents/oceanology2004_mida.pdf

O'Dea, L., Cummins, V., Wright, D., Dwyer, N., & Ameztoy, I. (2007). *Report on Coastal Mapping and Informatics Trans-Atlantic Workshop 1: Potentials and Limitations of Coastal Web Atlases*. University College Cork, Coastal & Marine Resources Centre: Cork, Ireland. Retrieved May 7, 2009, from the ICAN web site: http://ican.science.oregonstate.edu/node/47

O'Dea, L., Dwyer, N., Cummins, V., Dunne, D., & Tuama, Ó. É. (2004, May). *Improving Access to Coastal Information: Metadata in the Marine Irish Digital Atlas*. Paper presented at ECO-IMAGINE Conference, Seville, Spain.

Ocean Data and Information Network for Africa (ODINAFRICA). (2009). *African Marine Atlas*. Retrieved May 29, 2009, from African Marine Atlas website http://www.africanmarineatlas.net/

Oregon Coastal Management Program (OCMP), Oregon State University, & EcoTrust. (2009). *Oregon Coastal Atlas*. Retrieved May 7, 2009, from OCA website http://www.coastalatlas.net

Oregon State University. (2009). *North Coast Explorer*. Retrieved May 7, 2009, from North Coast Explorer website http://www.northcoastexplorer.info/

Purce, D., Wood, M., Kueck, J., Taylor, K., O'Dea, L., Wright, D., & Haddad, T. (2009). *Proceedings of the West Coast Coastal Atlas Workshop*, Seattle, WA. Retrieved Jul 1, 2009, from ICAN website http://ican.science.oregonstate.edu/westcoast/

Solutions, D. M., & Go, M. O. O. S. (2009). *Gulf of Maine Mapping Portal*. Retrieved May 7, 2009, from GOMMAP web site: http://www.gommap.org/.

Washington State Department of Ecology. (2009). *The Washington Coastal Atlas*. Retrieved May 7, 2009, from Washington State Department of Ecology website http://www.ecy.wa.gov/programs/sea/sma/atlas_home.html

Wright, D., Blongewicz, M. J., Halpin, P. N., & Breman, J. (2007). *Arc Marine: GIS for a Blue Planet*. Redlands, CA: ESRI Press.

KEY TERMS AND DEFINITIONS

Watermark: A semitransparent image of a logo or text that displays over the map, for the purpose of crediting the map source and preventing the misuse of the map image.

Overview Map: A small map which shows the full extent of the geographic area covered by the coastal web atlas. It is typically displayed on or next to the main map in the atlas, with a notation showing the current location that the user is viewing.

Scale Factor Limit: A high or low scale set by the atlas developer for specific datasets so that atlas users cannot view the datasets in the map above or below those scales when they zoom in and out of the map.

Volunteered Geographic Information: Geographic information which is voluntarily collected from or submitted by the general public and displayed in a web mapping application.

Identify Tool: A map tool that is used to identify map features (points, lines, polygons) and the tabular information behind them.

Query Tool: A map tool that is used to search attribute tables for specific contents, such as a place name or category of data.

Abstract Metadata: A minimal summary of a geographic dataset's metadata, which typically contains data name, data owner, year data was created and brief summary of data contents.

Discovery Metadata: A mid-level summary of a geographic dataset's metadata, providing more detail than abstract metadata. It may contain additional information such as keywords, scale at which the data was created, and bounding coordinates for the geographic extent of the dataset.

Full Metadata: Complete metadata documentation for a geographic dataset which meets a specific metadata standard.

Chapter 3
Coastal Web Atlas Implementation

Tanya Haddad
Oregon Ocean-Coastal Management Program, USA

Elizabeth O'Dea
Washington State, USA

Declan Dunne
University College Cork, Ireland

Kuuipo Walsh
Oregon State University, USA

ABSTRACT

Coastal Web Atlas (CWA) design may be based around interactive maps that provide users with access to rich information about the geography of the coastal zone. There are few firm rules about what elements should make up a CWA, or how those elements should be developed. Developers seeking to build a successful product for their audience should begin with knowledge of the needs and capabilities of their users, and design the elements of their CWA accordingly. In general, principles of good web design and usability should be adhered to at all times.

INTRODUCTION: GETTING STARTED

For those embarking on a coastal web atlas (CWA) project, there are many factors to consider in the early stages of design and planning. Primary among these is being knowledgeable about the needs and capabilities of the CWA's user community, and designing the contents of the CWA to meet those needs. The purpose of this chapter is to provide CWA developers and managers with a range of considerations for implementing various components of a coastal web atlas. Each CWA situation and audience will be different so there are few hard and fast rules to follow, but it is hoped that the range of design considerations presented here will improve awareness of the issues CWA designers face and provide pragmatic advice on potential paths to take.

In terms of getting started, the first recommendation to CWA project planners is to set aside time at the beginning of the project to thoroughly

DOI: 10.4018/978-1-61520-815-9.ch003

Copyright © 2011, IGI Global. Copying or distributing in print or electronic forms without written permission of IGI Global is prohibited.

consider the goals of the project, the nature of the audience it seeks to serve, and the long term plans for project support, both technical and financial. Discussion of these issues early in the project will improve the odds that major problems can be avoided.

O'Dea *et al.* (2007) have an excellent overview of critical considerations upon which project designers should reflect. These considerations (summarized below) are posed as questions that when answered and acted upon by a CWA project team should help form the basis of a successful project

MAPS

As stated in Chapter 2, a map area is at the heart of any coastal web atlas, as this is where the main geographic and thematic scope of the atlas is showcased to users. To some extent one of the early decisions a CWA design team must make is to determine how dominant the maps portion of their project will be. For some atlas projects the interactive map is the central functional focus of the website. In others, it is simply a portion of the atlas content offerings. Neither approach is faulty, and either can lead to a feature- and information-rich resource for CWA users. From a technical standpoint, there are many different methods by which a host can feature maps in a web atlas. These range in complexity of implementation and have different implications for amount of effort involved for atlas hosts. There is also a range of options for the distribution of computing workload between the atlas web server and the client browser of the atlas users. In this section, we cover the typical range of map implementations and lay out the pros and cons of each option.

Interface Interactivity (Static vs. Dynamic Interactive Maps)

Menno-Jan Kraak (2001) classified maps on the world wide web as being either "static" or "dynamic", where a static map is most like a paper map, with static extents and cartography, while a dynamic map is a map that is capable of moving, or being moved by the user and cartographic contents can be changed by the addition or subtraction of data layers. They further state that both static and dynamic map types can be either "view only" where users can only see or browse the map contents, or "interactive", where an interface is provided to allow users to click on the map or tools related to the map, that link to or calculate further information beyond what is readily visible in the map itself. These interactive maps are especially valuable in the context of a CWA, as they provide a mechanism for using maps as a gateway to detailed information about coastal areas.

What follows is a discussion of the various methods for implementing both static and dynamic maps and the typical issues associated with each type. It should be noted that no single approach below is ideal for all circumstances, rather each option is suited to specific situations, and designers are encouraged to consider which option is most appropriate for their particular setting. As always, the complexity of the map interface employed

Table 1. Questions for CWA project teams to consider

- What is the purpose of the atlas?
- Who is the atlas audience?
- What are their skills?
- What spatial data will it contain?
- Are there opportunities for data sharing?
- What functionality is necessary?
- What additional information will be included?
- Which software should be used?
- What operating system should be used?
- What technical and data standards should be met?
- What resources are available for development?
- How should the web site and navigation be organized?
- How will the content be managed?
- How should data be searched?
- How will the atlas be backed up?
- Is the design scalable and flexible?
- What institutional / financial support is available?
- How will the atlas be sustained and updated in the long term?

O'Dea et al. (2007)

should consider the browser, computer memory, and network connection speeds of the specific atlas user audience (O'Dea *et al.*, 2007).

Static Pre-Composed Maps

At the most basic level, simple static map images can be composed offline in any GIS or illustration software of the host's choosing, and embedded in the web page as any other type of image might be. This category would also include traditional PDF documents that contain static maps, which are linked into an atlas.

Pros: full control of cartography choices, low burden on atlas server, low burden on client browser.
Cons: large amount of cartographic work up front that might not scale well if geographic scope of the atlas changes over time. Maps are static rather than dynamic, so user interaction with the maps is minimal or limited to interactions inherent in HTML image map tags.

Static Maps Composed on the Fly

An enhancement from pre-composed static maps is maps generated by a mapping engine such as Minnesota Mapserver or ArcIMS. The maps are still static in the sense that the user is not given the ability to navigate the geography via a map interface, but by virtue of being drawn directly by the sever, there is a higher level of interaction possible for both atlas host and user. This map type option requires the ability to support a mapping engine on the atlas server, but even minimal implementation provides much increased scalability, due to the ability of the engine to respond to future changes in atlas geographic scope, scale or theme of map. Typically this approach is excellent for embedding maps in a report template such that as the scope of the report changes, the static maps embedding in the report can also change as appropriate.

Pros: Provides scalability in terms of ability to generate new future maps with less effort than pre-composed static maps. Maps can be drawn directly from latest available data. Permits higher level of customization of static map contents. Low burden on client browser.
Cons: Requires server support for a mapping engine. Requires up front effort to configure cartographic rules for the mapping engine to follow so that maps produced have a consistent look and feel. Often cartographic quality is lower than what can be obtained in traditional desktop software. Load on server CPU is higher than previous option, especially at times of web site congestion.

Dynamic Maps Composed on the Fly

Most first generation interactive map websites (for example, those made with ESRI's ArcIMS software) implement a mapping engine on the server that responds to map requests from an interface on the client's browser. The browser interface in turn allows the user to specify the composition of the next map to be displayed. Simple interface examples might include pan and zoom navigation tools, while more complex interface examples might include the ability to alter the make up of the map itself (e.g., switching layers on and off). Further utilities such as the classic "Info" tool are also possible if the map engine supports query responses. Since such interfaces have more coded complexity, they are a higher burden on the client's browser and server.

Pros: Highly scalable for new map creation because once the server, data, and cartography are set up with a chosen client interface, maps can be made at any allowable scale, extent and with any combination of data layers made available. Maps can be drawn directly from latest available data. Higher sense of user control over map display.
Cons: Highest burden on server of all possible configurations since every single user action results in a new request to the server. As a result, if

simultaneous web traffic is expected to be high, server hardware must be up to the task. Can also be a high burden on the client side browser if many complex user tools or widgets are included in the client interface.

Dynamic Pre-Composed Maps

This option is characteristic of second-generation commercial web mapping websites such as Google Maps, or open source interfaces such as Open-Layers. Typically such sites combine a dynamic user interface of varying feature complexity with pre-composed maps generated and cached in advance by a mapping engine, and are served as pre-cut map "tiles" at predetermined zoom levels. Such pre-composed tiles cannot be altered on the fly, but may be layered with other data on the fly if transparency and opacity is possible via the browser and provided interface.

Pros: Map navigation can feel extremely responsive to the user. Cartography is often improved over maps composed on the fly as scales are often fixed to specific zoom levels. This means more processor-intensive rendering can be employed during map generation when building a map cache. Once a cache has been created, this option is reasonably light on the server side burden.
Cons: Since the map is cut into both horizontal and vertical tiles that must be assembled on the client side by the map interface, this option is typically the heaviest load on the client's browser. Load times will be high the first time a user visits the map, especially if the user's computer is low on memory and CPU capacity. The user's browser cache can fill up if not set at a high enough level, and users must learn to empty their cache from time to time to ensure up to date content display.

Map Interface Features

Overview Map

Whenever the user has an option to pan and zoom into the contents of a map, the possibility emerges for the user to become "lost" within the map and lose his/her sense of place within the larger geography depicted in the whole map extent. This is a particular danger whenever the user is unfamiliar with the local terrain being displayed in the map. This situation emerges frequently in coastal areas in particular whenever users are exploring areas that are offshore, where the reference benefits of landmarks are not present. This problem of user disorientation can occur whenever large datasets are navigated. In the case of interactive maps, the most common solution is the provision of a "reference" or "overview map" which displays the entire extent of the navigable map in miniature with a highlight box that outlines the footprint of the current map extent being viewed in the main map viewer. Some designers choose to provide these overview maps as part of the permanent map interface layout, while others make it available via a clickable button to display the overview, and then dispatched when it is no longer needed.

Scale Bar

A completed maps depicts a scale bar (or other statement of map scale) at all times. In general, the statement of scale is not a feature that should be hidden from view behind a button, as users have no basis for interpreting the magnitude of visual map elements if scale is missing. Most map interfaces place scale in a corner of the map (often a lower corner) such that it does not obstruct any major portions of the map view.

Coordinate Display

Some CWA maps display "live" coordinate information relating to the position of the cursor on the map. Typically, since the cursor can be moved frequently, this information is usually calculated and provided on the client side. This implies fairly complex or "thick" client software and is therefore not recommended for CWAs that have audiences with limited browser/computer memory capabilities or network connection speeds. This type of feature can also be challenging to implement if the data in the map display are in a map projection different from geographic decimal degrees. Users typically expect to receive coordinate information in geographic decimal degrees of latitude and longitude, but projected data may not use decimal degrees and may be referenced to custom coordinates. In these instances, calculations must be made by the map interface to translate from custom coordinate space to geographic decimal degrees for display, further increasing the burden on the client capabilities of the user.

Watermarks and Copyright Information

When data owners require that a watermark or copyright statement be provided as part of any map containing their data, these elements are typically included as overlays on top of the composed map, often in a corner. While watermarks typically mark the map with the atlas name or logo and are static, copyright information may change depending on the geographic data and/or area displayed in the map. In some cases where multiple data owners require their name be displayed on the map, there may need to be multiple copyright lines. Atlas developers should be aware of when copyright statements are required and viewable on the map. Programming code can be generated to display correct copyright statements when necessary.

From a map server's perspective, a watermark or copyright statement is not much different technically from any other data layer, except where the position of the watermark or copyright statement is relative to the map frame rather than to coordinates in the real world. In the ArcIMS world, such layers are referred to as acetate layers, while in Minnesota MapServer, these layers are simply a form of annotation layer. In any case, implementation of watermarks and copyright information is relatively simple in most map server technologies.

Map Size on Page Area

Experts in information display and web and map usability alike refer to a concept of a ratio between the area of information content on a page to the area dedicated to navigation or interface controls and other layout mechanisms (Tufte, 1983; Nielsen, 2000; Harrower & Sheesley, 2005). In general the recommendation for a successful interface design is to maximize the ratio of information to other spaces on any given page or screen. In the case of an interactive map display, the primary information area is the map itself. Any design that can maximize the size of a map relative to the size of the user screen without eliminating access to navigation and other interface controls is likely to be more successful than one that ignores this recommendation.

GEOGRAPHIC DATA CONTENT AND DISPLAY

Having discussed the multiple options for displaying maps in a CWA, it is necessary to pay attention to the data that make up those maps. For the most part the most accessible data in a CWA are those which have been selected for display in the various maps embedded throughout the website. These data can be of various types, and organized in various ways that might make them easier for the user to interpret.

Data Types

As stated earlier, the geospatial data types that may be depicted in an online map span the full range of GIS file types. Depending on the map server software employed by a CWA, not all possible file types may be supported. Any map server software that supports the OGR (OpenGIS Simple Features Reference library) and GDAL (Geospatial Data Abstraction Library) libraries for vector and raster data respectively, likely has access to the largest range of data file types. For CWA developers not familiar with the concept of a "library," the following definition may be helpful:

"In computer science, a library is a collection of subroutines or classes used to develop software. Libraries contain code and data that provide services to independent programs." (Wikipedia, 2009)

The list of data formats supported by OGR (http://www.gdal.org/ogr/ogr_formats.html) and GDAL (http://www.gdal.org/formats_list.html) may be consulted online. If the server software employed by a CWA does not accommodate a specific desired file type, one option is to use GDAL or OGR binaries to transform the data into some other data type that is supported.

Data Availability to User

While some CWAs may have large archives of data available for download, there are challenges inherent in displaying large amounts of data via a single map interface in ways that do not detract from the utility of the map, or obscure useful information. As a result, CWAs address data availability in different ways:

Making all Data Available in the Map

Pros: The user can view any data that is available in the atlas, combined with any other data of their choice.
Cons: There is strong potential of too much data being displayed at one time, thus making the map difficult to interpret. The more layers added to the atlas, the more difficult they are to find in the list of layers.

Grouping Data by Theme

Pros: The user can more easily select the theme they are interested in, rather than hunting for specific layers.
Cons: Thematic grouping is done by the atlas developers, so the user is reliant on how the developers grouped the data. The user generally cannot add data from other themes to the map, although it is possible to design this flexibility into the system.

Limiting the Number of Layers Viewable at One Time

Pros: This minimizes cartographic confusion by avoiding too many layers displayed at once.
Cons: The limit set by the atlas developers may be too limiting for the user.

Showing High-Interest Data and Allowing All Other Data Sets Available in Downloadable Archive Format

Pros: The web map is kept simple, displaying the most popular layers in the map.
Cons: The user can only view a limited set of layers in the map. To view other data they must download and view them in a desktop mapping software, rather than viewing on the fly in the web map.

Enabling Users to Customize the Data Set Lists of What Can be Viewed in the Map

Pros: This option provides the user with the most control by searching the atlas database for data of interest and adding them to the list of high-interest data in the map. Such a solution could represent the most optimal combination of interface characteristics.
Cons: This requires more coding to enable the adding and removing of data. It can also have

implications for other parts of the map interface as newly added layers must also be added to legend and layer lists. This option also requires trained or savvy users to prevent confusion or frustration.

The choice of how to make data available to users is mostly a preference choice for the CWA designer, though certain practical matters do play a role. CWA map servers that compose each map on the fly on the server before sending the map to the browser generally have more latitude to allow large lists of data to be accessed via their map interface, while CWA maps designed to layer map data together on the client side (e.g., using tiled layers) may run into client memory limitations beyond a certain number of layers. A work around to the latter situation is to group thematically similar layers in to either aggregate layers (thus reducing the total number of layers of tiles needed for a map) or to group thematically similar layers into separate thematic maps, and provide the user with the ability to change map theme from within the map interface.

Data Sources and Increased Performance

Once decisions have been made as to what data will be displayed to atlas users via maps, the next decision involves how the data files themselves are arranged and provided to the map server. In terms of performance there are many things that can be done to improve the speed of map renderings for the Web. The first of these is moving the data as close to the server as possible. Unless a data layer is truly dynamic and changes often, the fastest option is simply to have a dedicated copy of the data on the server itself. The obvious exception to this principle is when a data set is being ingested from an outside service such as WMS, WFS, etc. Users interested in employing such services should consult the section at the end of this chapter for more information on that topic.

Data sets that are highly detailed and cover a large area can be optimized by building spatial indexes and cutting data sets into tiles to reduce file sizes. A tile index is then used to access the files so that as few files are opened as possible when only a small area is being viewed. More generalized overview layers can be used when the view is zoomed way out to cut down on the complexity of files that need to be opened when lots of detail will not be visible to the user. With regards to image files, while compressed files such as MrSID (Multi-resolution Seamless Image Database) and ECW (Enhanced Compressed Wavelet) can be used with acceptable performance in a desktop environment, these compression formats are typically ill-suited for use with interactive map servers due to the high amounts of memory required to open and uncompress the files each time a part of the image is requested. There also may be licensing issues with having such files located on a server. Therefore unless specialized image services are available for use, and since hard disk space is relatively cheap, it is preferable that images are stored in uncompressed formats, tiled together with indexes and swapped for overview images at higher zoom levels.

A final consideration regarding source data and performance relates to the subject of map projections. While desktop GIS users may have become accustomed to software programs that allow different datasets to be re-projected on the fly into a common coordinate space, relying on this sort of feature in an online setting is not recommended because it greatly reduces speed or performance. There is simply no reason to ask the server to perform this function in addition to all the other tasks that are required to serve a map to users.

Cartography

While a book on coastal web atlas implementation is no place to re-cap the lessons of centuries of cartographic principles, there are important considerations unique to online mapping of which experienced desktop cartographers should be aware. A primary issue is that in a dynamic map environment where users are allowed to pan and zoom at will, cartography ceases to be the composition of a single well-crafted map image. Rather, for every scale viewable to the user, cartographic rules specifying rendering information such as layer color, symbolization, and label conventions must be set up to dictate to the server the most aesthetic rendering of each data set viewable at that scale. When map images are rendered on the fly and do not limit available scales, inevitably some scales will have less optimal cartography than others. Map environments that do limit available scales do not have this problem. For example, if an interactive map has only 12 available scales, then 12 sets of cartographic rules each optimized to a specific zoom level will ensure that maps at every scale are as understandable as possible.

Some cartographic principles related to map performance have already been touched upon above. Besides the indexing and tiling of highly detailed datasets, CWA designers should also employ scale factor limits to constrain the scales below or above which certain data are viewable, and to substitute data versions that have more or less detail as appropriate to the scale being requested. In situations where map caches are being served, it is common to limit the total number of viewable scales, and matching scale dependency levels to cache scale levels is optimal. Typically cached map layers are groups of data sets that would be expected to receive high volumes of traffic and which do not contain data that change very often. This situation is common in base maps used for geographic reference.

MAP LEGENDS AND LAYER LISTS

Whenever a map is provided to a user, a legend or key is required to de-code any special symbols or colors used. In addition, users of traditional desktop GIS are familiar with the concept of a layer list that shows the drawing order of the layers that make up a map, and allows individual layers or groups of layers to be turned on and off as desired. In an online environment both these elements present challenges, in that a legend or layer list is yet another feature that will take up valuable screen real estate. While one obvious option is to combine these two elements into one interface feature, interactive map interface developers need to consider how to deal with this challenge in a way that works to the benefit of their specific audience of users. As the design of the legend and layer lists in a CWA are a key interactive component of the atlas, it is useful for atlas developers to carefully study existing examples and find an approach that works best for the targeted atlas audience. It may even be beneficial to provide a flexible design that gives users multiple options, such as including a merged legend and layer list as well as a button that opens a window or tab that displays only the legend.

To some degree, the decision of how to include a legend or layer list is related to the overall approach to maps that a CWA is taking. If the objective is to serve users that are familiar with a traditional desktop GIS then a traditional layer list may be acceptable. Layer lists are also inevitable if there is a need to make long lists of data layers available to users who need to be able to control what layers are viewable at any on time. In both cases, it may be optimal to make the list available in a toggle-on/toggle-off way that removes it from view when not being used, thus freeing up screen space for the main map view. On the other hand, if the objective of a map interface is to serve a wide spectrum of map users, few if any of which have any exposure to a GIS environment, designers need to reframe their thinking about how map layers are perceived by their users. CWA map interfaces that cater to the wider public may want to model the example set by most second generation commercial map websites (e.g., Google Maps). Such sites typically limit the total number of layers to a very small number that are critical to supporting the business purpose of the map, and separate these layers into two groups:

- Layers that are part of base maps, and
- Operational overlays or "informational" layers that sit on top of base maps and usually feature some sort of interactivity beyond simple panning and zooming.

Often such map frameworks limit the layers that can be turned on and off to the second group only, and limit the total number of operational layers to a very small number required to meet the objectives of an individual map application. While experienced GIS users may balk at the idea of limiting user control over the map contents in this way, millions of satisfied users of commercial web map sites are a testament to the fact that legends and layer lists are not always missed.

TOOLS FOR MAPS, SEARCH, AND DECISION SUPPORT

The concept of a "tool" in the software world is wide ranging and can take many forms, as discussed in Chapter 2. In the CWA web map world the word is most commonly used to refer to one of two concepts:

- Map interaction tools that are part of a web map interface, or
- Standalone tools that are built to perform standard operations on specific data sets and thereby streamline analytic inquiry or support a decision making process.

In this section we will explore the basics of each of these concepts of a tool.

Map Tools

Standard interface tools in any modern interactive map typically include at a minimum "Zoom in/out" and "Pan/re-center" tools. In interfaces that attempt to reproduce a desktop GIS type interface these functions are often provided as buttons in a toolbar, and are often accompanied by a "zoom to full extent" button. Map interfaces that are not descended from the desktop GIS universe commonly provide different mechanisms for the same map navigation functions. For example, vertical navigation (zooming) is often provided using a slider bar that depicts where in the range of scales the user is and allows movement to other scales via sliding or jumping to different places along the bar. Horizontal navigation (panning) is achieved through navigation arrows at the edges of a map, or assigned to the default cursor actions, where click and drag will result in a pan, and double click will result in a re-centering of the map.

According to Harrower & Sheesley (2005), the design of map interface elements (such as basic map interaction tools) can be assessed in two major areas: "*functionality* (what the tool is capable of) and *efficiency* (how well it supports those tasks, or how much work is required to achieve the same outcome)." They lay out four main functional criteria (see below), and two main efficiency criteria that relate to panning and zooming as bedrock interaction capabilities presented in most interactive map interfaces. They conclude by noting that research in human-computer interaction points to the idea that success in interface design is based on both the nature of the users and the nature of the tasks they need to accomplish via the design. Therefore different interaction methods work best when paired to particular users and map tasks. Based on this knowledge, CWA designers would be advised to consider their audience level of comfort with various map navigation options, before choosing a path.

Many web map interfaces incorporate additional functionality beyond the basic pan and zoom

Table 2. Map interface criteria

Map Interface Functional Criteria:
• Sequential versus non-sequential map browsing *(e.g., must users go through a sequence of pan or zoom levels or can they "jump" between levels?)*
• User defined or pre-defined map browsing precision *(e.g., confined zoom levels or infinite scale options)*
• Availability of local-global orientation cues *(e.g., "you are here" overviews, or reference maps)*
• Direct or indirect manipulation of map via interface controls *(e.g., how directly does the map respond to manipulation via the interface tools?)*
Map Interface Efficiency Criteria:
• Interface workload *(e.g., how much work does the user have to do?)*
• Information to Interface ratio *(e.g., how much space is dedicated to the map itself? See Chapter 18 for more on this topic.)*
Harrower & Sheesley (2005)

navigational functions of interactive maps. The next most common map tool type after navigation tools are feature identification tools. Such tools operate via a button or simple click on operational "info" layers and act to highlight the feature clicked upon and reveal further information about the selected location or map feature from the related attribute tables (discussed further below). In essence, the identification tool marries hyperlink functionality with online maps, thereby creating an incredibly versatile framework for distributing information about geographic locations. As such, an identification tool, however it is implemented, is a highly recommended addition to any CWA map interface. Typical implementations limit users to selecting map features one at a time or multiple features simultaneously from a single layer using the mouse pointer. Advanced implementations work in a drilldown mode, allowing selection of all visible map features directly below the point or area selected. Offering users an option for selecting a specific layer or multiple layers provides useful flexibility.

Beyond the functions described above, the array of other possible map interface tools begins to draw heavily on functions borrowed from the traditional desktop GIS world. Such functions include tools that allow search of map features, measurement of distances or area of map features, advanced spatial queries of map features, tools that incorporate tailored spatial analysis functions into a map interface, printing of custom maps, and others, as discussed in Chapter 2. While this list starts out with relatively simple map functionality, there is no question that as the level of complexity of a tools task increases, it becomes less appropriate for a wide audience. CWA designers targeting a wide audience of varying GIS experience are advised to keep complex analytic tools out of their main map interface to avoid user frustration.

Decision Support Tools

This brings us to the topic of tools focused on specific user needs, such as for decision support on a particular issue. O'Dea *et al.* (2007) state that "thought should be put into what [...] tools are necessary to meet the atlas purpose as well as users' skill levels", and that "the level of functionality may influence the software chosen for the atlas". Decision support tools are an area of considerable literature (e.g., Wright & Scholtz, 2005), the scope of which can only be touched upon here.

When a CWA design team sees the need for a specific type of analysis that they know will need to be repeated often by their audience, it is common to construct a tool so that it stands on it's own and is uncluttered by other analytic functions that do not pertain to the specific question at hand. Often such tools require custom coding since every decision process is somewhat unique. Typically, specialized tools can be expected to have a more limited audience than a standard map interface as analyses are often limited to sub-audiences with special interests. Also, since such tools often involve assumptions or require special instructions, they should be introduced with explanatory materials that give details about the issue the tool is intended to address, illustrate the sequence of steps a user can expect to encounter in an analysis, and prepare the user with background materials on how the tool functions analytically. In this way, stand-alone decision support tools can serve as both educational and utilitarian additions to a CWA interface.

Metadata Search

Once a CWA has amassed a data archive or network of data providers of some size, it becomes desirable to provide a metadata search tool, which enables users to search for various data sets, and learn about their qualities (e.g., title, abstract, purpose, date, area, etc.). This sort of tool is valuable for any CWA user seeking specific data, and becomes more valuable as the size of the archive grows. Having a search function typically requires that a database of metadata be designed and deployed. O'Dea *et al.* (2007) recommend that when implementing a searchable metadata database, attention should be given to what search elements would be the most useful to users (e.g., name, keyword, theme, date, region, etc.). Alternately CWA designers might opt to incorporate a product, such as GeoNetwork Open Source (http://geonetwork-opensource.org/) or ESRI's GIS Portal Toolkit, that is a catalog application specially designed to manage spatially referenced resources over the web. Such a catalog application allows single point management of both full and abstracted or "discovery" metadata and can simplify the many challenges that exist with incorporating metadata, into a CWA. Many participants in the ICAN prototype project have begun using this tool for its many metadata management features, as well as for the function it provides as a Catalog Service for the Web (CSW) feed, which allows external websites to also search the atlas data catalog.

ATTRIBUTE TABLES

Attribute tables associated with geographic data can provide a richer experience for CWA users; however there are a few usability concerns with providing access to data attributes that CWA designers should be aware of. Access to attribute information is frequently provided through an "Identify Tool". When a user selects a map feature with an Identify tool, the quality of information that appears in the resulting attribute table has an impact on the user experience. Confusion can result when attributes are displayed as codes instead of descriptive text. Generally, it is recommended that cryptic database codes be avoided in attribute display. Often these codes only make sense to specialists who are familiar with the data, but not to the general CWA user. Given that coastal atlases contain a very broad range of subject matter, it should not be presumed that even expert users will not be familiar with every subject in the atlas enough to know all possible codes and abbreviations. Tables that provide very little, if any, information can also frustrate the user. In other cases, tables can provide too many fields, which can overwhelm the user.

Atlases can address this problem in different ways. It is always useful for atlas developers to carefully select the most relevant and informative fields from the table and only display those. One approach to solving the problem of codes in attributes would be to provide a data dictionary or a link to metadata, which describes the codes in detail. Alternately a layer's attribute table could be modified by creating a new column with the code definitions, and this new field could then be displayed along with the code field to provide clarity. A third option is to simply use the Identify tool to query for a feature ID, and use this ID to drive a scripted template that provides a fully formatted HTML page of information relating to the feature clicked.

There are different schools of thought about where to display tabular results as described in Chapter 2. Some atlas designers choose to avoid pop-up windows as there are several usability problems associated with them. Pop-ups may be blocked by pop-up blockers, and users may be confused if nothing happens when they click on an item. Sometimes when they do appear, the pop-up fills the whole screen and can also cause confusion if the user wonders where the main map has gone. As long as CWA designers are aware of these issues, pop-ups may be used with the proper precautions. Alternatives include displaying the resulting table in a frame, pop-up bubble, or a map screen <div> overlay, on the main map page. Depending on the size of the chosen display location, CWA designers can be creative with the display of attribute data. Small locations such as pop-up bubbles are limited in what they can contain, but are nice in that they remain connected to the map, highlighting the feature to which the information pertains. Large spaces such as pop-up windows and screen overlay <div>s can provide more flexibility in design and additional space for links to metadata, a data dictionary, and a map which shows the highlighted feature or features.

Additional Non-Spatial Information

Apart from providing access to GIS attribute tables, a CWA may also provide users with coastal information that is not necessarily linked to a map. This is particularly the case if a CWA has a number of focus topic areas and there are background materials such as text, images, multimedia or informational charts and statistics relating to coastal topics that could improve visitors understanding about coastal resources or issues, and otherwise enhance the learning experience. In general, the hyperlinked, feature-rich nature of the web is well suited for communicating such informational extras to users, with a few standard caveats to CWA designers to respect the needs and technical capabilities of users.

In the case where information may logically be tied to a map, users who are map-centric often prefer to have access to informational extras from within the map. This can be achieved using the same suggested alternatives described above for displaying GIS attribute data from within a map (e.g., a frame, pop-up bubble or overlay on the main map page). Alternatively, when informational extras are lengthier than can logically be provided via a map, a nice alternative is to house them as stand-alone web pages that can be cross-linked from within the map page and vice versa. This approach has an added advantage in that informational material will be more likely to indexed by a search engine, and this in turn may drive new users to a CWA who might not otherwise have known it was there.

CONTENT MANAGEMENT

When a CWA is first being assembled, there is web content comprised of text and images that provides structure to the CWA and compliments the already discussed geospatial data holdings and maps. A simple way to deal with this information is to organize it into HTML pages. However much like the composition of static maps, the production of static HTML pages does not scale well as the web site grows large over time. When this happens or is expected to happen, the management of the web content typically needs to move to another level of organization.

This can be achieved via use of a formal content management system (CMS), which is typically a database framework that simplifies administration of web assets and provides a high level of control over what portions of a collection are published at any one time. CWA administrators that are unfamiliar with the concept of a CMS may have experience with them under a different name. CMS systems come in many varieties, from simple blog software (such as the popular WordPress), to bulletin board packages (such as phpBB), to full on portal systems (such as Drupal or Joomla). The packages mentioned here are all open source and widely used across the internet, and users interested in learning more about the range of CMS options may visit http://www.opensourcecms.com to view and test a very comprehensive list.

In any case, when an atlas moves to a formal CMS it is an opportunity to inventory the available web content and begin to introduce logical groupings into the collection if they are not already present. The CMS system employed will typically allow the appointing of specific content stewards who can then take charge of their portion of the atlas content and be responsible for keeping it up to date. In this way, use of a CMS can distribute the burden of content maintenance across a broad range of stewards and lessen the load on the core CWA staff without compromising security or quality. A CMS also provides easy remote access for system administrators so that content can be published or un-published from anywhere Internet access is available, providing valuable flexibility to atlas maintainers.

SOFTWARE AND HARDWARE TECHNOLOGY

One of the initial tasks facing a CWA development team is the choice of software and hardware technologies. Issues such as hardware and software platform, together with institutional and financial support, and longer-term sustainability and maintenance are important in determining appropriate technology. In general a web-mapping portal can be setup on a minimal hardware and software specification. After this, it is really a question of scalability, projected public usage of the website, and maintainability in determining your required specification.

Hardware

Disk space, physical memory, central processing unit (CPU), and Internet bandwidth are key hardware components. If the datasets required by the website are large, then you will need multi-disk storage and potentially fast SCSI-based disks with a configuration such as RAID 0. This configuration spreads or "strips" the data load over all disks in the array. However, if one disk in the RAID 0 configuration fails, then all data is lost and the website will potentially be down until the faulty disk is replaced and the website rebuilt from a backup. In contrast, a RAID 1 disk array configuration duplicates or "mirrors" data across paired disks. So if one disk in a pair fails, then the other disk in that pair continues to function as a single drive. However, RAID 1 does not have the performance benefit of RAID 0. A common compromise is RAID 5, which benefits from most of the striping performance of RAID 0 and the fault tolerance of RAID 1. For RAID 5, if one disk fails then the array still operates without data loss. If two disks fail, then data is lost.

Physical memory or RAM (random access memory) is also an important consideration for the website's hardware specification. If the physical memory required to render a map is too small, then temporary disk-based memory is used which significantly reduces performance. Temporary disk-based memory may also be used if there are multiple applications and users simultaneously competing for physical memory on the server. Therefore, sufficient RAM is advised. For memory greater that 4GB, a 64-bit operating system (or a 32-bit operating system which supports a physical address extension) is required. A CWA team is advised to consult operating system documentation regarding its memory support. The next hardware issue is peak traffic levels anticipated on the website. If the website will have many concurrent hits, then the server may not cope with this load. This is especially true for maps that are dynamically composed on the fly. In such cases, multiple servers with multiple CPUs, together with load balancing and proxy server caching techniques are required to help manage large loads. If large loads are anticipated, then pre-composed map tiles is a good approach to help lighten the workload on the server. A high bandwidth connection to the Internet is also advised. If your server is connected to the Internet via a low bandwidth connection, then your website may appear as slow and potentially frustrating to end users. This is especially true if there are many concurrent users and there is a high multimedia content on the website. Finally, a web hosting service can also be considered as an alternative to purchasing hardware and configuring the software environment. Some specialist hosting service providers support web mapping environments.

Web Servers and Operating Systems

After hardware comes the configuration of the software environment. At the lowest level is the operating system, which is typically either Windows or a version of UNIX (e.g., Linux, Solaris, Mac OS X, etc.). After this the foundation for setting up any website is a web server. Apache HTTP server and Microsoft Internet Information Services (IIS) are the most deployed web servers today. Apache takes up almost half the web server market, with IIS taking almost a quarter of the market (Netcraft, 2009). Apache is open source, and is supported on Windows and all major UNIX platforms, including Linux, Solaris, and Mac OS X. IIS is bundled with and works on the Windows platform. A standard web server is capable of delivering static HTML content. This is sufficient for delivering static maps. However if the atlas must deliver dynamic maps, then the web server must also be capable of generating HTML content and maps on the fly. Dynamic web frameworks include JavaServer Pages (JSP), Java Servlets, PHP, ASP, ASP.NET, and CGI. Many of these frameworks and environments must be downloaded, configured and directly plugged into the web server as an extension. However an exception is JSP/Servlet engines such as Apache Tomcat and Jetty. These can be directly used as a standalone web server. While this is acceptable in a development environment, it is better practice to have static content severed by the specialist web server, and JSP/Servlet content severed via a JSP/Servlet engine. A "connector" (such as JK or a proxy) can make these two servers appear as a single virtual server to the outside world.

Map Software

For dynamic map support, the CWA implementing team must select a web mapping software package. The software package can be either proprietary or open source. Both have advantages and disadvantages. Commercial software typically provides generic 'out of box' web mapping solutions, user manuals, plus options for consultancy support. However, disadvantages include the cost, proprietary technology, and the requirement of programming skills to customize the interface. On the other hand, the advantages of open source are it is free from commercial licensing fees, and is non-proprietary technology. Disadvantages include that it does not generally provide an 'out of box' web mapping solution, documentation can be inadequate in some cases, and it often requires HTML editing and scripting skills. ESRI ArcGIS Server and MapServer are good examples of proprietary and open source solutions respectively. The proprietary ArcGIS Server provides an 'out of box' solution where it is easy to publish maps. It is supported on Windows, Linux and Solaris. ArcGIS web browser viewers support standard JavaScript and also Rich Internet Application (RIA) technology such as Flex and Microsoft Silverlight. RIA technologies are proprietary and generally require a web browser plug-in, which is free. On the other hand, the open source MapServer only provides an engine for rendering dynamic maps, which is supported on Windows and all major UNIX systems. The CWA developer must then build a web site around this map rendering engine. However some independent frameworks and viewers exist that help to build MapServer applications on top of this engine. Examples include p.mapper and cartoweb.

Database

Web mapping software requires underlying datasets in order to render maps. Datasets can either be stored as files (e.g., shapefile for vector data and GeoTiff for raster data) or within a geodatabase or spatial database. A geodatabase takes advantage of RDBMS capabilities, where it is possible to generate dynamic and integrated queries on the spatial data. A geodatabase approach is usually required for data centers, but in many cases a file-based approach is sufficient for many basic CWA applications. ESRI supports all major file formats. It also supports geodatabase technology for vector and raster datasets using the ArcSDE component of ArcGIS Server. Underneath, ArcSDE supports true RDBMS including Oracle, SQL Server, and PostgreSQL. MapServer supports major file formats using the OGR and GDAL libraries. It supports geodatabase technology for vector data using RDBMS such as PostGIS/PostGreSQL, MySQL and Oracle Spatial. To help performance, data should be indexed and/or tiled. For example, MapServer documents optimization and performance improving techniques at (http://mapserver.org/optimization/).

Standards

An atlas should be designed in line with latest international standards. For example, the web content should be build to World Wide Web Consortium (W3C) standards for HTML, and CSS. For metadata, standards include ISO 19115/19139. ISO 19115 defines the general-purpose metadata standard for geographic information, while ISO 19139 defines how it is implemented in XML. In order for an atlas to remotely access maps, datasets and metadata hosted by third parties on the fly, OGC standards are important. OGC standards include Web Map Service (WMS), Web Feature Service (WFS), Web Coverage Service (WCS), and Catalogue Service for Web (CSW). WMS enables remote access to maps as raster images. Both WFS and WCS enable remote access to the actual data. WFS enables access to feature data, which includes the vector geometry (i.e. points, lines, polygons) plus its attributes. WCS enables access to gridded data (e.g., GeoTIFF containing data). In addition, the CSW standard enables discovery and access to remote metadata catalogues. Metadata standards and OGC services are described in more detail in Chapter 4.

Design and Testing

The CWA developers should test and refine their website against the current release of different web browsers such as Internet Explorer, Firefox, Safari, Opera, and Chrome. Also testing against the previous release of web browsers, and with browsers on different platforms (e.g. Windows, Linux, Mac and popular PDA or cell phone browsers) may be considered depending on how prevalent they are within the target audience of CWA users. In practice the implementation of W3C standards between web browsers can differ slightly, so testing and tuning of the CWA website is important before public release.

Another important consideration is the potential network speed of the target audience. People in developed countries have become accustomed to fast broadband Internet access. However, the rollout of broadband is less progressed in developing countries and even rural areas of developed countries. If the atlas audience includes non-broadband users then this should be taken into consideration for the atlas design.

Monitor size and resolution are other factors. A 640 x 480 resolution, followed by an 800 x 600 resolution, are the lowest common denominators for a computer monitor. PDA and cell phones displays have even lower resolutions. The general trend with computer monitors is the support of larger resolutions such as 1024 x 768 and higher. This trend will vary from developed to developing countries, so this should be considered in atlas design.

INTERFACING WITH WEB SERVICES AND THIRD-PARTY APPLICATION PROGRAMMING INTERFACES (APIS)

OGC Services

While local files and geodatabases are a reliable source of data, it is also possible for a CWA to interface with and visualize remote datasets on the fly using Open Geospatial Consortium (OGC) technologies. OGC services include Web Map Service (WMS), Web Feature Service (WFS) and Web Coverage Service (WCS). The main difference between these OGC standards is that WMS enables remote access to maps as raster images, while WFS and WCS enables remote access to the actual data. In addition, metadata that describes these OGC services and their datasets can be remotely accessed via the Catalogue Service for Web (CSW) standard. An example use case for a WMS service is a metrological center that makes their weather forecast maps available in real-time using the WMS protocol (i.e., as a WMS server). An independent website can then access, visualize and overlay this real-time map using this WMS standard (i.e., as a WMS client). Such interoperable and distributed maps are very important as they open up many possibilities and opportunities for web mapping applications and decision support systems to access specialized maps and products from third parties. While WMS enables access to map imagery, other OGC standards such as WFS and WCS can be used to remotely access the actual data. WFS enables access to feature data, which includes the vector geometry (i.e. points, lines, polygons) plus its attributes. WCS enables access to gridded data (e.g., GeoTIFF containing data). Web mapping software applications and viewers including ESRI ArcGIS Server and MapServer support access to and visualization of remote maps and data from OGC services, in addition to accessing maps and data from local files or a local database. Also, specialist web mapping frameworks exist for viewing OGC services. Examples include Mapbender and Internet Mapping Framework (IMF). Even virtual globe viewers such as Google Earth and NASA World Wind can connect to and visualize WMS services directly from the desktop. Chapter 4 discusses OGC technology in more detail.

Commercial Map APIs

CWA designers that do not have the capability to deploy an in-house map server have the option of embedding maps from one of the many commercial map application programming interfaces (APIs) that have proliferated since 2005. Companies such as Google and Microsoft allow the embedding of their base maps and other data layers and map controls into the home interface of any established web site with relatively simple code. These options are particularly attractive to CWA hosts who may have limited server resources, as the bulk of the burden of serving the maps is carried by the commercial API provider. For the most part the interfaces provided by these commercial services are relatively cross-browser friendly, have benefited from user testing by large numbers of users across the Internet, and are easy to use. Drawbacks include limited geospatial analysis capabilities for CWAs who are interested in more than just simple map browsing and simple map interactions.

Other concerns to be aware of regarding the use of commercial map APIs in Coastal Web Atlases can be summarized by examining the specific case of Google Maps and Google Earth, two of the most popular platforms. Issues that might be of concern include:

- In Google Maps there is an upper limit to the size of the KML files that can be displayed. This is both a Megabytes limit, as well as a "number of features" per file limit. This is a technical barrier basically imposed intentionally to avoid browser overload.
- There are printing restrictions on the Google Maps that contain imagery. In many places it is not possible to print a Google Map that contains imagery (controlled on Google end).
- In Google Earth (GE), there is a concern to do with licensing. To use GE for business use (including government business) projects and users are required to buy a license from Google Earth Pro.
- Google Maps and Earth also have a restriction that says, "You may not use the [product] in a manner which gives you or any other person access to mass downloads or bulk feeds of numerical latitude and longitude coordinates."
- Any data plotted by the user on top of and with reference to imagery that is not public domain may not be data to which the user can claim ownership (as it has been derived from a proprietary base).
- Similarly, in Google Maps, there is a concern regarding plotting data within the API based on the underlying street and base map data. Specifically there may be a problem if a CWA wants to allow a user to plot points, lines or polygons on a map from within the Google interface (as opposed to data from a layer that exported from a regular GIS).

In Europe, where data in the public domain is in shorter supply than in the United States, CWA projects that are tempted to use Google Maps or Earth or other commercial APIs as a base to display their data to the public should be careful to read the Google terms and determine if they are compatible with the licenses that exist for the desired overlay data. For example, the base data displayed in Google Maps comes from many private providers that may have license restrictions on what a user can do "on top" of their data, even if the data is displayed via Google Maps. See the Google maps site (http://www.google.com/intl/en_ALL/help/legalnotices_maps.html) for a list of countries that have special restrictions.

In some countries, the restrictions are narrow (e.g., in Denmark, one cannot make applications that are specifically like telephone books) while in other places the restrictions are so broad as to pose a huge problem (e.g., in Great Britain, the base data displayed is ultimately from the Ordnance Survey, therefore the license essentially means that anything plotted on top by a user could be considered a derived product, theoretically owned by the Ordinance Survey). Some local data providers only require the local copyright holder acknowledged on the map, which is relatively simple to satisfy.

In summary, if a CWA is in a position where base map data is readily available and the CWA stewards have license concerns about overlay data, then using a non-commercial API such as the OpenLayers API (http://www.openlayers.org) with internal data is the best choice to emulate a Google Maps experience. If a Google Earth experience is desired, the only real open source alternative is NASA World Wind, a sophisticated product that has analysis capabilities not found in Google Earth, presenting an attractive alternative for projects that are interested in applications that go beyond pure visualization.

For CWA projects that are lucky enough not to have concerns about any of the above restrictions, then the freely available commercial products are a great option for incorporating interactive maps into an atlas.

CONCLUSION

When discussing the elements of websites in general, and maps and interactive map interfaces as a main ingredient of a CWA in particular, this chapter is not specific to CWA implementation. Rather it pertains more generally to good design principles as might be sought in any web site. There are few firm rules about what elements should make up a map interface, or how those elements should be deployed. CWA developers seeking to build a successful product for their audience should begin with knowledge of the needs and capabilities of their users, and design the elements of their CWA accordingly.

REFERENCES

Cartoweb. (2009). Retrieved October 15th, 2009 from Cartoweb - Advanced Geographical Information System for the Web web site: http://www.cartoweb.org/

Drupal. (2009). Retrieved October 15th, 2009 from Drupal.org – Community Plumbing web site: http://www.drupal.org.

Environmental Systems Research Institute. (2009). Retrieved October 15th, 2009 from ESRI: GIS Software that Gives You The Geographic Advantage Web site: http://www.esri.com.

ESRI. (2009). Retrieved October 15th, 2009 from Wikipedia: http://en.wikipedia.org/wiki/Esri.

GDAL. (2009). Retrieved October 15, 2009, from GDAL - Geospatial Data Abstraction Library web site: http://www.gdal.org.

GeoNetwork Open Source. (2009). Retrieved October 15, 2009, from GeoNetwork Open Source web site: http://geonetwork-opensource.org.

Google Earth. (2009). Retrieved October 15th, 2009 from Google Earth web site: http://earth.google.com.

Google Maps. (2009). Retrieved October 15th, 2009 from Google Maps web site: http://maps.google.com/

Google Maps Legal Notices. (2009). Retrieved October 15, 2009, from Google Maps Legal Notices web site: http://www.google.com/intl/en_ALL/help/legalnotices_maps.html.

Harrower, M., & Sheesley, B. (2005). Designing better map interfaces: A framework for panning and zooming. *Transactions in GIS*, *9*(2), 77–89. doi:10.1111/j.1467-9671.2005.00207.x

Internet Mapping Framework. (2009). Retrieved October 15th, 2009 from Moxi Media Inc. - Internet Mapping Framework web site: http://www.moximedia.com/imf.html

Joomla! Content Management System. (2009). Retrieved October 15, 2009, from Joomla! - the dynamic portal engine and content management system web site: http://www.joomla.org.

Kraak, M. J. (2001). Setting and needs for web cartography . In Kraak, M. J., & Brown, A. (Eds.), *Web Cartography: Developments and Prospectus* (pp. 1–7). London: Taylor & Francis.

Library (Computing). (2009). Retrieved October 15, 2009 from Wikipedia web site: http://en.wikipedia.org/wiki/Library_(computing).

MapBender. (2009). Retrieved October 15th, 2009 from MapbenderWiki web site: http://www.mapbender.org.

p*mapper*. (2009). Retrieved October 15th, 2009 from p.mapper - a MapServer PHP/MapScript Framework web site: http://www.pmapper.net

MapServer Optimization. (2009) Retrieved October 15, 2009, from MapServer Optimization web site: http://mapserver.org/optimization.

National Aeronautics and Space Administration World Wind. (2009). Retrieved October 15th, 2009 from World Wind JAVA SDK web site: http://worldwind.arc.nasa.gov.

Netcraft. (2009). Retrieved October 15, 2009, from Netcraft – Internet Research web site: http://news.netcraft.com/

Nielsen, J. (2000). *Designing Web Usability.* Berkeley, CA: New Riders.

Nielsen, J., & Loranger, H. (2006). *Prioritizing Web Usability.* Berkeley, CA: New Riders.

O'Dea, L., Cummins, V., Wright, D., Dwyer, N., & Ameztoy, I. (2007). *Report on Coastal Mapping and Informatics Trans-Atlantic Workshop 1: Potentials and Limitations of Coastal Web Atlases.* University College Cork, Ireland, Coastal & Marine Resources Centre. Retrieved October 15, 2009, from ICAN web site: http://dusk.geo.orst.edu/cmrc/Cork06_CoastalWebAtlas_FinalReport_web.pdf

OGR. (2009). Retrieved October 15, 2009, from OGR Simple Feature Library web site: http://www.gdal.org/ogr/

Open Source Content Management Systems. (2009). Retrieved October 15, 2009, from Open Source Content Management Systems web site: http://www.opensourcecms.com

OpenLayers. (2009). Retrieved October 15, 2009, from OpenLayers: Free Maps for the Web web site: http://openlayers.org.

phpBB. (2009). Retrieved October 15th, 2009 from phpBB - Creating Communities Worldwide web site: http://www.phpbb.com

PostGIS. (2009). Retrieved October 15th, 2009 from PostGIS - Home web site: http://postgis.refractions.net

Postgre, S. Q. *L.* (2009). Retrieved October 15th, 2009 from PostgreSQL: The world's most advanced open source database web site: http://www.postgresql.org

Tufte, E. R. (1983). *The Visual Display of Quantitative Information.* Cheshire, CT: Graphics Press.

Van Elzakker, C. P. J. M. (2001). Use of maps on the Web . In Kraak, M. J., & Brown, A. (Eds.), *Web Cartography: Developments and Prospectus.* London: Taylor & Francis.

WordPress. (2009). Retrieved October 15th, 2009 from WordPress - Blog Tool and Publishing Platform web site: http://www.wordpress.org

Wright, D., & Scholz, A. (Eds.). (2005). *Place Matters: Geospatial Tools for Marine Science Conservation and Management in the Pacific Northwest.* Corvallis, OR: Oregon State University Press.

KEY TERMS AND DEFINITIONS

Cache: A location on disk where resources (such as images) that are expected to be used often, are stored for easy retrieval. Many programs (such as web browsers) employ caches in order to improve the perceived speed or responsiveness of a tool or interface. In an interactive map server environment, a map cache or tile cache may be employed for map compositions involving many data layers, or for single data layers. Using such a cache simplifies life for a map server because if a map tile has already been rendered for an area, the server need only retrieve the tile and send it to the user.

Scale Factor Limit: A cartographic technique that can also be employed to improve map rendering speed in an online environment. The cartographer specifies a scale factor (e.g. 1 to 10,000 scale) above or below which a certain dataset will not draw. Reasons to prevent drawing of a layer could include the inability of the user to see the level of detail in a layer (user is zoomed too far out), or the inappropriateness of a coarse layer for detailed viewing (user is zoomed too far in). In both cases, the cartographer may choose to specify an alternate layer that is more appropriate for the given zoom level.

Spatial Index: A tool used within a spatial database to speed a spatial query. Essentially a spatial index makes it easier for the computer to skip to the relevant portion of a large spatial data set based on geography of the query and the physical storage of the data set on disk.

Thick Client: Refers to interfaces that operate on the client (user) side of a user-server interaction, that have complex coding and are relatively large in terms of bytes of code loaded into the browser on launch. A thick client interface will generally be capable of handling more processing on the "client side" of a transaction than a "thin-client" interface, but the burden on the users computer (in terms of memory or CPU speed needed) will be higher. Some thick clients require the installation of plug-ins to a browser.

Tile Index: Refers to a spatial data layer that records how several separate data files fit together to form a seamless wide area data set. Tiles may be rectangular, but this is not required. An example of non-rectangular tiling would be the tiling of a statewide road data set by county or province.

Chapter 4
Coastal Atlas Interoperability

Yassine Lassoued
University College Cork, Ireland

Trung T. Pham
University College Cork, Ireland

Luis Bermudez
Southeastern University Research Association, USA

Karen Stocks
University of California San Diego, USA

Eoin O'Grady
Marine Institute, Ireland

Anthony Isenor
Defense R&D Canada – Atlantic, Canada

Paul Alexander
Marine Metadata Interoperability Initiative & Stanford Center for Biomedical Informatics Research, USA

ABSTRACT

This chapter defines the coastal web atlases interoperability problem, introduces interoperability standards, and describes the development of a semantic mediator prototype to provide a common access point to coastal data, maps and information from distributed coastal web atlases. The prototype showcases how ontologies and ontology mappings can be used to integrate different heterogeneous and autonomous atlases (or information systems), using international standards such as ISO-19139 for metadata encoding and the Open Geospatial Consortium's Catalogue Service for the Web specification. Lessons learned from this prototype will help build regional atlases and improve decision support systems as part of a new International Coastal Atlas Network (ICAN).

DOI: 10.4018/978-1-61520-815-9.ch004

Copyright © 2011, IGI Global. Copying or distributing in print or electronic forms without written permission of IGI Global is prohibited.

INTRODUCTION

Advances achieved in Web mapping technologies are allowing a much wider audience to develop and use Web geographic information systems and, in particular, coastal web atlases (CWAs), which were discussed in Chapter 1. While multiple benefits are derived from these tailor-made atlases (e.g., speedy access to multiple sources of coastal data and information, economic use of time by avoiding individual contact with different data holders), the potential exists to derive added value from the interoperability of disparate CWAs, to optimize decision-making at a variety of levels and across themes.

Within the context of coastal Web atlases, interoperability can be briefly defined as the ability of several autonomous, heterogeneous and distributed CSWs to communicate and exchange resources (information, metadata, data or maps) or be used together despite their differences. In a wider context, interoperability has been defined in ISO 2382-1 (ISO/IEC, 1993) as the "capability to communicate, execute programs, or transfer data among various functional units in a manner that requires the user to have little or no knowledge of the unique characteristics of those units."

Interoperability of coastal web atlases will help to unify the discovery and access of information to users (e.g., scientist, coastal resource managers) and categorize results using a convenient vocabulary. Data discovery relies on documentation provided as part of metadata (notably discovery metadata), such as the dataset title, abstract, extent, keywords, etc. In the context of distributed resources, this information is present in different heterogeneous formats, and systems, according to several existing metadata models and standards. While various efforts exist to provide a unified common form to harmonize different standards (see section 4.3), problems still arise with metadata semantics.

Metadata terms may be agreed upon within an organization or a region, but not necessarily between or among other organizations or regions. Use of the word "seabed" in Europe versus use of the word "seafloor" to describe the same feature in North America is a good example of this scenario. From both human and computational standpoints, users need assurance that the concepts, terminology, even the abbreviations that are shared between two or more individuals, systems, or organizations are understood by all to mean the same thing. In this way the quality of data retrieval and subsequent data integration are greatly increased, as they are based on meaning rather than on mere keywords.

Several communities including the database, artificial intelligence and geosciences communities have studied interoperability methodologies in the past two decades. These include database integration (e.g., Yétongnon *et al.*, 2006) and mediation (e.g., Wiederhold, 1992) approaches. Several mediation systems and prototypes have also been developed: examples of such systems are TSIMMIS (Garcia-Molina *et al.*, 1997), PICSEL (Goasdoue *et al.*, 2000), Information Manifold (Kirk *et al.*, 1995), and AGORA (Manolescu *et al.*, 2001).

This chapter discusses standards and tools, defines and proposes solutions on the use of controlled vocabularies and ontologies, and provides an overview of the International Coastal Atlas Network (ICAN) prototype. The solution proposed here is based on ontology mediation and on information that is shared via web services.

INTEROPERABILITY: WHY IT MATTERS

In recent years significant momentum has occurred in the development of Internet resources for decision makers, scientists and the general public who are interested in the coast, where worldwide, 20% of humanity live within a 25 km range, and 39%, or 2.2 billion people, live within a 100 km range (e.g., Wright 2009). Given that no CWA functions alone as an island, and is often part of a larger universe of resources that is needed for effective marine spatial planning, resource management, and emergency planning, CWAs must build a common approach toward managing and disseminating the coastal data, maps and information that they contain (Wright et al., 2007). Sometimes more than one CWA may be needed in order to address regional problems such as hazard mitigation, climate change, intergovernmental marine spatial planning, etc. As an example, if there is a dataset missing in one atlas, it may be immediately located within another. Often, similar datasets from different atlases cover different areas and can be combined in order to build a dataset that covers a wider area. For instance local high-resolution bathymetry data from different atlases one can participate in building a regional high-resolution bathymetry dataset.

LEVELS OF INTEROPERABILITY

The levels of interoperability required for CWAs are not less than those required for interoperating amongst two or more distributed heterogeneous systems in the wider context of information systems. Because information systems heterogeneity may occur at several levels, interoperability, also, can be defined at several levels. Bishr (1998) provides an overall discussion of GIS interoperability and a definition of interoperability at various abstraction levels: system, syntactic, structure and semantic.

- The system level involves protocols for networks, Web services and operating systems, e.g., FTP and HTTP protocols. At the system level, users may connect to the host or transfer data files between systems. Users may also download data files in some standard format.
- The syntactic level involves standard formats and languages for transporting and possibly archiving data, e.g., Geography Markup Language (GML).
- The structure level involves standard data structures, e.g., a GML application schema or a database schema.
- The semantic level deals with meaning and data standards (Asch *et al.*, 2004). Semantic interoperability aims to solve semantic conflicts between different systems, generally using standards and mapping (correspondence) rules.

STANDARDS

Efforts to create standards for the exchange of geospatial information over the past years have produced a number of national and international standards related to data and metadata formats and models, and to web services. Standards supporting geospatial interoperability have been developed by a number of national and international standardization organizations and consortiums, which will be discussed in this section.

Objectives

An important aspect of interoperability consists of the design an implementation of web services that support data and metadata exchange. Consequently, one way to facilitate interoperability is to standardize these three components. Such standardization will help harmonize data and metadata access, delivery and interpretation. It is difficult or practically impossible to achieve absolute interoperability, i.e. interoperability where a system can communicate with any system using any data and exchange protocol standards. Therefore, it is crucial to set one's feet on the ground first, before developing interoperable or interoperability systems, by setting the standards to cope with. Such standards will define the ground basis for interoperability and the extent to which mediation or translation efforts will be spent. As a matter of fact, interoperability typically relies on two things: standardization and mediation. While standardization aims to harmonize data formats and models, query languages and exchange protocols, mediation aims to translate from one model, language or format to another and integrate responses from different systems. It is obvious then that the more data formats and models and query languages are standardized the less effort we need to spend in mediating between systems; hence the need for standards as a way to set a ground basis for interoperability and to minimize translation and mediation efforts.

Standardization Bodies

Several national and international organizations and consortiums have been active in developing interoperability standards over the past decades. Each organization or consortium tackles some levels of interoperability within one more or less broad application area or domain (Web resources in general, geographic information, marine information, etc.). For instance, while W3C defines standards for data exchange on the Web in general, OGC focuses on geospatial information.

The aim of this section is to introduce a number of national and international standardization organizations and consortiums of relevance to Web GIS or CWAs interoperability. The list is not meant to be exhaustive; rather it focuses on the key organizations.

World Wide Web Consortium

W3C (http://www.w3c.org) is an open international organization whose goal is to lead the World Wide Web to its full potential by developing common protocols. W3C open standards support the development of technologies for advancing the Internet by ensuring interoperability, such as:

- eXtensible Markup Language (XML) (Ray, 2003), which provides a syntactic infrastructure for structured documents, thus facilitating syntactic interoperability;
- Resource Description Framework (RDF) (Powers, 2003), a data model for referring to objects ("resources") and how they are related to each other; such a data model underpins the Ontology Web Languages (c.f. next point);
- Ontology Web Language (OWL) (Lacy, 2005), which facilitates formal description of concepts, terms, and relationships within a given domain; OWL and RDF facilitate structural semantic interoperability by linking concepts and vocabularies from different domains or organizations;
- Web services (Cerami, 2002), which are software systems intended to support interoperable interactions between systems or machines over a network; web services support system and syntactic interoperability by defining common protocols and formats for data exchange.

International Organization for Standardization

ISO [http://www.iso.org] is the most established and recognized developer and publisher of International Standards. ISO is composed of national standards institutes of 160 countries, one member per country.

ISO/TC211 (ISO/T211, 2009) is the ISO technical committee specialized in geographic information and geomatics. ISO/TC211 develops specifications for the methods, tools and services for managing, acquiring, processing, analyzing, accessing and presenting geospatial data. ISO/TC211 also provides a framework for developing interoperable spatial applications allowing geospatial data to be shared and exchanged between different systems.

ISO/TC211 is responsible for the 191XX series of standards for representing and exchanging geographic information. The committee currently has approximately 40 standards in development including the following:

- ISO 19103 - Conceptual schema language
- ISO 19104 – Terminology
- ISO 19107 - Spatial Schema
- ISO 19108 - Temporal Schema
- ISO 19111 - Spatial referencing for coordinates
- ISO 19112 - Spatial referencing by geographic identifies
- ISO 19113 - Quality principles
- ISO 19114 - Quality evaluation procedures
- ISO 19115 - Metadata (ISO/TC211, 2003)
- ISO 19119 – Services
- ISO 19127 - Geodetic codes and parameters
- ISO 19128 - Web Map Server Interface
- ISO 19136 - Geography Markup Language (ISO/TC211, 2007)
- ISO 19138 - Data quality measures
- ISO 19139 - Metadata - Implementation Specification (ISO/TC211, 2006)

A complete list of these standards can be accessed through the ISO/TC211 web site: www.isotc211.org.

Open Geospatial Consortium

OGC (OGC, 2009) defines standards, both for data models and for Web services. It promotes interoperability in the fields of geographic data by defining standard interfaces and implementation guidelines for geospatial data and geo-processing services. The OGC web services specifications are compliant with the W3C Web service standards.

European Committee for Standardization

CEN (CEN, 2009) is a business facilitator in Europe, aiming to provide a platform for the development of European Standards and other technical specifications. CEN is composed of thirty national members working together to develop European standards, which are also adopted as national standards in each of the thirty member states.

CEN/TC 287 is the CEN technical committee specialized in geographic information. CEN/TC 287 developed various standards for supporting interoperability in the fields of geographic information, such as the metadata, services and encoding standards amongst others.

INfrastructure for Spatial InfoRmation in Europe (INSPIRE)

INSPIRE (INSPIRE, 2009) is an EU initiative aiming to develop a spatial information infrastructure for Europe. It is envisaged that such an infrastructure will deliver to the users integrated spatial information services. Such services will facilitate identification of, and access to, spatial or geographical information from a wide range of sources ranging from the local to the global levels, in an interoperable way and for a variety of uses. INSPIRE envisages "a distributed network of databases, linked by common standards and protocols to ensure compatibility and interoperability of data and services" (Smits *et al.*, 2002). INSPIRE aims to achieve its goals by defining a set of common rules and an architectural framework for implementing them. Such an architectural framework is based on the deployment of interoperable spatial information services over the Internet. The role of these services is to facilitate (1) the production and publication, (2) the discovery and delivery of, and ultimately, (3) the use and understanding of geographic information.

Geographic Metadata Standards

Geographic metadata standards harmonize geographic metadata formats, models, structures and semantics with the aim to facilitate metadata sharing and interpretation and data discovery and access. Several national and international metadata standards have been developed over the past few years. Below is an introduction to the key ones.

Dublin Core Metadata Element Set (DCMES)

The Dublin Core Metadata element Set (ISO/TC 46/SC 4, 2003) is a vocabulary of fifteen properties for use in resource description. The name "Dublin" is due to its origin at a 1995 invitational workshop in Dublin, Ohio; core because its elements are broad and generic, usable for describing a wide range of resources.

DCMES consists of the following fifteen comprehensive metadata elements, organized alphabetically: contributor, coverage, creator, date, description, format, identifier, language, publisher, relation, rights, source, subject, title, and type.

DCMES trades richness for wide visibility. Its simplicity, as pointed out in (ISO/TC 46/SC 4, 2003), can be both a strength and a weakness. On the one hand, simplicity facilitates metadata creation and improves interoperability. On the other hand, it does not accommodate the semantic and functional richness and precision supported by complex metadata schemas. Because of such simplicity, DCMES is not intended to replace any other metadata standard. Rather it should co-exist with metadata standards that offer other or richer semantics and functionalities.

ISO-19115 and ISO-19139

The ISO-19115 (ISO/TC211, 2003) international standard defines general-purpose metadata for geographic information. It defines metadata elements, provides a schema and establishes a common set of metadata terminology, definitions and extension procedures.

ISO-19115 defines thirteen top-level metadata elements, the main being called the Metadata entity set information. This element encapsulates information about metadata, such as the language, standard name, version, etc.; and is an aggregate of the twelve remaining elements listed and defined below as per (ISO/TC211, 2003).

- Identification information contains information to uniquely identify the data, e.g., citation for the resource, an abstract, keywords, spatial and temporal extents, points of contact, etc.
- Constraint information contains information pertaining to the restrictions placed on data.
- Data quality information contains a general assessment of the quality of the dataset. It also contains the scope of the quality assessment and is an aggregate of lineage and data quality elements: completeness, thematic accuracy, logical consistency, temporal consistency, and positional accuracy.
- Maintenance information contains information about the scope and data updating frequency.
- Spatial representation information defines the mechanisms used to represent spatial information in a dataset.
- Reference system information defines the spatial and temporal reference systems used in the dataset.
- Content information contains information identifying the features catalogue used and/or information describing the content of a coverage dataset.
- Portrayal catalogue information contains information identifying the portraying catalogue used.
- Distribution information contains information about the distributor of, and options for obtaining, a resource.
- Metadata extension information contains information about user specified extensions.
- Application schema information contains information about the application schema used to structure the dataset.

The ISO-19115 standard specifies a set of metadata elements, their types and relationships among themselves. What it defines is a metadata model, which facilitates structural and semantic interoperability. However, since various implementations are possible, it does not guarantee syntactic interoperability. ISO-19115 has been implemented in several ways since its publication, and the GIS community ended up with a number of heterogeneous implementations, which raised the problem of interoperability again. A standard implementation was then needed, and ISO-19139 (ISO/TC211, 2006) was the answer.

ISO-19139 is an XML implementation of ISO-19115. The choice of XML (Ray, 2003) as standard metadata format is not random as XML is the data exchange format for the Web. It consists of a set of XML schemas (XSD) (van der Vlist, 2002) implementing the metadata model specified by ISO-19115 and prescribing the formats of the metadata records.

Geographic Data Standards

Extensible Markup Language (XML)

XML (Ray, 2003) is a general-purpose specification that provides a syntactic infrastructure for structured documents. As a markup language, XML uses a set of annotations to text in order to define its structure or display. XML is recommended by W3C and is widely used as the Web data exchange format.

The structure of an XML document can be defined using an XML schema language. Several XML schema languages exist. XML Schema, or XML Schema Description, or XSD (van der Vlist, 2002) is one of such several schema languages, recommended by W3C. An XML schema expresses a set of rules to which an XML document must conform in order to be considered valid. XSD rules include element and attribute types, cardinality and element structures.

Various XML grammars have been developed over the past years in order to standardize data structures in several disciplines and domains such as geography, mathematics, music, etc. The geography Markup Language or GML (Lake *et al.*, 2004) is one of such XML grammars designed to describe geographic data.

Geography Markup Language (GML)

GML (Lake *et al.*, 2004) is the XML grammar defined by OGC to express geographic features. GML is used as an exchange data format for the delivery of geographic data on the Web, by Web Feature Services (WFS) (Vretanos, 2005). The GML schemas are the set of XML schemas that define the GML grammar. They offer a set of predefined elements for the encoding of geographic feature classes, geometries, temporal properties, topologies, etc. GML defines features distinct from geometry objects. A feature is an application object that represents a physical entity, e.g., a building, a river, or a person. A feature may or may not have geometric aspects. A geometry object defines a location or region instead of a physical entity, and hence is different from a feature.

In order to expose an application's geographic data with GML, a community or organization creates a *GML application schema* specific to the application (the application schema). This *GML Application Schema* is an XML schema description (van der Vlist, 2002) that builds on (extends) the GML schemas in order to describe the object types whose data are exposed by the application. By doing so, organizations or communities define the common structure (feature classes, properties, associations, data types, etc.) for their data being exchanged by WFS servers. For instance, a community could define that a station is a feature, and that it has properties location, name and observation data as properties. Examples of common GML application schemas are the Climate Science Modeling Language (CSML, 2009) for atmo-

Figure 1. Principle of a client interaction with an OGC web service

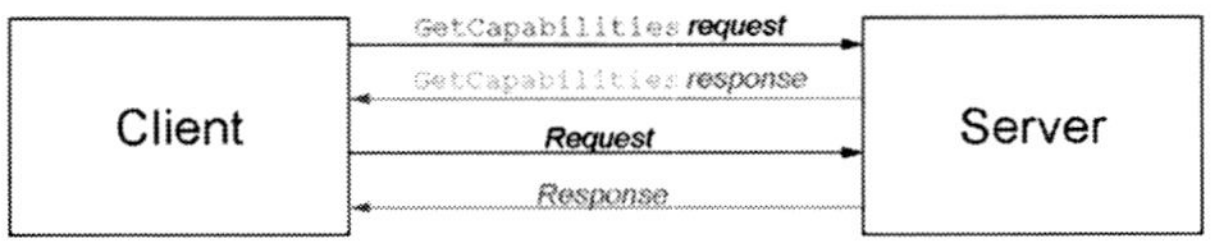

spheric and oceanographic data and GeoSciML (GeoSciML, 2009) for geoscientific data.

GML is widely accepted by industry, government and academia and it has been endorsed by ISO-19136 standard (ISO/TC211, 2007); however, because of the specialization of the schema by each community and because GML is very general, it is often difficult to achieve interoperability (Bermudez, 2009). Each schema that is created will require the development of new tools (software, visualization capabilities etc.) and also translation between one GML extension to another.

Information Exchange Standards

The major information exchange standards are developed around Web services (Cerami, 2002). Web services are an emerging technology that allows different Web-based applications to interact in order to exchange data and software. Such interaction relies on the use of XML as a data format, Simple Object Access Protocol (SOAP) (Snell *et al.*, 2001), Web Service Description Language (WSDL) and Universal Distribution Discovery and Interoperability (UDDI) open standards (Cerami, 2002) over an Internet protocol backbone. Through Web Services, distributed and heterogeneous data sources or systems can be shared and interoperated within an organization or across the Internet.

OGC has driven the development of, and recommended, several geographic Web services designed specifically for the purpose of delivering geographic resources on the web. OGC web services are the foundation of an interoperable framework for web-based discovery, access, integration, exploitation and visualization of distributed geospatial data sources. Several web services have been recommended by OGC, such as Web Map Service (WMS), Web Feature Service (WFS), Web Coverage Service (WCS), Catalogue Service for Web (CSW). The main of these services will be introduced in Section 4.4.

OPEN GEOSPATIAL SERVICES

The typical client interaction with an OGC Web service can be summarized in the follow four steps as illustrated in Figure 1.

1. The client contacts the server and queries it about its capabilities, i.e. supported operations and resources, using a GetCapabilities request;
2. The server sends back to the client an XML document containing the functionalities supported by the service; such a document is called the GetCapabilities response;
3. The client requests data possibly using operations within the capabilities of the server;
4. The server provides the user with the requested data.

Such processing is identical for the various OGC services. However request types and syntaxes vary depending on the type of service in question. In the following subsections we introduce three key OGC services (CSW, WMS and WFS) and their main request types.

Catalogue Service for the Web (CSW)

Catalogue services are the main technology for locating, managing and maintaining distributed geospatial data and services. The OGC CSW (Nebert and Whiteside, 2005), is a key international standard for interoperable distributed catalogue services. It defines a common interface for discovering, retrieving and querying metadata about geographic data and services. An OGC catalogue service is an online repository for storing, indexing, publishing and searching metadata records.

CSW supports the ability to publish and search collections of data and services metadata by means of various operations or request types. The operations supported by CSW are introduced below. Mandatory operations are followed by the star symbol (*). Others are optional.

- GetCapabilities* – for retrieving service metadata from a server;
- DescribeRecord* – for discovering elements of the information model supported by the service;
- GetDomain – for retrieving information about the valid values of one or more named metadata properties.
- GetRecordById* – for extracting a metadata record by its unique identifier;
- GetRecords* – for searching metadata records, possibly using filters, such as keyword, location and time search;
- Transaction – for creating, modifying and deleting catalogue records;
- Harvest – for retrieving an accessible resource from a specified location, possibly periodically in order to refresh the information in the catalogue.

Web Map Service (WMS)

The OGC WMS specification (de la Beaujardière, 2004) defines a common interface for disseminating digital maps, rendered from spatial data, across the Internet. WMS supports the networked interchange of web based map layers, which are generally rendered in a digital image such as JPEG, GIF, PNG, etc. Typically, a client submits an HTTP request to a WMS server including a number of standard parameters. The server then returns the responses based on those parameters. The WMS specification defines the three operations defined below. Mandatory operations are followed by the star symbol (*), others are optional.

- GetCapabilities* – returns an XML document that contains information about the server and the data that are available;
- GetMap* – returns a map image;
- GetFeatureInfo – returns thematic information of a specific feature on the map.

As per the common functioning of an OGC service, the GetCapabilities request allows a WMS client to retrieve metadata about the WMS including supported operations and data holdings (spatial extent, thematic layers, styles available, etc.). Knowing what is available from the WMS, the user can now request a specific map for a specific area using a GetMap request. The server will return a map image in JPEG, GIF, PNG, or other format. Given a map, the client can request data about the feature represented by a given point (x, y) on that map, using a GetFeatureInfo request. The response from the server can be a text, XML/GML, HTML, or Word file containing attribute information for the selected feature(s).

Web Feature Service (WFS)

WFS (Vretanos, 2005) is an OGC specification defining a common interface for delivering geospatial features on the Web. Geospatial features delivered by WFS are encoded in GML (c.f. Section 4.3.4). WFS supports several operations, which are introduced below.

- GetCapabilities – returns the capabilities of the WFS, indicating which feature types the WFS can service and what operations are supported on each feature type.
- DescribeFeatureType – describes using an XML schema, the structure of a feature type the WFS can service.
- GetFeature – retrieves feature instances, with the possibility for the user to specify which feature properties to fetch and to constrain the query both spatially and non-spatially.
- GetGmlObject – retrieves element instances (GML objects) given their identifiers (XLinks that refer to their XML IDs), with the possibility for the client to specify whether nested XLinks embedded in returned element data should also be retrieved.
- Transaction – allows a client to modify features using create, update, and delete operations on geographic features.
- LockFeature – allows an authorized client to lock one or more instances of a feature type for the duration of a transaction. This ensures that serializable transactions are supported.

As one might notice, it is not specified here which operations are mandatory and which are optional. Rather, three types of WFS are defined depending on the operations they support:

1. **Basic WFS** – implements the GetCapabilities, DescribeFeatureType and GetFeature operations, therefore is considered as a READ-ONLY WFS.
2. **XLink WFS** – supports all the operations of a basic web feature service and in addition implements the GetGmlObject operation for local and/or remote XLinks, and offers the option for the GetGmlObject operation to be performed during GetFeature operations.
3. **Transactional WFS** – supports all the operations of a basic web feature service and in addition it would implement the Transaction operation. The GetGmlObject and LockFeature operations remain optional for a transaction WFS.

Web feature services constitute the basis for system and syntactic interoperability of geospatial data services by harmonizing data exchange protocols on the Web, query languages and data formats.

CONTROLLED VOCABULARIES AND ONTOLOGIES

Why are They Needed?

The standards presented in this chapter and the technologies used to implement these standards, are oriented toward achieving interoperability between CWAs. These standards address the structure and information requirements, which provide CWAs with the necessary syntactic and structure level interoperability (see section 4.2). To move toward semantic interoperability, the systems need to understand the actual data content being passed within these standard structures.

In many cases, the data content passed between systems is based on numeric data values. These values are supported through the use of words or terminology, which identifies the data values to the community-of-interest. In many professional communities, the terminology used to describe the data is often discipline-specific. In many cases, the different communities describe similar data differently.

As a simple example, consider the use of the word "altitude" (Miller *et al.*, 2009). Altitude refers to the height of something (such as an airplane in flight) above a reference point (like ground level). Although altitude is a common term, architects would not consider 'altitude' as proper terminology for describing the level of a rooftop. Most likely, the architect would use the word "height". Similarly, if someone were in a boat looking downward into the water, they would describe the distance to the bottom as "depth".

The three words – altitude, height, and depth – are all similar in that they represent measures of distance relative to specified levels; but they are all used differently. As well, they are associated with different communities-of-interest. The collective set of terms used by a community is called a vocabulary.

Definition of Controlled Vocabulary

Vocabularies provide a mechanism for communication- be it written, oral or electronic- because the meaning of the terms are known and agreed upon by the community members. When a vocabulary is formally managed, it becomes a controlled vocabulary. In this case, "managed" means the terms are stored and maintained using agreed upon procedures. Procedures should exist for adding terms, modifying terms and, more rarely, deprecating terms from a controlled vocabulary.

Controlled vocabularies are very powerful when combined with formal metadata standards (e.g., ISO 19115). This is because the terms within the controlled vocabulary can be used as the content for specific metadata elements that make up the standard. In many cases, the controlled vocabulary terms completely define the allowable content for a particular metadata element. This control helps avoid misspellings and inconsistencies in the metadata content. Moreover, in the world of computers, the controlled vocabulary offers enhanced capabilities because it can be incorporated into automated procedures. For example, in a data system a controlled vocabulary can simplify system input and contribute to quality control of that input. Input is simplified by providing users or other systems with a list of allowed entries for the specific metadata elements. Similarly, the controlled vocabulary can be used to check existing or imported metadata descriptions for consistency and correctness, including things like spelling and hyphenation.

Controlled vocabularies may be defined internal or external to a metadata standard (Isenor and Spears, 2007). An internally controlled vocabulary often represents the allowed content for a metadata attribute described by the standard. As an example, consider ISO 19115. This standard allows for different kinds of reference dates to describe the data resource. The specific kind of date is noted by using a metadata attribute called "*datetype*". The allowed content of *datetype* is made up of the set {creation | publication | revision}. These three terms are the only permissible content for this metadata attribute. These three terms represent a controlled vocabulary, defined internal to the ISO 19115 standard.

A metadata standard is intended to have broad support across many communities. Thus, the controlled vocabularies defined internal to the standard are intended to have broad applicability. Controlled vocabularies defined external to the standard typically have support within smaller communities-of-interest. These externally defined controlled vocabularies provide the details required by the community-of-interest to distinguish the subtleties of the data, which the vocabulary describes.

In terms of a CWA, the externally defined controlled vocabulary will allow an agreed upon meaning for the exchanged data. This means the thematic data, feature data, and the supporting numerical data will be described using terminology contained within the vocabulary, and accessible to the larger community. As well, community members will be able to use the controlled vocabulary in mappings to other controlled vocabularies. This is very powerful, as it allows the community to establish a semantic-based link to other similar communities.

Types of Controlled Vocabularies

Various forms of controlled vocabularies exist, but all can help improve data understanding by restricting and better defining data content. To provide the type of functionality noted above, the controlled vocabulary must be in a form that allows computer processing. The exact form of the controlled vocabulary will vary, but in general there are three broad categories of controlled vocabularies (Isenor and Neiswender, 2009) extracted from (Neiswender, 2009): "*flat, multi-level and relational vocabularies.*

- *Flat controlled vocabularies provide a set of used terms. Some flat controlled vocabularies will provide additional information about each term (e.g., glossary).*
- *Multi-level controlled vocabularies build upon a flat controlled vocabulary by assigning each term to a category (e.g., taxonomy).*
- *Relational controlled vocabularies provide a set of terms, and capture how they are associated with each other (e.g., ontology).*"

ONTOLOGIES

As one progresses through the categories of controlled vocabularies, the functionality of the vocabulary improves. At the apex is the ontology. The ontology provides a means for describing the complex internal relationships between the terms in the vocabulary and declaring machine-processable definitions for vocabulary terms that can be reasoned with. A coastal and marine controlled vocabulary in the form of an ontology has been identified as a critical component for the integration (or interoperability) of CWAs (O'Dea *et al.*, 2007).

Definition of Ontology

The term "*ontology*" has many definitions depending on the area of discourse. In computing, an ontology is defined in (Sowa, 1999) as follows.

"The subject of ontology is the study of the categories of things that exist or may exist in some domain. The product of such a study, called an ontology, is a catalog of the types of things that are assumed to exist in a domain of interest D from the perspective of a person who uses a language L for the purpose of talking about D. The types in the ontology represent the predicates, word senses, or concept and relation types of the language L when used to discuss topics in the domain D. An informal ontology may be specified by a catalog of types that are either undefined or defined only by statements in a natural language. A formal ontology is specified by a collection of names for concept and relation types organized in a partial ordering by the type-subtype relation."

In some views, an ontology is simply a controlled vocabulary that is expressed in a particular structured format called OWL (Web Ontology Language). Most sources, however, require that an ontology explicitly defines concepts, and expresses relationships between concepts in class and subclass relationships (Gruber, 1993).

Depending on which perspective one adheres to, ontologies can include some of all of the following:

- Classes (general concepts),
- Instances of concepts,
- Relationships among concepts,
- Properties of the concepts or relationships, including functions, processes, constraints, and rules.

In summary, an ontology is a set of concepts in some domain of study, their properties and the relationships between them. As an example, con-

Figure 2. Marine habitat ontology

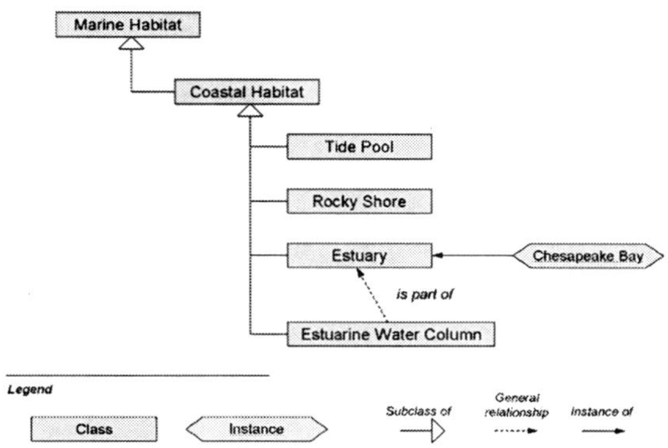

sider a hypothetical ontology that defines marine habitat types (c.f. Figure 2). Estuary, estuarine water column, tide pool, rocky shore, and coastal habitat might be classes of habitats within this ontology. The Chesapeake Bay would be an instance of an estuary. One could define the relationship between a tide pool and a coastal habitat by saying that a tide pool "is a" coastal habitat – this defines tide pools as a subclass of coastal habitats. Or, to define a different kind of relationship, one could say that the estuarine water column "is part of" an estuary. An example of a property might be the requirement that, by definition, estuaries must have an average salinity of less than full seawater, or and that a rocky shore must be intertidal, and it must have a hard substrate. Another kind of property, a rule, can define how concepts in one ontology map to those in another ontology, for example a rule might indicate that "habitat type" in one ontology maps to the multiple fields of "geoform" and "biotope" in another ontology.

Ontologies are important for interoperability because they allow complex concepts to be expressed, described, and related in a structured and, ideally, machine-readable way. The ability to express relationships between concepts also makes them critically useful for creating interoperability between systems that each uses a different vocabulary to cover the same domain area. The participating vocabularies can be mapped to each other within an ontology, allowing data connected to one set of vocabulary terms to be searched, accessed or otherwise related using the other vocabulary's terms.

A suite of tools exist to develop/define ontologies, express them in a standard structure, and work with them once developed (Alexander, 2009). Servers also exist for describing and sharing ontologies so that they can be discoverable by the larger community. Some of the ontology registries and libraries that are relevant to the coastal data community include the MMI Ontology Registry and Repository (ORR, 2009), the OntoSelect Ontology Library (Buitelaar, 2004) and the Protégé Ontology Library (POL, 2009). Search engines like Swoogle can also help with finding ontologies; the W3C SWEO Community Project on Linking Open Data maintains a list of ontology search engines (SWSE, 2009).

Ontology Languages

An ontology language can be briefly defined as a formal language for building and encoding ontologies. The ontology community has specified several ontology languages over the past few years. Efforts of the Web Ontology Working Group of W3C focused on the specification of standard web-based ontology languages. This led to a series of XML-based languages and models the most important of which are Resource Description Framework (RDF), RDF Schema (RDFS), Web Ontology Language (OWL), and Simple Knowledge Organization System (SKOS), which we will discuss in this section.

Resource Description Framework (RDF)

The Resource Description Framework (RDF, 2009) is, as defined by W3C, "*a general-purpose language for representing information in the Web*". RDF was designed to describe resources on the web by means of *statements*. A *statement* in RDF

Figure 3. Illustrative example of a generic RDF resource description

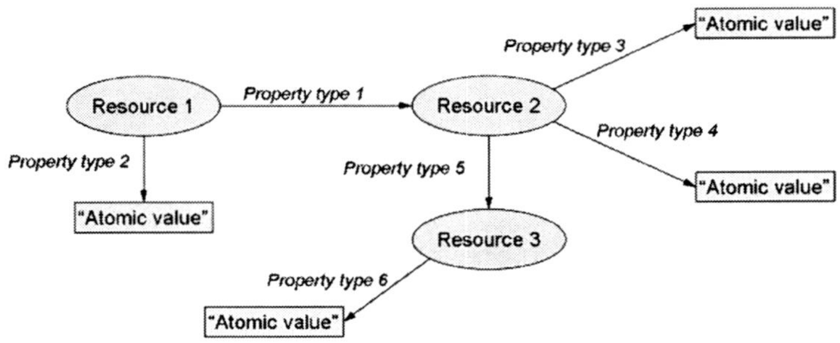

describes a *resource* using *properties* and *property values*. A *resource* is a web page or a real world object like a publication, a person or an institution. A *property* is a relationship that associates to a resource some "value", for example *name*, *author*, etc. The associated value is called *property value*, e.g., "Seán Murphy" or "http://www.w3schools.com". In RDF, values may be atomic in nature (text strings, numbers, etc.) or other resources, which in turn may have their own properties. A collection of these properties that refer to the same resource is called a *description*. An RDF statement is defined as a triple (*subject, predicate, object*). The *subject* refers to the resource being described, the *predicate* to the property and the *object* to the property value.

RDF provides a syntax-independent model for representing resources and descriptions. An RDF description can be represented as a directed labeled graph. The so-called RDF graph has nodes and labeled directed arcs that link pairs of nodes and this is represented as a set of RDF triples, where each triple contains a subject node, predicate and object node. Nodes are RDF URI (Uniform Resource Identifier) references, RDF literals, or blank nodes. Blank nodes may be given a document-local, non-RDF URI references identifier called a blank node identifier.

Predicates are RDF URI references and can be interpreted as either a relationship between the two nodes or as defining an attribute value (object node) for some subject node. An example of a generic RDF graph is illustrated in Figure 3.

A more realistic and intuitive example is given in the graph of Figure 4, which represents the following statements:

- The *title* of Painting 1 is "Mona Lisa";
- The *author* of Painting 1 is Artist 1;
- The *name* of Artist 1 is "Leonardo da Vinci";
- The *birth date* of Artist 1 is 15 April 1452.

An XML syntax for RDF graphs, called RDF/XML (Beckett, 2004), has been developed by W3C. The purpose of it is to provide a common

Figure 4. Illustrative example of a resource description for a painting resource

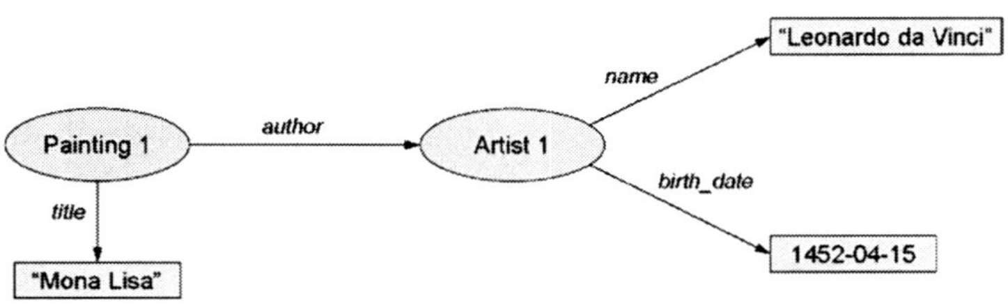

machine processable and extensible encoding for RDF graphs. In RDF/XML, nodes and predicates are represented in XML terms: element names, attribute names, element contents and attribute values. For instance, an RDF/XML encoding of the previous painting example would be something similar to the following:

RDF Schema (RDFS)

RDF provides a way to express statements about resources. However, RDF user communities need also the ability to define the vocabularies (terms) they intend to use in those statements, specifically, to indicate that they are describing specific kinds or classes of resources, and will use specific properties in describing those resources. For instance, a given organization http://www.example.com (referred to as "ex") would want to describe the classes ex:painting and ex:artist from the example above, and define properties such as ex:title, ex:author, ex:name and ex:birthdate to describe them. RDF itself provides no means for defining such application-specific classes and properties. Instead, such classes and properties can be described as an RDF vocabulary, using extensions to RDF provided by the RDF Vocabulary Description Language, referred to as RDF Schema (RDF-S or RDFS) (Brickley, 2004). RDF Schema further extends RDF by adding more modeling primitives often found in ontology languages like classes, class inheritance, property inheritance, domain, range restriction.

Simple Knowledge Organization System (SKOS)

SKOS (Miles, 2009) is a common data model for knowledge organization systems such as thesauri, classification schemes, taxonomies, subject-heading systems, and taxonomies within the framework of the Semantic Web. SKOS provides a standard way to represent knowledge organization systems using the Resource Description Framework (RDF). Encoding this information in RDF allows it to be passed between computer applications in an interoperable way. Using RDF also allows knowledge organization systems to be used in distributed, decentralized metadata applications. Decentralized metadata is becoming a typical scenario, where service providers want to add value to metadata harvested from multiple sources.

SKOS Data Model

The SKOS data model views a knowledge organization system as a *concept scheme* comprising a set of *concepts*. These SKOS concept schemes and SKOS concepts are identified by URIs, enabling anyone to refer to them unambiguously from any context, and making them a part of the World Wide Web. SKOS concepts can be:

- Labeled with zero or many lexical (UNICODE) labels in different natural languages (e.g., English, French, etc.); in any language, one of these labels can be specified as the preferred label;
- Assigned one or more lexical codes, called notations, used to uniquely identify the concept within the scope of a given concept scheme;
- Documented using notes of various types such as scope notes, definitions, editorial notes, etc.;
- Linked to each other using hierarchical and associative semantic relationships such as Related Term, Broader Terms, Narrower Term, etc.;
- Grouped into collections which can be labeled and/or ordered;
- Mapped to other SKOS concepts in different concept schemes, using four basic types of mapping links: hierarchical, associative, close equivalent and exact equivalent.

An example of SKOS data expressed as an RDF graph is illustrated below. The graph is expressed in TURTLE, the Terse RDF Triple Language (Beckett, 2008), a textual syntax that allows RDF graphs to be completely written in a compact and natural text form, with abbreviations for common usage patterns and data types.

```
<A> rdf:type skos:Concept ;
    skos:prefLabel "Gravel"@en ;
    skos:altLabel "G"@en ;
    skos:broader <B> ;
    skos:inScheme <S> .
<B> rdf:type skos:Concept ;
    skos:prefLabel "Folk Class"@
en ;
    skos:altLabel "Folk Sediment
Class"@en ;
    skos:topConceptOf <S> .
<S> rdf:type skos:ConceptScheme
;
    dct:title "Marine Geology
Thesaurus" ;
    skos:hasTopConcept <B> .
```

In this example, it is expressed that *A* is a concept called, in English, "Gravel" (preferred label) or alternatively "G". Concept *A* has a broader concept *B*, called "Folk Sediment Class" or preferably "Folk Class". Both concepts *A* and *B* belong to a concept scheme *S* called "Marine Geology Thesaurus", concept *B* being a top level concept of this scheme, i.e. is not narrower than any other concept in *S*.

Usage

SKOS may be used on its own, or in combination with formal knowledge representation languages such as the Web Ontology language (OWL) to express and exchange knowledge about a domain. However, SKOS is not a formal knowledge representation language.

Web Ontology Language (OWL)

OWL, the Web Ontology Language (McGuinness, 2004), is used to explicitly represent ontologies, i.e. the meaning of terms in vocabularies and the relationships between those terms. It has more facilities for expressing meaning and semantics than XML, RDF, and RDF-S, and thus OWL goes beyond these languages in its ability to represent machine interpretable content on the Web. OWL adds more vocabulary for describing properties and classes: among others, relations between classes (e.g., disjointedness), cardinality (e.g., "exactly one"), equivalence, richer typing of properties, characteristics of properties (e.g., symmetry, transitivity), and enumerated classes.

OWL provides three increasingly expressive sublanguages (McGuinness, 2004): OWL Lite, OWL DL and OWL Full, OWL OWL Full being an extension of OWL DL which is in its turn an extension of OWL Lite (c.f. Figure 5). Only OWL Lite and OWL DL guarantee computational completeness, i.e., all computations performed on the statements of an OWL Lite or DL ontology will finish in finite time.

- OWL Lite is intended to support applications with basic expressiveness requirements. It supports the basic constructs, including hierarchy of classes and simple cardinality constraints (values 0 and 1).
- OWL DL is intended to support applications requiring the maximum expressiveness while retaining computational completeness (i.e., all conclusions are guaranteed to be computable) and decidability (all computations will finish in finite time). Its name is due to its correspondence with description logics, a field of research that has studied the logics that form the formal foundation of OWL. OWL DL includes all OWL language constructs, but they can be used only under certain restrictions. For

Figure 5. The OWL sublanguages

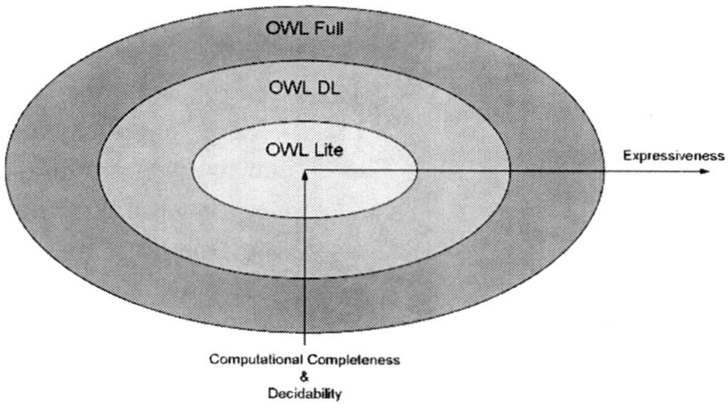

instance, a class cannot be an instance of another class in OWL DL.

- OWL Full practically allows "to say anything". For example, in OWL Full, a concept could be a class and an individual at the same type. It is intended for applications requiring maximum expressiveness with no computational guarantees.

ONTOLOGIES FOR INTEROPERABILITY

In the literature, several approaches propose ontology-based solutions for interoperability both in the general context of information systems and databases (Bianchini *et al.*, 2003 & Maedche *et al.*, 2003) and in the context of geographic information systems (Heydari, 2009 & Kemp 2007). Such approaches use ontologies as a way to express the structure and semantics of the information being exchanged and/or as a way to map between heterogeneous ontologies/data schemas (structures & semantics) or between controlled vocabularies or also between categories/hierarchies.

Figure 6 illustrates an ontology-based interoperability scenario. In this architecture, distributed heterogeneous databases, DB_1, …, DB_n, use different structures and semantics. The structure and semantics of one of these databases, DB_i, are expressed using an ontology O_i, called local ontology. A mediator allows transparent access to the distributed databases, by offering users a common virtual ontology, G, called global ontology. The global ontology is linked to the local ontologies using mappings. Users pose queries over the global ontology. The mediator uses the ontology mappings to translate the a user query into queries over DB_1, …, DB_n. Responses are then collected, restructured and translated according to ontology G, then integrated and returned to the user.

This scenario is quite common in interoperability and mediation problems. The International Coastal Atlas Network (ICAN) community developed a coastal web atlas mediator that uses a similar approach, which will be described below.

THE ICAN INTEROPERABILITY APPROACH

The ICAN community developed an ontology-based mediator for coastal web atlases. The current version of the ICAN prototype (ICAN, 2009) supports the OGC Catalogue Service for the Web (CSW). Each atlas delivers metadata records through a CSW.

Figure 6. Interoperability scenario with ontology mappings. A user query is posed over a virtual common ontology, called global ontology. A mediator uses mappings between the global ontology and the local ones (O_1, ..., O_n) to rewrite the user's query into queries supported by the local databases (DB_1, ..., DB_n). Responses are integrated by the mediator and returned to the user conforming to the global ontology.

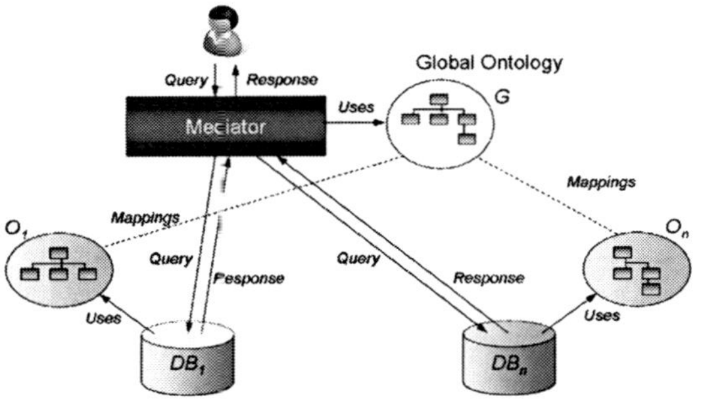

Figure 7. ICAN global theme ontology. Terms highlighted in Blue are those selected by the user

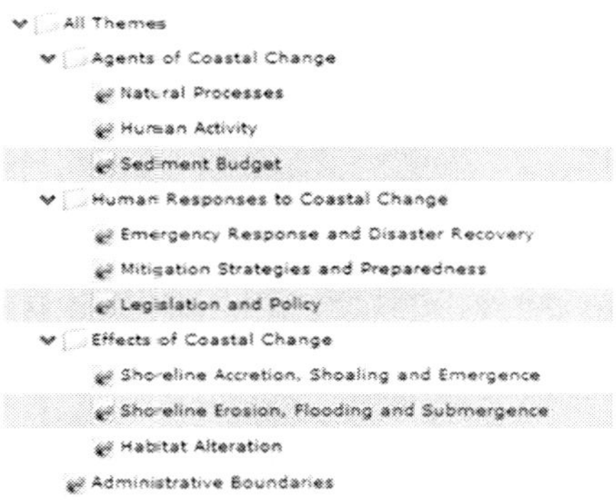

Figure 8. ICAN ontology-based CSW mediation architecture

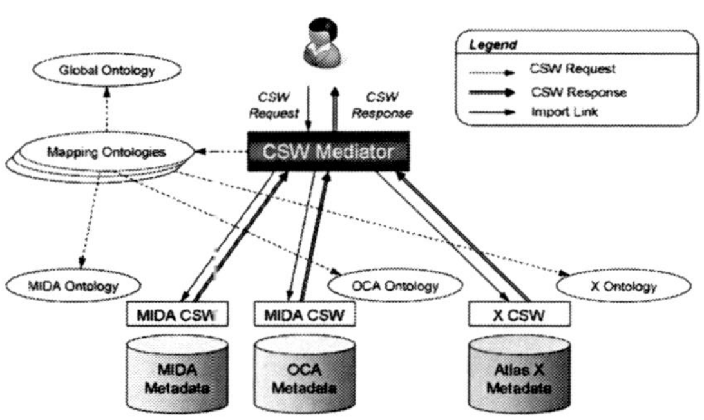

Architecture

The ICAN approach uses ontologies for both facilitating interoperability and semantic data discovery. Metadata records for a given atlas use the terms of an ontology, called local ontology. For instance the Marine Irish Digital Atlas uses its own ontology; the Oregon Coastal Atlas uses another one. The approach relies on a common ontology (global ontology) that defines common terms (keywords) for users of the ICAN mediator (c.f. Figure 7). Mappings between the global ontology and the local ones allow the mediator to translate terms from the global ontology to a local one.

The ICAN solution is based on a global atlas that offers a virtual catalogue service, called global catalogue, which acts as a CSW mediator and which offers unified and transparent access to the atlases' CSWs. As illustrated in Figure 8, users of the global CSW are provided with the global ontology. The user formulates a CSW request using an area of interest and keywords defined in the global ontology. The global CSW rewrites the user's request into CSW requests over the local atlases' CSWs using their local ontology terms, executes the so-obtained requests, and collects metadata records (responses) from local CWAs.

This architecture facilitates extendibility as new catalogue services can be added and removed at any time without affecting the global CSW, provided that they come with the ontologies for the terms used by their metadata records and that mappings between these terms and the global ontology's terms are provided.

A user of the ICAN mediator may select one or more keywords and an area of interest (bounding box) and submit a search request to the mediator. The mediator translates the terms selected by the user into local terms and submits a request to the local atlases using terms from their ontologies. Results of such requests are then gathered and returned to the user (c.f. Figure 9).

Figure 9. ICAN mediator interface. Results from MIDA and OCA corresponding to the keywords selected in Figure 8

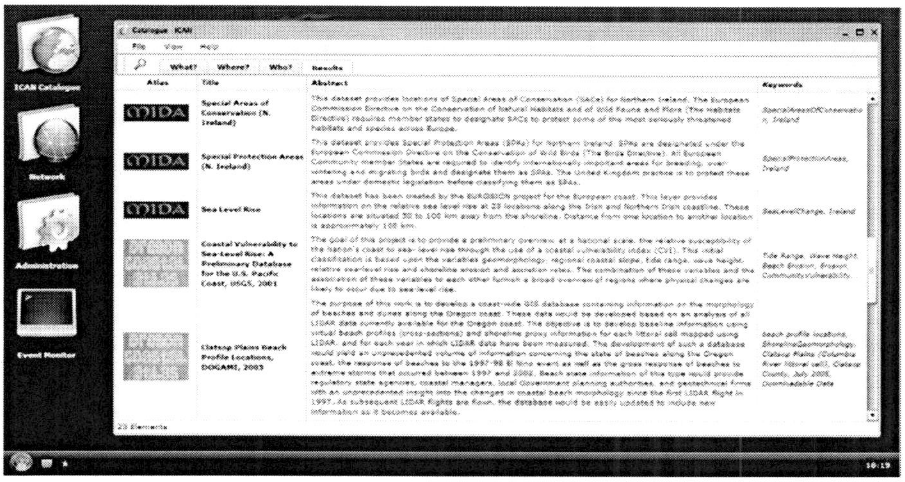

Global and Local Ontologies

A (global or local) CSW uses an ontology, which defines the terms used as values in its metadata records (for example thematic keywords, places, etc.). In the initial CSW mediator prototype, the ontologies define the terms for keywords provided as part of metadata. Conforming to the ISO-19115 standard, five types of keywords are defined: discipline, theme, place, temporal, and stratum. For each atlas, an ontology of terms related to these five keyword types is defined. Such an ontology is called local ontology. Relationships between the terms contained within one ontology are provided as part of the same ontology. Figure 8 shows the ICAN global ontology for theme keywords.

Ontology Mappings

Ontology mappings link the global ontology to the local ontologies. This link is crucial as it is the only means to allow the CSW mediator rewrite user requests expressed with terms from the global ontology into requests over the local CSWs using their own terms (i.e., terms from local ontologies). Therefore they act as semantic translators.

As OWL comes with a powerful mechanism for linking concepts both within and across ontologies, mappings are defined in separate OWL ontologies. For each local ontology, a mapping ontology defines the mappings between the local ontology and the global one. A mapping ontology imports both a local and the global ontology and defines relationships between their concepts. An example showing extracts of the MIDA and the OCA mapping ontologies is illustrated in Figures 10 and 11.

In Figures 10 and 11 terms preceded by prefix "global" are from the global ontology. Those preceded by prefixes "mida" and "oca" are respectively from the MIDA and OCA ontologies. Relationships represented with thin lines are defined as part of the local ontologies. Those represented with thick lines are defined as part of the mapping ontologies. For example, Coastal Protection and Shore Stabilization are defined in the MIDA and OCA mapping ontologies as narrower terms than Human Responses to Coastal Change.

Figure 10. Extracts of the MIDA mapping ontology

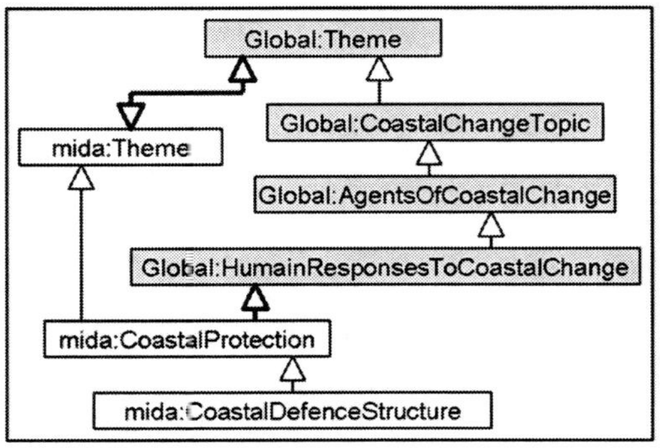

Figure 11. The OCA mapping ontology

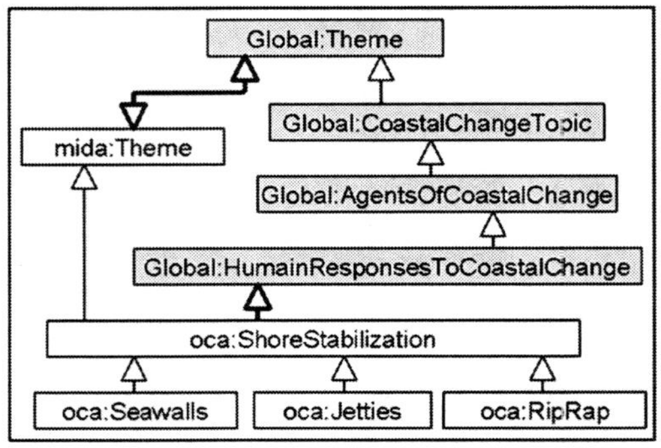

Query Rewriting and Execution

Query rewriting is the most important task of the mediator as it refers to the process of rewriting a user query posed over a global schema or ontology into queries over local data sources (in this context CSWs). We mainly focus on the GetRecords CSW requests in this section. Consider a human or software user that poses a GetRecords request, searching for metadata records available through the global CSW. The user might be interested only in records corresponding to data within a given extent and related to some given keywords. He or she would formulate his request using a filter. For instance, the following is a GetRecords request formulated by a user interested in metadata records about data covering any region all over the world and related to *Human Responses to Coastal Change*.

```
http://ican.ucc.ie/srv/en/csw?
request=GetRecords&service=CSW&v
ersion=2.0.1&resultType=results
&namespace=csw:http://www.open-
gis.net/cat/csw&maxRecords=1000
&elementSetName=summary&constrai
ntLanguage=FILTER
&constraint_language_ver-
sion=1.1.0
&constraint=
<?xml version="1.0"
encoding="UTF-8"?>
<Filter xmlns=http://www.open-
gis.net/ogc xmlns:gml="http://
www.opengis.net/gml"
 xmlns:csw="http://www.opengis.
net/cat/csw/2.0.2">
      <And>
            <PropertyIsLike wild-
Card="%" singleChar="_" es-
cape="\">

<PropertyName>keyword</Property-
Name>
                        <Literal>Hu
manResponsesToCoastalChange%</
Literal>
                        </Property-
IsLike>
            <BBOX>
                  <PropertyName>/
csw:Record/ows:BoundingBox</
PropertyName>
                  <gml:Envelope
                  srsName="http://
www.opengis.net/gml/srs/epsg.
xml#4326">

<gml:lowerCorner>-180 -90</
gml:lowerCorner>

<gml:upperCorner>180 90</
```

```
gml:upperCorner>
              </gml:Envelope>
          </BBOX>
     </And>
</Filter>
```

Query rewriting consists in translating the global keywords contained in the user's queries into local keywords and rewriting the initial request using the so-obtained terms. In order to do so, the mediator starts by parsing the user's request. It identifies the clauses related to keywords, and extracts the corresponding keyword literals. For instance, in the example above, the only clause related to keywords is the following, and the corresponding keyword is Human Responses to Coastal Change.

```
<PropertyIsLike wildCard="%"
singleChar="_" escape="\">
     <PropertyName>keyword</
PropertyName>
     <Literal>HumanResponsesToCo
astalChange%</Literal>
</PropertyIsLike>
```

For each local atlas, the CSW mediator uses its inference engine to obtain all the local atlas' terms that are equivalent to, or narrower than, the keyword literal considered. Next, the initial clause is replaced by a disjunction of clauses, each containing a keyword literal corresponding to one of the so-obtained local keywords. For instance, in the example above, the CSW mediator will translate keyword Human Responses to Coastal Change into the MIDA keywords Coastal Protection and Coastal Defence Structure. Thus the corresponding clause will be rewritten according to MIDA as follows.

```
<Or>
     <PropertyIsLike wild-
Card="%" singleChar="_" es-
cape="\">

<PropertyName>keyword</Property-
Name>

<Literal>CoastalProtection%</
Literal>
     </PropertyIsLike>
     <PropertyIsLike wild-
Card="%" singleChar="_" es-
cape="\">

<PropertyName>keyword</Property-
Name>
          <Literal>CoastalDefenc
eStructure%</Literal>
     </PropertyIsLike>
</Or>
```

This process is repeated for each clause in the request's filter, of course by avoiding repetition of keywords in a disjunction of clauses. Each so-obtained final GetRecords request is sent to the corresponding local CSW. Records obtained as results from the local CSWs are then collected and sent back to the user.

Implementation and Future Work

An initial version of the global coastal atlas prototype has been implemented in Java, using the Jena 2 framework (Jena, 2009) for inferring and is available at http://ican.ucc.ie. The prototype allows the user to:

- Select keywords from a list of "global" keywords;
- Select an area of interest by dragging a bounding box in a map area;
- Submit their query.

The CSW mediator will consult a registry of atlases and identify the atlases that may have data within the bounding box selected, as a bounding box is associated with each atlas representing its geographic extent in order to optimize query execution by avoiding rewriting queries over CSWs with no data covering the area of interest. Next, the atlas mediator will translate the user's request according to the process described in subsection 4.8.3, collect the responses and send them back to the user through the graphical interface.

Current work is focusing on integrating Web Map Services for overlaying maps from different CWAs. Future work will integrate web feature services in order to integrate the datasets available through the CWAs.

CONCLUSION

The atlas mediator prototype described in this chapter is a first step towards atlas interoperability.

The prototype showcases how ontologies and ontology mappings can be used to integrate different heterogeneous and autonomous atlases (or information systems), particularly OGC CSWs. It will be tested further by members of ICAN (30 organizations from over a dozen nations). The next step of the ICAN initiative is to integrate web map services and web feature services in order to define a more complete approach for integrating maps, data and metadata. Also, thematic information will be considered and interfaces will be specified for sharing this type of information, which is highly important in atlases. The aim is to define a complete solution for integrating CWAs.

ACKNOWLEDGMENT

This work has been produced as part of the Geoscientific Data Integration (GeoDI), the Griffith Geomatics for Geosciences, and the Marine Metadata Interoperability (MMI) projects. GeoDI is a three-year project which started in February 2008 and which is funded by the Irish National Development Plan (NDP) under the SeaChange programme. The three-year Griffith Geomatics for Geosciences project started in June 2008 and is based on research grant-aided by the Department of Communications, Energy and Natural Resources under the National Geoscience Programme 2007-2013. The views expressed in this study are the authors' own and do not necessarily reflect the views and opinions of the Irish Minister for Communications, Energy and Natural Resources. MMI is funded by the National Science Foundation (OCE0607372, OCE0606959).

REFERENCES

Alexander, P. (2009). *Core Technologies for Ontologies*. Retrieved April 09th, 2009 from The MMI Guides: Navigating the World of Marine Metadata website http://marinemetadata.org/guides/vocabs/ont/coretech

Asch, K., Brodaric, B., Laxton, J. L., & Robida, F. (2004). *An International Initiative for Data Harmonization in Geology. In 10th EC GI & GIS Workshop*. Warsaw, Poland: ESDI State of the Art.

Beckett, D. (2004). *RDF/XML Syntax Specification, W3C Recommendation 10 February 2004*. Retrieved July 25th, 2009 from W3C web site: http://www.w3.org/TR/rdf-syntax-grammar/

Beckett, D., & Berners-Lee, T. (2008). *Turtle – Terse RDF Triple Langauge, W3C Team Submission 14 January 2008*. Retrieved July 24th, 2009 from W3C web site http://www.w3.org/TeamSubmission/turtle/

Bermudez, L. E., Bogden, P., Bridger, E., Galvarino, C., Graybeal, J., Forest, D., et al. (2009). Web Feature Service (WFS) and Sensor Observation Service (SOS) comparison to publish time series data. *International Symposium on Collaborative Technologies and Systems - Workshop on Sensor Web Enablement*, Baltimore, MD.

Bianchini, D., De Antonellis, V., & Melchiori, M. (2003). Domain ontologies for virtual knowledge sharing and service composition in virtual district. In *Proceeding of the International Workshop on Web Semantics* (WebS2003), Prague, Czech Republic, September 1-5 2003.

Bishr, Y. (1998). Overcoming the semantic and other barriers to GIS interoperability. *International Journal of Geographical Information Science*, *12*(4), 299–314. doi:10.1080/136588198241806

Brickley, D., & Guha, R. V. (2004). *RDF Vocabulary Description Language 1.0: RDF Schema, W3C Recommendation, 10 February 2004*. Retrieved July 16th, 2009 from W3C website http://www.w3.org/TR/rdf-schema/

Buitelaar, P. Eigner Thomas, & Declerck T. (2004). OntoSelect: A Dynamic Ontology Library with Support for Ontology Selection. In *Proceedings of the Demo Session at the International Semantic Web Conference*, Hiroshima, Japan.

CEN. (2009). *European Committee for Standardisation*. Retrieved Septembr 24th, 2009 from the website http://www.cen.eu

Cerami, E. (2002). *Web Services Essentials - Distributed Applications with XML-RPC, SOAP, UDDI & WSDL*. Sebastopol, CA: O'Reilly Media.

CSML. (2009). *Climate Science Modelling Language*. Retrieved September 20th, 2009 from the website http://ndg.nerc.ac.uk/csml/

de la Beaujardière, J. (2004). *OGC Web Map Service Interface (Version 1.3.0), Open Geospatial Consortium Inc*. Retrieved May 18th, 2009 from OGC portal http://portal.opengeospatial.org/files/?artifact_id=4756

Garcia-Molina, H., Papakonstantinou, Y., Quass, D., Rajaman, A., Sagir, Y., & Ullman, J. (1997). The TSIMMIS approach to mediation: Data models and Languages. *Journal of Intelligent Information Systems*, *8*, 117–132. doi:10.1023/A:1008683107812

GeoSciML. (2009). Retrieved September 20th, 2009 from the web site http://www.geosciml.org/

Goasdoue, F., Lattes, V., & Rousset, M.-C. (2000). The Use of CARIN Language and Algorithms for Information Integration: The PICSEL System. *International Journal of Cooperative Information Systems*, *9*(4), 383–401. doi:10.1142/S0218843000000181

Gruber, T. (1993). A Translation Approach to Portable Ontology Specification. *Knowledge Acquisition*, *5*(2), 199–220. doi:10.1006/knac.1993.1008

Heydari, N., Mansourian, A., Taleai, M., & Fallahi, G. R. (2009). Ontology-based GIS web service for increasing semantic interoperability among organizations involving drilling in city of Tehran. In *GSDI 11 World Conference,* Rotterdam, The Netherlands.

ICAN. (2009). International Coastal Atlas Networks. Retrieved October 21th, 2009 from the website http://ican.ucc.ie/

INSPIRE. (2009). Retrieved September 24th, 2009 from the website http://inspire.jrc.ec.europa.eu

Isenor, A., & Neiswender, C. (2009). Categories of Controlled Vocabularies. In *The MMI Guides: Navigating the World of Marine Metadata*. Retrieved September 15th, 2009 from Marine Metadata Interoperability website http://marinemetadata.org/guides/vocabs/voctypes/voccat

Isenor, A. W., & Spears, T. W. (2007). A conceptual model for service-oriented discovery of marine metadata descriptions. *CMOS Bulletin*, *35*(3), 85–91.

ISO/IEC. (1993). ISO 2382-1: Information technology – Vocabulary – Part 1: Fundamental terms. *International Organization for Standardization, 3.0 edition.*

ISO/TC211. (2003). [– *Geographic Information – Metadata*. Amsterdam: International Organization for Standardization.]. *ISO*, *19115*, 2003.

ISO/TC211. (2006). *ISO/PRF TS 19139 – Geographic information – Metadata – XML schema implementation*. Amsterdam: International Organization for Standardization.

ISO/TC211. (2007). [*Geographic information – Geography Markup Language][GML] [*. Amsterdam: International Organization for Standardization.]. *ISO*, *19136*, 2007.

ISO/TC211 Geographic information/Geomatics. (2009). Retrieved September 24th, 2009 from the website http://www.isotc211.org

ISO/TC 46/SC 4. (2003). *ISO 15836:2003: Information and documentation – The Dublin Core metadata element set*. Amsterdam: International Organization for Standardization.

Jena. (2009). *Jena 2 Inference support*. Retrieved September 14th, 2009 from the website http://jena.sourceforge.net/inference/

Kemp, Z., Tan, L., & Whalley, J. (2007). Interoperability for Geospatial Analysis: a Semantics and Ontology-based Approach. In J. Bailey & A. Fekete, (Eds.), *Proc. Eighteenth Australasian Database Conference (ADC 2007),* Ballarat, Australia, (CRPIT, Vol. 63, pp. 83-92). ACS.

Kirk, T., Levy, A. Y., Sagiv, Y., & Srivastava, D. (1995). The Information Manifold . In Knoblock, C., & Levy, A (Eds.), *Information Gathering from Heterogeneous, Distributed Environments*. Stanford, CA: Stanford University.

Lacy, L. W. (2005). *OWL: Representing Information Using the Web Ontology Language*. Victoria, Canada: Trafford Publishers.

Lake, R., Burggraf, D. S., Trnini'c, M., & Rae, L. (2004). *Geography Mark-Up Language (GML) – Foundation for the Geo-Web*. New York: John Wiley.

Maedche, A., Motik, B., Stojanovic, L., Studer, R., & Volz, R. (2003). Ontologies for enterprise knowledge management. *IEEE Intelligent Systems*, *18*(2), 26–33. doi:10.1109/MIS.2003.1193654

Manolescu, I., Florescu, D., & Kossmann, D. (2001). Answering XML queries over heterogeneous data cources. In *Proceedings of VLDB*, (pp. 241–250).

McGuinness, D. L., & van Harmelen, F. (2004). OWL Web Ontology Language – Overview, W3C Recommendation 10 February 2004. Retrieved January 20th, 2009 from W3C website http://www.w3.org/TR/owl-features/

Miles, A., & Bechhofer, S. (2009). *SKOS Simple Knowledge Organization System Reference, W3C Recommendation 18 August 2009*. Retrieved September 10th, 2009 from W3C website http://www.w3.org/TR/2009/REC-skos-reference-20090818/

Miller, S. P., Bermudez, L., Montgomery, E., Isenor, A., & Neiswender, C. (2009). Vocabularies: Dictionaries, Ontologies, and More. *The MMI Guides: Navigating the World of Marine Metadata*. Retrieved September 17th, 2009 from The Marine Metadata Interoperability website http://marinemetadata.org/guides/vocabs

Nebert, D., & Whiteside, A. (2005). *OGC Catalogue Services Specification*. Open Geospatial Consortium Inc. Retrieved May 6th, 2009 from OGC website http://portal.opengeospatial.org/files/?artifact_id=5929

Neiswender, C. (2009). What is a Controlled Vocabulary? *The MMI Guides: Navigating the World of Marine Metadata*. Retrieved September 17th, 2009 from Marine Metadata Interoperability website http://marinemetadata.org/guides/vocabs/vocdef

O'Dea, L., Dwyer, N., Cummins, V., & Wright, D. (2007). Potentials and limitations of coastal web atlases: Outcomes of a transatlantic coastal mapping workshop. In *Proceedings of CoastGIS 2007*, Santander, Spain.

OGC. (2009). Retrieved September 24th, 2009 from the website http://www.opengeospatial.org

ORR. (2009). Pntology Registry and Repository. Retrieved September 10th, 2009 from the website http://mmisw.org/or

POL. (2009). *Protege Ontology Library*. Retrieved from the website http://protegewiki.stanford.edu/index.php/Protege_Ontology_Library

Powers, S. (2003). *Practical RDF*. Sebastopol, CA: O'Reilly Media.

Ray, E. T. (2003). *Learning XML* (2nd ed.). Sebastopol, CA: O'Reilly Media.

RDF. (2009). *Resource Description Framework (RDF)*. Retrieved September 14th, 2009 from the website http://www.w3.org/RDF/

Smits, P., Düren, U., Østensen, O., Murre, L., Gould, M., & Sandgren, U. (2002). *INSPIRE Architecture and Standards Position Paper*. Ispra, Italy: JRC-Institute for Environment and Sustainability.

Snell, J., Tidwell, D., & Kulchenko, P. (2001). *Programming Web Services with SOAP*. Sebastopol, CA: O'Reilly Media.

Sowa, J. F. (1999). *Knowledge Representation: Logical, Philosophical, and Computational Foundations*. Pacific Grove, CA: Brooks Cole Publishing Co.

SWSE. (2009). *Semantic Web Search Engines*. Retrieved September 14th, 2009 from the website http://esw.w3.org/topic/TaskForces/CommunityProjects/LinkingOpenData/SemanticWebSearchEngines

van der Vlist, E. (2002). *XML Schema - The W3C's Object-Oriented Description for XML*. Sebastopol, CA: O'Reilly Media.

Vretanos, P. A. (2002). *Web Feature Service Implementation Specification*. Open Geospatial Consortium Inc. Retrieved May 6th, 2009 from OGC website http://portal.opengeospatial.org/files/?artifact_id=7176

Wiederhold, G. (1992). Mediators in the Architecture of Future Information Systems. *IEEE Computer*, *25*(3), 38–49.

Wright, D. J. (2009). Spatial data infrastructures for coastal environments. In X. Yang (Ed.), *Remote Sensing and Geospatial Technologies for Coastal Ecosystem Assessment and Management*, (LNGC, pp. 91-112). Berlin: Springer-Verlag.

Wright, D. J., Watson, S., Bermudez, L., Cummins, V., & Dwyer, N. O'Dea, L., Nyerges, T., Benoit, G., Berman, M., Helly J., & Uhel, R. (2007). *Report on Coastal Mapping and Informatics Trans-Atlantic Workshop 2: Coastal Atlas Interoperability*. Oregon State University, Corvallis, Oregon, USA. Retrieved July 29, 2009 from http://workshop1.science.oregonstate.edu/final_rpt2.

Yétongnon, K., Suwanmanee, S., Benslimane, D., & Champin, P.-A. (2006). A Web-Centric Semantic Mediation Approach for Spatial Information Systems. *Journal of Visual Languages and Computing*, *17*(1), 1–24. doi:10.1016/j.jvlc.2005.04.002

KEY TERMS AND DEFINITIONS

Interoperability: The ability of several autonomous, heterogeneous and distributed information systems (e.g., coastal web atlases) to communicate and exchange resources (information, metadata, data, maps, etc.) or be used together despite their differences.

Interoperability Standard: A standard aiming to facilitate interoperability by providing common formats, languages, exchange mechanisms or protocols for communicating information between systems.

Semantic Interoperability: The condition where two or more computer systems are able to exchange information and have the meaning of that information accurately and automatically interpreted by the receiving system.

Catalogue Service for the Web: An interface for managing and delivering geographic metadata on the web.

Controlled Vocabulary: A consistent set of terms pertaining to a given domain or application with explicit meanings and relationships among them.

Ontology: A type of knowledge organization system that defines a set of concepts with meanings, relationships among them, and logical rules on these meanings and/or relationships.

Mediation: Mediation is a virtual data integration approach that relies on a mediator as a common interface for accessing a set of heterogeneous, autonomous and possibly distributed data bases or information systems.

Query Rewriting: In mediation, query rewriting is the process of translating a user's query over the common data structure and model into queries supported by the heterogeneous databases or information systems.

Ontology Mappings: Rules or relationships that define semantic links between terms from different ontologies, e.g., equivalence, broader term, narrower term, etc.

Section 2
Coastal Web Atlas Case Studies around the World

Chapter 5
Overview of Coastal Atlases

Dawn J. Wright
Oregon State University, USA

Gabe Sataloff
NOAA Coastal Services Center, USA

Tony LaVoi
NOAA Coastal Services Center, USA

Andrus Meiner
European Environment Agency, Denmark

Ronan Uhel
European Environment Agency, Denmark

ABSTRACT

This chapter provides a brief overview of various coastal web atlas projects around the world, providing a contextual bridge to the atlas case studies of Chapters 6-14. A summary of the policy context within which many European atlases operate is followed by a summary of other efforts emerging in Australia, the Western Pacific, Africa, and the Caribbean (as facilitated mainly by the Ocean Data and Information Network of the United Nations Educational, Scientific and Cultural Organization's International Oceanographic Data and Information Exchange). Atlas projects in the U.S. are summarized mainly via the results of a recent national survey of coastal managers reporting on the deployment and content of their atlases, with concluding thoughts on where there might be opportunities to develop approaches for a federated coastal atlas of the U.S.

INTRODUCTION

Having defined and explained the principles of coastal web atlas design, implementation, and interoperability, it is the task of Section 2 of the book to present various case studies of actual projects in action (Chapters 6-14). The aim of these case study chapters is to appreciate the diversity in approaches (e.g., lead agency or organization, intended audiences, software tech-

DOI: 10.4018/978-1-61520-815-9.ch005

Copyright © 2011, IGI Global. Copying or distributing in print or electronic forms without written permission of IGI Global is prohibited.

nologies used, degree of interactivity, available datasets, geographic extent of atlas, etc.) but also to understand, as pointed out in Hills *et al.* (2006) that certain aspects of good practice transcend these various categories, regardless of user, local or regional situations and culture. For example, a recent U.S. workshop hosted by the Washington State Department of Ecology and National Oceanic and Atmospheric Administration (NOAA) Coastal Services Center, brought together over thirty participants from Alaska, British Columbia, Washington, Oregon and California, representing twelve different coastal web atlas (CWA) projects (including a legislative atlas and digital multipurpose marine cadastre for the entire U.S.). These atlases were presented in detail, but there were also discussions and a data table drawn up on important comparisons between the atlases from which to learn (e.g., distinguishing features, data included, challenges encountered, future directions, textual vs. map content, geographic extent, inland vs. marine extent, limits to number of displayed layers, public access to layers, and more; Purce, 2009), and how atlases may ultimately work together for regional ocean governance.

This chapter provides a gateway to the case studies of the chapters that follow, by first summarizing very briefly the policy context within which the European atlases are developing, four of which are described in detail in Chapters 7, 10, 13, and 14. Next, a summary of similar efforts in Australia, the Western Pacific, Africa, and the Caribbean are included, two of which are detailed in Chapters 11 and 12, but the rest are either nascent efforts not yet resulting in mature atlases or for which full case studies are not yet available. Finally, an overview of atlas efforts in the U.S. is given from the standpoint of a recent national survey of coastal managers who reported on the characteristics of their atlases, the audiences served, and the various technical issues. This provides some context for the U.S. case study examples presented in Chapters 6, 8, and 9.

OVERVIEW OF EUROPEAN UNION ATLASES

Recently a number of important policy decisions and developments have taken place in the European Union (EU) regarding the management of coastal, marine and maritime resources. These will shape the design, functioning and sharing of coastal and marine information services in the coming years, including the design and implementation of CWAs. One critical need is to streamline monitoring and reporting activities that support the production of policy-relevant assessments of the marine and coastal environment, including an emphasis on ecosystem-based management issues. Several activities at the EU level have been initiated and/or further developed in 2007-2009 to prepare the production of regular indicator-based assessments and the delivery of information services over the period 2009 to 2012 in support to new policies adopted by the EU.

Interoperable coastal information systems and useful operational services are becoming of more use to a large community of practitioners and users at the coastal zones across the world. In Europe, this is in particular important in light of the emerging Integrated Maritime Policy (IMP) of the EU, where initiatives such as the web-based European Atlas of the Seas, broad-scale seabed habitat mapping and promotion of maritime spatial planning have prominent roles (European Commission, 2007; Vivero *et al.* 2007; see also Chapter 16).

A 2008 EU marine environmental law (Marine Strategy Framework Directive or MSFD) aims at applying an ecosystem-based approach to the management of human activities. There is work going on towards the establishment of a limited set of pan-European common indicators for the marine environment by 2010. It should be noted that "pan-European" includes the whole continent from the Urals to Portugal, but at this time only the EU member states will be participating.

Within the framework of the EU initiative on global monitoring for environment and security (GMES; joint initiative of the European Commission and the European Space Agency), the EU is developing operational oceanography services in line with Global Ocean Observing System community needs. Operational oceanography has been identified as a GMES fast track service. These core and derivative web services services are to be integrated into European Marine Observation and Data Network (EMODNET) and a future European marine information system and thus be made easily available to the European user community; development of coastal information systems also facilitates the coordination of on-ground monitoring data in the context of GMES.

As part of the implementation of the EU Recommendation on Integrated Coastal Zone Management (ICZM) the Member states together with EU organisations have identified core sets of indicators for sustainable development and ICZM effectiveness. Regional and national authorities have tested this set, with new activities planned in the Mediterranean. Also, the identification and mapping of important biotopes, ecosystems and protected areas - under EU nature directives - is being gradually extended to coastal and marine areas with the view to create a coherent view of ecological networks. These activities include a focus on coastal zone use potentials, vulnerabilities and adaptations to environmental change.

European cooperation in coastal atlas development is beneficial for the integrated assessment of coastal zones and marine environment and wider implementation of Shared Environmental Information System (SEIS) initiative and the INfrastructure for SPatial InfoRmation in Europe (INSPIRE) directive (e.g., Masser, 2007). Key objectives of such cooperation are:

- Achieving the atlases interoperability at the operational level (proof-of-concept) and validating first outcomes;
- Informing and attracting a larger population of potential stakeholders of the activity, and promoting an exchange of best practice in coastal and ocean information services;
- Developing a long-term strategy for effective governance of coastal atlases and design of further applications for International Coastal Atlas Network or ICAN (Dwyer and Wright 2008; see also Chapter 15);
- To offer information support for integrated maritime governance of coastal regions and implementation of maritime spatial planning.

This European experience can be shared via international cooperation by way of specific projects contributing to global monitoring and assessment activities, especially under the Global Earth Observation System of Systems (GEOSS). GEOSS promotes scientific, data-centric connections networks of observational systems (hence the "system of systems"), while also promoting applications across nine societal benefit areas (water, climate, weather, ecosystems, energy, health, agriculture, disasters, biodiversity; Craglia *et al.*, 2008). Cooperation should also support the UN initiative on "Assessment of Assessments" (Global Regular Assessment of Marine Environments or GRAME). GRAME aims to evaluate existing marine assessments and to propose a framework for a regular reporting process on the state of the marine environment globally.

ATLASES IN AUSTRALIA, THE PACIFIC, AFRICA AND THE CARIBBEAN

There are undoubtedly scores of CWAs around the world that we have yet to learn about, and hence are beyond the scope of this chapter and this book. But this section briefly touches on efforts that were presented at a third workshop of the International Coastal Atlas Network (ICAN) in Copenhagen, Denmark (Dwyer and Wright, 2008). With regard to Australia, there was a coastal atlas project launched in 1998, originally conceived of as a network of Australian government and state/territory nodes using a variety of interactive mapping tools. The nodes were largely managed by the Commonwealth of Australia, as well as by various states and territories. The commonwealth node has now been incorporated into an Australian Natural Resources Atlas, which is primarily focused on terrestrial data. Coastal data can be found primarily in the New South Wales Coastal Atlas.

Another major effort is the Australian Marine Spatial Information System developed by Geoscience Australia, which contains over 80 layers of information including maritime boundaries, bathymetry, physical and environmental information, legal interests, fisheries and shipping, extending beyond the coastal realm to the Australian Exclusive Economic Zone and in the Antarctic.

In the Pacific region the Regional Sub-Commission for the Western Pacific (WESTPAC), within Intergovernmental Oceanographic Commission or IOC of UNESCO, focuses on small Pacific island nation states and their vulnerabilities to climate change. IOC's International Oceanographic Data and Information Exchange (IODE) is in the process of establishing an Ocean Data and Information Network (ODIN)-WESTPAC to strengthen regional networking of marine related libraries and information resources in the region. Similarly, the South Pacific Applied Geoscience Commission (SOPAC), administered and funded by member counties, as well as the EU, the Office of US Foreign Disaster Assistance, and several UN agencies, has set up MapServer installations on fourteen Pacific islands.

The IODE ODIN program in general links training, equipment, and operational support (i.e., how to set up data centers) within regional contexts focused on Africa, South America, the Caribbean, and Pacific. Two major ODIN efforts are the African Marine Atlas and the Caribbean Marine Atlas (see Chapters 11 and 12). Figure 1 also diagrams related efforts throughout Africa.

Figure 1. Example of the many partners of, contributors to and linkages with the African Marine Atlas project at global, continental, regional, and national levels from Dwyer and Wright (2008). Acronym definitions available online at: http://ican.science.oregonstate.edu/acronyms. (© 2009, Oregon State University & University College Cork. Used with permission.)

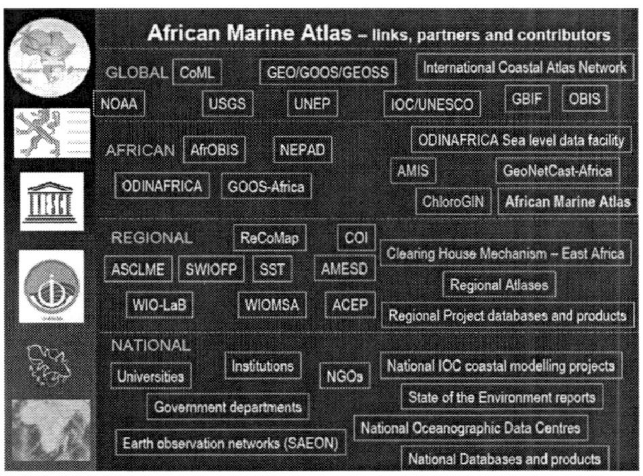

OVERVIEW OF U.S. ATLASES: A TALE OF TWO SURVEYS

The recent reports of the Pew Oceans Commission and the U.S. Commission on Ocean Policy (Pew Oceans Commission, 2003; Juda, 2005) have clearly shown that coastal communities are critical to the economy of the U.S., and to its overall health and well-being as a nation, and further that geographic technologies will be a fundamental, critical tool to address the threats of climate change, coastal hazards, overpopulation, and more. In a 2006 survey of coastal managers, the National Oceanic and Atmospheric Administration (NOAA) Coastal Services Center (Center) found that almost 75% of the respondents considered the geographic information system (GIS) to be a highly useful tool in their jobs (NOAA Coastal Services Center, 2006). According to Crossett *et al.* (2004), coastal counties in the U.S. are home to over 53% of the population, while these counties consist of only 17% of the land mass. High population densities and the coastal regions' need for geographic information make coastal atlases an essential resource for a range of coastal management activities. In addition to providing geographic information to coastal managers, the atlases can also provide a valuable educational resource to the local communities on ecosystems, coastal hazards, and a vast number of other topics.

Though almost all coastal states and territories provide access to geographic information and related tools and materials in atlases, there is very little coordination between neighboring states or regions on the format of the information provided. Groups such as ICAN seek to document best practices for CWAs (see Chapter 15), but there are no requirements or compelling business drivers for states or other groups to follow the best practices or coordinate across state boundaries. This may change as various regional ocean governance bodies are formed within the United States to address ocean and coastal management issues on a regional basis.

In October of 2008, ICAN and the Center performed an informal survey of the coastal atlas sites available for each coastal state and territory to better understand what type of information is being provided, in what manner, and where there might be opportunities to develop approaches for a federated coastal atlas of the U.S. The survey consisted of linking to atlas sites provided by the Office of Ocean and Coastal Resource Management's partner states website (OCRM, 2009). This was then followed by further Internet searches using keyword combinations, including individual state names, "atlas," "coastal atlas," "coastal zone management," "GIS," and "coastal maps." This Internet-based search allowed for the collection of information on the types of geographic mapping resources available (static vs. dynamic), main data themes, and availability of other resources (tools, outreach materials, permitting information, etc). The results were recorded in Microsoft Excel, and some basic statistics were tabulated (See Appendix).

Of the thirty-four coastal states and territories in the U.S., all but one had at least one site with a mapping component (Figure 2 and see Appendix). Fifteen of the thirty-four (44%) coastal states and territories had multiple atlases. Of the fifty-five total atlas sites surveyed, forty-four (80%) had a mapping component available, and, of those forty-four, thirty (68%) had interactive maps. The interactive maps themselves were quite varied. Some sites had clickable PDF or JPEG images, while others contained interactive mapping applications that use ESRI's ArcIMS software, open source map servers, or mashups (web applications that combines data or functions from two or more sources – e.g., Google Maps with weather forecasting information – to create a new service). Mapping tools were fairly consistent and primarily included the ability to zoom in and out, pan around the map, turn map layers on and off, and identify features. Some atlases had additional

Figure 2. Map of the thirty-four coastal states of the U.S., including the Great Lakes region. (© 2009, under the auspices of the GNU Free Documentation License, Creative Commons Attribution ShareAlike 3.0, http://creativecommons.org/licenses/by-sa/3.0/. Used with permission.)

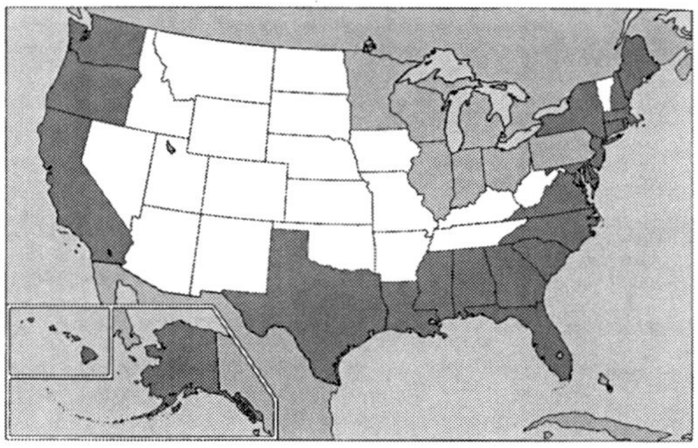

map tools, such as Alaska's dynamic nearshore fish query and Washington State's Percent Land Cover tool.

The data themes available were disparate and varied according to the principal coastal management issues for each given state. In addition, the geographic scope of the data was very different for many of the atlases. Some atlases were focused solely on the coastal region, while others were statewide geographic clearinghouses that contained coastal data. The Center found in its prior survey of coastal managers (NOAA Coastal Services Center, 2006) that the most common datasets used for coastal management were current shoreline, coastal land use and land cover, bathymetry, elevation and topography, sensitive habitats, protected areas, and public access areas. These data types, though not quantified in the current survey, were among the most commonly seen.

In addition to maps, most of the atlases had online technical assistance resources. Two of the most common resources were the ability to fill out or search building or dock permits in the coastal area, and the listing of laws and regulations specific to the coastal region. Many sites had educational materials available, ranging from short overviews of the coastal region ecosystem to full scientific white papers. Only a few of the atlas sites provided additional tools to aid coastal managers with decision-making. Examples of decision-making tools include the Oregon Coastal Atlas' Coastal Risk and Vulnerability Assessment Tool (documented also in Haddad *et al.*, 2005), the Illinois Ecological Compliance Assessment Tool (EcoCAT), the Gulf of Maine's Northeast Regional Ecosystem Assessment, and the Michigan Coastal Management Program's Lake Superior Decision Support Tool.

CONCLUDING THOUGHTS

CWAs are an important and growing resource for coastal managers, scientists, and the public living within those areas. Though most U.S. coastal states and territories have a site to provide information about their coastal regions, there is little consistency in how the information is presented. This is still also the case in Europe and other regions of the world. Survey results show that coastal areas each have their own issues to address, which is one reason why there is little similarity between the types of information provided. Larger-scale issues such as climate change will undoubtedly provide an outlet for different states and regions to address them in a unified manner, through an interconnected network of atlases (see Chapters 15 and 16).

Mapping interfaces of atlases will be another area for future growth within a federated approach. With the advent of mashup mapping technologies, it has become easier for novice developers to create informative maps for little to no cost. With more Web map applications appearing across the Internet, it is likely that coastal managers and the public have become more aware of and more comfortable with the use of geographic mapping services within their websites. CWA developers should continue to take advantage of these technologies to better serve their constituents in the coastal zone.

To comply with federal and industry standards for geographic data management in the U.S., federal agencies and state and local governments must meet the Office of Management and Budget (OMB) requirements outlined in the OMB Circular A-16 (OMB, 2009) mandate by documenting all GIS and geographic data using the Federal Geographic Data Committee's (FGDC) Content Standard for Digital Geospatial Metadata (CSDGM; FGDC, 2009), and make their metadata publicly accessible via the FGDC Clearinghouse's E-Government Geospatial One-Stop (GOS) portal (GOS, 2009), as well as the Data.gov portal (Data.gov, 2009). Adherence to OMB and other federal and industry standards (such as the Open Geospatial Consortium, or OGC, and International Standards Organization or ISO), for geographic data and metadata is a critical component of data quality, data management, data distribution and access, and data interoperability. As described in Section 1 of the book, data layers and maps used in CWAs should adhere to FGDC or ISO metadata standards to ensure that all data and maps are properly documented and made accessible to the public for use.

And finally, though many states and countries have several CWAs with different foci (e.g., ecology, hazards, etc.), the atlases do not always link to one another, making it still difficult for a user to find information (solutions for which were proposed in Chapter 4). Therefore, states and nations should seek neighboringneighbouring partnerships to better provide relevant coastal information for regional ocean governance (as evidenced by the recent west coast coastal atlas workshop mentioned at the outset of this chapter).

REFERENCES

Craglia, M., Goodchild, M. F., Annoni, A., Camara, G., Gould, M., & Kuhn, W. (2008). Next-generation Digital Earth: A position paper from the Vespucci Initiative for the Advancement of Geographic Information Science. *International Journal of Spatial Data Infrastructures Research*, *3*, 146–167.

Crossett, K. M., Culliton, T. J., Wiley, P. C., & Goodspeed, T. R. (2004). *Population Trends along the Coastal United States: 1980-2008. Coastal Trends Report Series*. Silver Spring, MD: U.S. Department of Commerce, National Oceanic and Atmospheric Administration.

Data.gov. (2009). Retrieved October 31, 2009 from the Data.gov web site: http://www.data.gov.

Dwyer, N., & Wright, D. J. (2008). *Report of International Coastal Atlas Network Workshop 3 on Federated Coastal Atlases: Building on the Interoperable Approach*, European Environment Agency, Copenhagen, Denmark. Retrieved July 29, 2009 from http://ican.science.oregonstate.edu/ican3_final_rpt

European Commission. (2007). *An Integrated Maritime Policy for the European Union*. Commission of the European Communities. Brussels, 10 October 2007, COM (2007) 575 final. Retrieved July 29, 2009 from the European Commssion website http://eur-lex.europa.eu/LexUriServ/site/en/com/2007/com2007_0575en01.pdf

Federal Geographic Data Committee. (2009). *FGDC Content Standard for Digital Geospatial Metadata*. Retrieved October 31, 2009 from the Federal Geographic Data Committee website http://www.fgdc.gov/metadata/csdgm.

Geospatial One-Stop. (2009). Retrieved October 31, 2009 from the Geospatial One-Stop website http://gos2.geodata.gov/wps/portal/gos

Geospatial One-Stop. (2009). Retrieved October 31, 2009 from the Geospatial One-Stop website http://gos2.geodata.gov/wps/portal/gos

Haddad, T. C., Wright, D. J., Dailey, M., Klarin, P., Marra, J., Dana, R., & Revell, D. (2005). The tools of the Oregon Coastal Atlas. In Wright, D. J., & Scholz, A. J. (Eds.), *Place Matters: Geospatial Tools for Marine Science, Conservation and Management in the Pacific Northwest* (pp. 134–151). Corvallis, OR: Oregon State University Press.

Hills, J. M., Alcock, D., Higham, T., Kirkman, H., Le Tissier, M., & Pagdilao, C. (2006). Capacity building for integrated coastal management in Asia-Pacific: The case for case studies. *Coastal Management*, *34*(3), 323–337. doi:10.1080/08920750600686703

Juda, L. (2005). The report of the U.S. Commission on Ocean Policy: State perspectives. *Coastal Management*, *34*(1), 1–16. .doi:10.1080/08920750500364930

Masser, I. (2007). *Building European Spatial Data Infrastructures*. Redlands, CA: ESRI Press.

NOAA Coastal Services Center. (2006). *2006 Coastal Resource Management Customer Survey. U.S. Department of Commerce (DOC), National Oceanic and Atmospheric Administration (NOAA), National Ocean Service (NOS), Coastal Services Center*. CSC.

Ocean and Coastal Rsource Management. NOAA. (2009). *States and Territories Working With NOAA on Ocean and Coastal Management*. Retrieved October 31, 2009 from the Ocean and Coastal Resource Management web site: http://coastalmanagement.noaa.gov/mystate/welcome.html

Office of Management and Budget, Executive Office of the President of the United States. (2009). *Circular No. A16 Revised.* Retrieved October 31, 2009 from the Office of Management and Budget web site: http://www.whitehouse.gov/omb/rewrite/circulars/a016/a016_rev.html.

Pew Oceans Commission. (2003). *America's Living Oceans: Charting a Course for Sea Change. A Report to the Nation*. Arlington, VA: Pew Oceans Commission.

Purce, D., Wood, M., Kueck, J., Taylor, K., O'Dea, L., Wright, D., & Haddad, T. (2009). *Proceedings of the West Coast Coastal Atlas Workshop, Seattle, Washington, USA*.

Vivero, J. S., & Mateos, J. R. C. (2007). *The Atlas of European Seas and Oceans - marine jurisdictions, sea uses and governance*. Ediciones del Serbal. 146 pp. Retrieved July 29, 2009 from the EurOcean web site: http://www.eurocean.org/np4/file/18/Atlas_of_the_European_Seas.pdf

KEY TERMS AND DEFINITIONS

Coastal Web Atlas: A collection of digital maps and datasets with supplementary tables, illustrations and information that systematically illustrate the coast, oftentimes with cartographic and decision support tools, all of which are accessible via the Internet. Also known as web atlas, digital atlas, digital coastal atlas.

Content Standard: A document that fully outlines all the vital information pertaining to a data set's source, content, format, accuracy, and lineage (i.e., what processing changes the data set has gone through over time). A national and international content standards for geographic metadata

Federated: A condition in which it is possible to simultaneously search multiple sources (such as coastal web atlases or other online databases or resources) and to place them on a single map. A federated search often provides a single search interface, as well as single map canvas on which to place data layers.

Information Management: The means by which an organization, agency, or individual collects, documents, shares, and uses information. It often involves creating or identifying the appropriate resources to find quality information in order to fill gaps in knowledge. Coastal web atlases are therefore a key component of information management as applied to coast region of the world.

Metadata: Information that fully describes or documents a dataset, such as its geographic coverage, quality, completeness, accuracy, etc. are a "pedigree" of sorts for a data set and helps you to judge its "fitness for use" or reliability, thereby helping one to use it more appropriately and efficiently. Metadata allow a potential user, for comparative purposes, to understand how the data were collected. They also provide the all-important details of how you can actually obtain the data in question, or who best to contact. Data that do not have accompanying metadata are often hard to find, difficult to access, troublesome to integrate, and perplexing to understand or interpret.

Spatial Data Infrastructure or SDI: A framework via an organization of people or government agencies, via the Internet, or via a series of guiding policies or standards to assist people with acquiring, processing, using, and preserving spatial data. The spatial data are often in geographic information system (GIS) format, are not, but not limited to this.

Web GIS: Geographic information systems functions as deployed via a web site or series of web sites rather than just on the desktop.

CSDGM: Content Standard for Digital Geospatial Metadata.

EcoCat: Ecological Compliance Assessment Tool.

EMODNET: European Marine Observation and Data Network.

FGDC: Federal Geographic Data Committee.

GEOSS: Global Earth Observation System of Systems.

GMES: Global Monitoring for Environment and Security.

GRAME: Global Regular Assessment of Marine Environments.

ICAN: International Coastal Atlas Network.

ISO: International Standards Organization.

INSPIRE: INfrastructure for SPatial InfoRmation in Europe.

MSFD: Marine Strategy Framework Directive.

OGC: Open Geospatial Consortium.

SEIS: Shared Environmental Information System

APPENDIX

Table 1. Coastal Atlases of the U.S. Resulting from 2008 Survey. (All are official partners in the U.S. Coastal Zone Management Program.)

State	Organization	Maps	URL
AK	Alaska Coastal Management Program	yes	http://www.alaskacoast.state.ak.us/
AL	Outdoor Alabama	yes	http://www.outdooralabama.com/
AL	Alabama Coastal Area Management Program	no	http://www.adem.state.al.us/fieldops/coastal/coastal.htm
AS	Environment Division	no	http://www.asdoc.info/CZM/1CZMGT.htm
CA	Central Coast Joint Data Committee	no	http://www.ccjdc.org/index.htm
CA	California Coastal Commission	yes	http://www.coastal.ca.gov/
CT	Connecticut's Changing Landscape	yes	http://clear.uconn.edu/projects/landscape/analysis/calcap.htm
CT	Connecticut Coastal Management Program	no	http://www.ct.gov/dep/cwp/view.asp?a=2705&q=323536&depNav_GID=1622
DE	Delaware Coastal Programs	yes	http://www.swc.dnrec.delaware.gov/coastal/Pages/CoastalPrograms.aspx
FL	Florida Blueways	yes	http://research.myfwc.com/features/view_article.asp?id=3108
FL	Florida Coastal Management Program	no	http://www.dep.state.fl.us/cmp/
GA	Georgia Coastal Comprehensive Plan	yes	http://www.georgiaplanning.com/coastal.htm
GA	Coastal Resources Division	yes	http://crd.dnr.state.ga.us/content/displaynavigation.asp?TopCategory=6
Guam	Guam Bureau of Statistics and Planning	yes	http://www.bsp.guam.gov/content/category/6/15/37/
HI	Hawaii Coastal Zone Management	yes	http://hawaii.gov/dbedt/czm/
IL	Coastal Management Program	yes	http://dnr.state.il.us/owr/CMP/
IN	Lake Rim GIS	yes	http://igs.indiana.edu/arcims/lrim/index.html
IN	Indiana Lake Michigan Coastal Program	no	http://www.in.gov/dnr/lakemich/
LA	Louisiana Coastal Area	no	http://www.lca.gov/index.aspx
LA	Coastal Louisiana Resource Information System	yes	http://www.ladigitalcoast.uno.edu/index.html
LA	Coastal Management Division	no	http://dnr.louisiana.gov/crm/coastmgt/coastmgt.asp
MA	Mass Office of Geographic and Env. Information	yes	http://www.mass.gov/mgis/massgis.htm
MD	Eyes on the Bay	yes	http://mddnr.chesapeakebay.net/eyesonthebay/index.cfm
MD	Maryland Shorelines Online	yes	http://shorelines.dnr.state.md.us/default.asp
MD	Chesapeake and Coastal Program	no	http://www.dnr.state.md.us/bay/czm/index.html
ME	Maine Coastal Program	Yes	http://maine.gov/spo/coastal/resources/dataandmaps.htm
ME	Department of Marine Resources	yes	http://www.maine.gov/dmr/maps/mapindex.html
ME	Gulf of Maine Coastal Program	no	http://www.fws.gov/r5gomp/gisindex.htm
ME	Gulf of Maine Council	yes	http://www.gulfofmaine.org/
ME	Earth Systems Data Collaborative	yes	http://www.datacollaborative.unh.edu/gulfofmaineportal.shtml
MI	Coastal Management	yes	http://www.michigan.gov/deq/0,1607,7-135-3313_3677_3696---,00.html
MN	Minnesota Coastal GIS	yes	http://www.nrri.umn.edu/coastalGIS/default.html

continued on following page

Table 1. continued

State	*Organization*	*Maps*	*URL*
MN	Minnesota Lake Superior Coastal Program	yes	http://www.dnr.state.mn.us/waters/lakesuperior/index.html
MS	Coastal Management and Planning Office	yes	http://www.dmr.state.ms.us/
NC	Division of Coastal Management	yes	http://dcm2.ehnr.state.nc.us/
NC	North Carolina Coastal Reserve & NERR	yes	http://www.nccoastalreserve.net/
NH	New Hampshire Dept. of Environmental Service	yes	http://www2.des.state.nh.us/gis/onestop/
NJ	Jacques Cousteau NERR	yes	http://www.jcnerr.org/
NJ	New Jersey Coastal Management Program	yes	http://www.state.nj.us/dep/cmp/czm_data.html
NJ	New Jersey Dept. of Environmental Protection	yes	http://www.state.nj.us/dep/gis/depsplash.htm
NMI	Coastal Resources Management Office	yes	http://www.crm.gov.mp/default.asp
NY	Long Island South Shore Estuary Reserve Council	no	http://www.estuary.cog.ny.us/
NY	New York State Division of Coastal Resources	yes	http://www.nyswaterfronts.com/maps_relief.asp
OH	Ohio Department of Natural Resources	yes	http://www.dnr.state.oh.us/AtlasGIS/tabid/19562/Default.aspx
OR	Oregon Coastal Atlas	yes	http://www.coastalatlas.net/
PR	Puerto Rico Coastal Zone Management Program	yes	http://www.drna.gobierno.pr/oficinas/arn/recursosvivientes/costasreservasrefugios/pmzc/coastal-zone-management-program
RI	Coastal Resources Management Council	yes	http://www.crmc.ri.gov/
SC	South Carolina Office of Ocean and Coastal Resource Management	yes	http://www.scdhec.net/environment/ocrm/
TX	Texas Coastal Management Program	yes	http://www.glo.state.tx.us/coastal/cmp.html
VA	Center for Coastal Research Management	yes	http://ccrm.vims.edu/index.html
VA	Virginia Coastal Zone Management Program	yes	http://www.deq.state.va.us/coastal/
VI	Division of Coastal Zone Management	yes	http://www.czm.dpnr.gov.vi/
WA	Washington State Department of Ecology	yes	http://www.ecy.wa.gov/programs/sea/sma/atlas_home.html
WI	Wisconsin Lake Superior Coastal Mapping Portal	yes	http://maps.aqua.wisc.edu/lscmp/wlscmp_index.htm
WI	Wisconsin Coastal Management Program	yes	http://www.doa.state.wi.us/section.asp?linkid=65&locid=9

Chapter 6
Oregon, USA

Tanya Haddad
Oregon Coastal Management Program, USA

Robert J. Bailey
Oregon Coastal Management Program, USA

Dawn J. Wright
Oregon State University, USA

ABSTRACT

This coastal web atlas case study for Oregon, USA, featuring the Oregon Coastal Atlas (OCA), describes a web site where people interested in coastal and marine information for the Oregon coast can find, learn about, and utilize data about the various resources and management issues in the Oregon coastal zone. The OCA currently hosts an archive of over 3,700 individual coastal data layers (including orthophotographs) and serves a variety of static and dynamic map products as well as informational tools focused on coastal places and topics.

INTRODUCTION AND OVERVIEW

The Oregon Coastal Atlas (OCA) is multi-group project established by a partnership between the Oregon Coastal Management Program (OCMP), Oregon State University (OSU) Department of Geosciences, and Ecotrust, a non-governmental organization based in Portland, Oregon. The project group formed in the year 2000, received funding in the fall of 2001, and launched the Atlas on the Internet in December 2002. A major revision of the website interface was undertaken in 2007 resulting in the current version of the OCA which launched in March 2008 and is visible at http://www.coastalatlas.net. As a relatively mature Coastal Web Atlas (CWA) the Oregon Coastal Atlas project has encountered and overcome many technical and institutional challenges and has incorporated many lessons learned into its current design and approach. The Atlas currently hosts an archive of over 3,700 individual coastal data layers (including orthophotographs) and serves a variety of static and dynamic map products as well as informational tools focused on coastal places and topics.

DOI: 10.4018/978-1-61520-815-9.ch006

Copyright © 2011, IGI Global. Copying or distributing in print or electronic forms without written permission of IGI Global is prohibited.

Original Motivation

The original motivation for the OCA was to make it easy to search for and retrieve datasets and information products produced by management projects in the Oregon Coastal Zone. Such data may have been collected by any of the many partner local, state, or federal agencies that participate in the federally-funded OCMP. Prior to the debut of the OCA, such data and all the related project supporting information accumulated largely in the hard copy and disk (actually often floppy disk, CD or tape) archives of the OCMP or partner agencies and as such were virtually inaccessible to the intended audience of program partners, other key coastal zone constituents, and the generally interested public. Thus the project was initially designed to address the problem of finding and sharing data products from past projects and also to ensure that data stored on older forms of digital media were migrated into current formats so as not to become technologically trapped in the past. In the process of addressing this original need, the project team realized that the archive of data would be an asset to a wider audience if paired with informational products and tools that could make the data contained in the archive more accessible to non-experts. As a result, the design of the OCA (Figure 1) evolved to address both the needs of GIS specialists, as well as non-specialists seeking to explore a special topic of coastal management concern (e.g., a local planner or resource manager) or simply learn about the Oregon coast (e.g., a coastal visitor).

Funding

The project received initial seed funding from National Oceanic and Atmospheric Administration (NOAA) Coastal Services Center, which afforded the OCMP the opportunity to form partnerships with OSU Geosciences and Ecotrust to develop the OCA project idea and apply for further funding to implement the concept on a coastal zone wide scale. In late 2001, the project secured key funding from the National Science Foundation (NSF) to fund the project's three-way partnership for three years. This base funding was augmented by various special project funding from other sources such as Federal Geographic Data Com-

Figure 1. Home page of the Oregon Coastal Atlas as visible at http://www.coastalatlas.net/

mittee (FGDC), e.g., for metadata development, and early adoption of Web Map Service (WMS) services. When the initial development grants expired in 2005, the project underwent a transition in which the OCMP eventually assumed the primary stewardship role and incorporated the OCA project into its standard ongoing portfolio of services to coastal management constituents. This assured that the project would continue to be maintained and grown over time.

Project Scope

The OCA covers the entire Oregon Coastal Zone, which encompasses all coastal watersheds except those of the Columbia, Umpqua and Rogue rivers on the terrestrial side, and the 3-nautical-mile territorial sea plus an ocean stewardship area that extends seaward to the toe of the continental slope, on the marine side. Data are not cut at these boundaries however, so if a data set is larger, it is allowed to extend outside of this zone. For regional context, the extent from (-132°W, 39 °N) to (-116 °W, 53 °N) is the full footprint of the displayed area of interest in the OCA online map interface.

The target audiences are the decision makers and constituents of the Oregon coastal zone. These are comprised of local planners, state and federal agency staff, tribal governments, researchers, consultants, and interested citizens.

DATA COLLECTION

The initial focus for the OCA data collection was to "rescue" data sets that had accumulated and were mostly unused in the OCMP archives. In most cases these data were archived in paper file folders that consisted of hard copy reports and project notes (the project metadata), often with digital data attached via a floppy disk or data tape. In many cases, there was a fear that as hardware technology changed, even the OCMP would be unable to access the digital files due to lack of access to appropriate media/drive hardware.

All data that had been acquired in full or in part by federal Coastal Zone Management (CZM) funds passed through OCMP were potential candidates for inclusion. Digital data that fit these criteria existed in the OCMP archives dating back to the mid-late 1980s. In some cases hard copy data in photo print or table format existed back into the early 1970s. In addition, data continued to be accumulated through the 1990s and first decade of the 21st century in various digital formats.

In all the cases, the data in question were collected as part of the normal course of implementing the Oregon CZM program and answering the interdisciplinary management-based questions that CZM entails. This was the "bar of relevance" that a data set had to meet in order to be considered for migration into the OCA archive. Since various management issues wax and wane in importance over time, the total archive ranges from data relating to broad topics such as estuary or ocean planning, to data for more site-specific topics such as coastal erosion and habitat restoration. In all, the archive now comprises over 3,700 individual data layers of both vector and raster data types.

In terms of data origin, some data were collected directly by OCMP, while other data were collected by partner agencies using CZM (and other) funds. In some projects, data collected were entirely new, while in other projects, complimentary data were assembled from accessible public sources (usually federal sources such as NOAA or USGS). In all cases, the data in the archives were considered to be publicly funded and thus eligible to be included in a more publicly accessible venue, with the important caveat that the distribution format includes detailed metadata to convey the original project credits, conditions, background information, and intent.

SYSTEM DESIGN

When it came time to design the architecture of the OCA, the first question the project team had to answer was "how to organize all this data and information so that it makes sense to our audience?" This question quickly led to another: "who are our audiences, and how will they need to use this information?" It was answering this second question that really helped formulate the overall design of the OCA. The resulting design has, for the most part, been well received and proved durable, having remained relatively unchanged through system upgrades over the years.

As the project developed in the early stages, it became apparent that the data archive itself was an opportunity to illustrate the resources of the coastal zone and to raise awareness of some of the management challenges common in coastal areas. Since GIS was not a particularly common tool in the intended audience, providing the data in downloadable GIS file formats would likely address only the needs of specialized users. For users without access to a GIS, it was important to make the data sets available via a map interface that would not be too difficult to use, and allow for personalized maps to be made and exported for offline use. For certain data sets that were expected to be of high interest due to wide public applicability (such as beach access locations and beach water quality results), the team felt that they would be more frequently used if a simple tool were developed that could be used to easily repeat common lines of inquiry. Locations of high interest along the coast could be profiled with texts and photographs and grouped with informational items about common coastal management topics into a section that could inform casual browsers of coastal information. These ideas for different modes of accessing the coastal data in the archive eventually played a strong role in the system design of the OCA.

As stated in the project goals, the intended audiences for the OCA were the "decision makers and various constituents" of the Oregon coastal zone. This is obviously a broad cross section of potential users, some of whom could be expected to be more technology-savvy than others. The project was starting with a product, geospatial data, that had potential to be useful to many people, but which was typically packaged in GIS data formats that were inaccessible to non-specialists. This distinction between GIS specialist and non-specialist was a fundamental division in the OCA audience types. Further, the non-specialist audience consisted of several significant sub-types such as planners/resource managers and researchers, in addition to the more general public. In order to serve the specialist customers, it was important to organize and document the archive data and make it easy to retrieve. The strategy to serve the various non-specialist customers was to derive useful information products from the data, and to offer various mechanisms (tools) for accessing certain data types that would need to be repeatedly accessed to answer specific types of questions.

In order to meet the needs described above, the project team developed a four-pronged information architecture organized around functions that the project team determined would best meet the needs of the various levels of audience the OCA would need to serve. The four functional branches of the architecture named Maps, Tools, Learn, and Search, were used to shape the user interface and navigation scheme of the site design. An early project diagram from 2002 (Figure 2) illustrates the anticipated types of content that could eventually fill the architecture. It is interesting to compare this diagram with the direction the project subsequently took.

Search

Once the conceptual system architecture was devised, the first task was to inventory the data in the archives, standardize it, and create a searchable database so that a catalogue of the OCA data holdings could be published. This entailed: (1) sifting

Figure 2. Original OCA information architecture diagram from 2002, showing the four main branches of the site and potential content. The architecture is interesting to look at in comparison to the actual direction the project has taken.

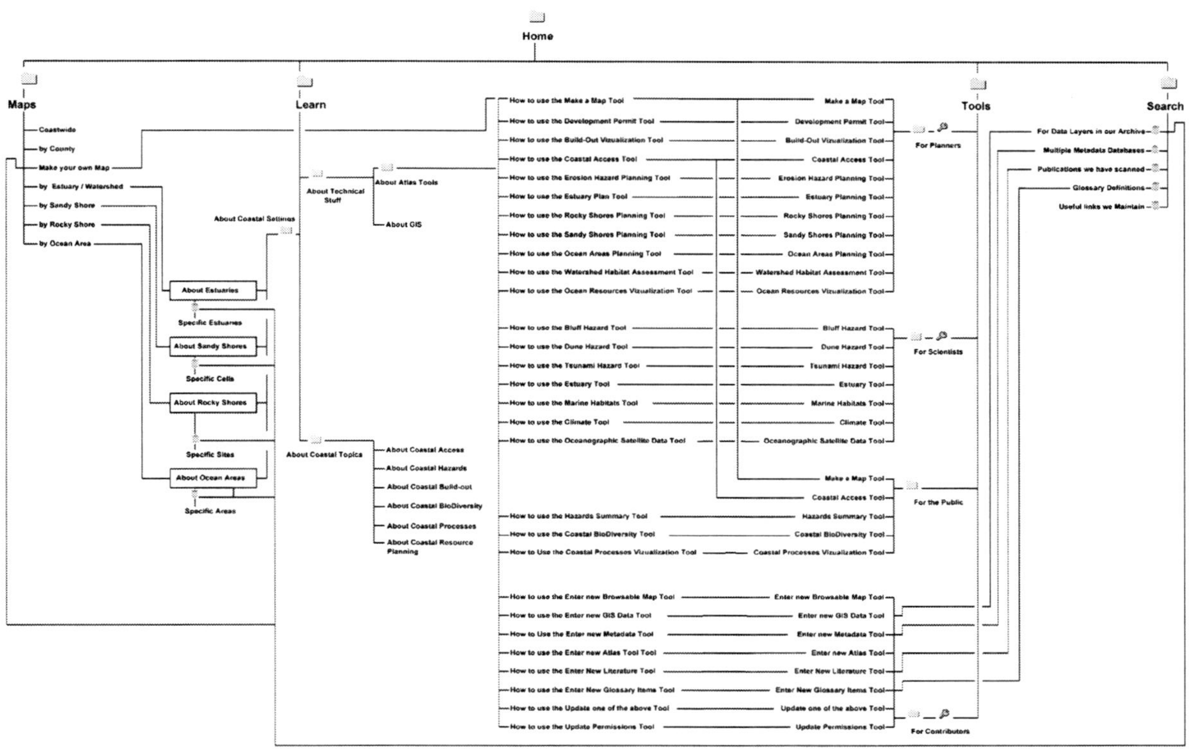

through paper archives, (2) accumulating digital data from the various forms of aging media, (3) transforming the data into a consistent geographic projection so that the various data layers could be used together easily, and (4) meticulously organizing the original project information from the paper archives into FGDC-compliant metadata to accompany the GIS data files.

This question of metadata became a central early focus of the OCA project. A majority of the initial data in the archive did not have any type of consistent documentation and therefore such documentation needed to be created before the data could be widely shared. Metadata is important because it documents not only the origins of a data set, in terms of authorship and institutional stewardship, but also the methods of data creation, original intended use of the data, and the possible appropriateness of the data for future uses. In the United States the standard for geospatial metadata is established by the Federal Geographic Data Committee (FGDC), and so this was the standard to which all of the OCA metadata were composed. A "Don't Duck Metadata" grant from the FGDC enabled the OCA partnership to train and employ a cadre of metadata interns to work through the metadata backlog that was preventing the data in the OCA archive from being shared. Gradually the collection was made available as the documentation was completed.

Each data set collected and documented was zipped along with metadata, and basic information metrics were entered into a simple database from which a searchable interface was ultimately created. Early internal versions of the archive database were constructed in Microsoft Access, but the web accessible version of the archive is built with the MySQL database. Vector data were shared as shapefiles, and raster data (which in OCA are primarily orthophoto images and base maps) were shared as GeoTiffs. In Oregon, a state Geographic Information Council (OGIC) sets state-wide data standards for framework spatial data and, as a result, all data in the archive are shared in the Oregon Geographic Information Council's standard Lambert projection.

The search section visible on the current version of the OCA remains largely unchanged from the initial implementation of the function. As such, it is the oldest portion of the current interface and upgrades in its functionality are being considered. When the OCA debuted in 2002 there were around 300 individual data layers in the searchable archive. In late 2009 the number of data layers has grown to over 3,700. This order of magnitude increase has led to a need to enhance the performance of the Search functions available to users to ensure that the most relevant results are always returned. Several mechanisms are being explored to improve the Search function, among them the possible implementation of GeoNetwork Open Source. Among other benefits, the switch to GeoNetwork Open Source would allow for search to be implemented against complete XML metadata records, as opposed to the simple "discovery" metadata that is currently utilized when a search is performed.

Maps

The second step in the OCA project implementation was to build the most obvious of the information products: the Maps section. Maps created with the main Map tool are of use to any user who does not have a ready way to access raw GIS files. In the first version of the OCA, the maps section was composed of an interactive map powered by Minnesota MapServer that allowed the user to browse maps of the Oregon coast composed from a predefined set of background and overlay layers. The pre-packaged standard data could be supplemented with data from the searchable archive, and the map user could compose a custom PDF of their area of interest for download and later use. The second version of the OCA maps, implemented in 2007, extended the map interface options to include a thematic map selection function. This allowed for the presentation of collections of data sets organized around informational themes or focused landscape settings. Currently there are nine focused collections in place with three more in the current planning stages. For technical reasons (primarily speed) it is preferable to have a larger number of data collections with a smaller number of data sets per collection, than the reverse.

The first version of the maps interface was based on portions of the GMap PHP MapScript sample application interface created by DM Solutions Group, and employed a relatively static HTML interface with minimal JavaScript to ensure the widest possible cross-browser compatibility and cater to the slow web-connection speeds that were prevalent on the Oregon coast at the time. The second version of the OCA maps (Figure 3) maintained the majority of the original mapping functions but wider adoption of web standards by browser manufacturers and improved connections speeds made it possible to significantly upgrade the map interface itself. Implementing a JavaScript-based interface that utilized cached tiles to provide a more interactive user experience did this. This second version of the OCA map interface is based on ka-Map, an early JavaScript API for developing highly interactive web-mapping interfaces also developed by DM Solutions. The ka-Map API has been superseded by the highly popular OpenLayers JavaScript API for map interfaces that has attracted many talented open source developers. Within OCA, the OpenLayers API is being used for several Learn and Tool features that require embedded interactive maps and the main OCA map interface is planned to be revised around an interface that incorporates OpenLayers. At present,

Figure 3. The main OCA Map interface is based on the Ka-Explorer interface, which was developed for the Food and Agriculture Organization of the United Nations (FAO-UN) by Lorenzo Becchi and Andrea Cappugi (www.ominiverdi.org)

however, the OCA main map page continues to operate with the Ka-Explorer interface because it provides the basic tools needed for map navigating, is easy for users, and benefits from wide cross-browser compatibility.

Learn

The third step of OCA construction involved assembling the beginnings of the Learn section, which was intended to be targeted towards visitors who would be exploring the contents of the Atlas via texts and pictures. The intention was that these users would be best served by providing them with basic information that introduced them to the various topics and locations that the OCA covered, thereby providing an alternative pathway to the contents of the data archive. Initially, the contents of the Learn section were fairly sparse, limited mainly to discussion, pictures and maps of the main geographic settings covered: estuaries, sandy shores, rocky shores and ocean areas. However as the project team anticipated, the Learn section has grown over time to cover a variety of interesting data sets.

Significant additions to the Learn section have come from a variety of sources. Over the years the OCMP has been fortunate to host several NOAA Coastal Management Fellows and the products of Fellowship projects have helped to fill in the information contained in the OCA Learn section. Examples include Estuaries (Chad Nelson, Tanya Haddad), Sandy Shores (David Revell), Rocky Shores (Laurel Hillmann) and various Ocean topics (Andy Lanier). Additionally, special programs of the OCMP, such as the coastal hazards program, have posted information in the Learn section about topics such as coastal shore protective structures. Lastly, additional useful data sets have come from other agencies that have programs of wide interest to the coast-visiting public. These agencies have approached OCMP to host products in the Learn section as a way to receive wider attention and distribution for their information. Examples of such hosted information include: beach water quality information from the Department of Human Services BEACH program and rapidly moving landslide hazard risk information from the Department of Geology and Mineral Industries. Both these data sets are extremely popular amongst OCA users. The beach water quality information gets more traffic in the summer months, while the landslide risk information is most popular during the wet months of December and January.

Tools

Last but not least, the Tools section was assembled. Each of the various tools featured in the Tools section was constructed to perform a specific task, usually on a specific type of coastal data. In the first version of the OCA the tools section contained links to a variety of tools around the Internet that were of interest to users working on the Oregon coast. In addition, a number of tools were constructed by the project team to address a variety of common coastal zone management questions. This generation of tools was partially documented in Haddad *et al.* (2005). The tools ranged in complexity from very basic reporting tools to more complex tools designed with geospatial analyses performed by ArcIMS. The experience of the OCA project with this first generation of tools was mixed with many lessons learned. Primary among these lessons was that simple tools that were easy to construct (e.g., reporting tools) often were the most popular with users and received the most traffic while more specialized and complex tools, the most difficult to construct and maintain, had limited audiences due to content and interface issues. Eventually, as the first iteration of the OCA project morphed into the second, many of these more complex experimental tools were simply not migrated into the new framework.

In the current (2009) OCA Tools section, only a few tools were actually constructed by the OCA project team. Most were simply documented and linked to by the OCA in order to give OCA users a catalogue of coastal-specific geospatial applications. Like the Learn section, the Tools section started out fairly sparsely populated with an expectation that both the OCA constructed tools and the wider links to tools created by other entities would grow over time. This has largely been the case. In recent years a wide variety of interesting geospatial tools relating to coastal management issues have been constructed by various federal and academic entities, and links to such products are added to the OCA Tools sections as they become available.

An additional important feature that the Tool section provides is the option for the user to filter the listed tools by the primary audience type for which the tool is appropriate. The four main audience types are: the Public, Planners, Researchers and Contributors. Tools listed in the catalogue are tagged to fall into one or more of these bins by the OCA administrator based on internal OCMP user evaluations. Users of the OCA see all available tools, or can select the appropriate tool based on the audience type to which they feel they belong.

Technologies Used

The current version of the OCA is built mostly on an open source set of software (Figure 4). The web server employed is the popular Apache. The primary database is MySQL, accessed via PHP scripts. The majority of the non-geospatial content is organized by the open source content management system (CMS) Joomla! A CMS is employed because it is a simple mechanism for controlling what content is publicly published on the website. Information can be brought online or taken offline fairly quickly with a CMS and permission can be granted to multiple content creators, editors, publishers and administrators so that the task of maintaining a project can be distributed among a wide variety of partners. When the first phase of the OCA project came to a close facing an uncertain funding future, the transition to an off-the-shelf CMS allowed the project staff to contract gracefully without impairing the site administration and allowed for a framework where distributed contributors could remain active.

From a technical perspective, Joomla is a CMS framework written in PHP that organizes web content utilizing the MySQL database that can be extended in functionality via user-created components, modules and plug-ins. In the OCA, the Maps section of the OCA is implemented as an instance of the PHP and JavaScript open source ka-Map ka-Explorer interface, embedded into Joomla. The various maps served in the map interface are simply a collection of MapServer text-based map files that are displayed via ka-Map. Similarly, the discovery metadata that are part of the primary search function are examples of content managed via a user-created extension to Joomla. In contrast the materials in the Learn and Tools sections of the OCA are managed almost exclusively by the Joomla "core" functions.

Figure 4. The software behind the various branches of the Oregon Coastal Atlas architecture

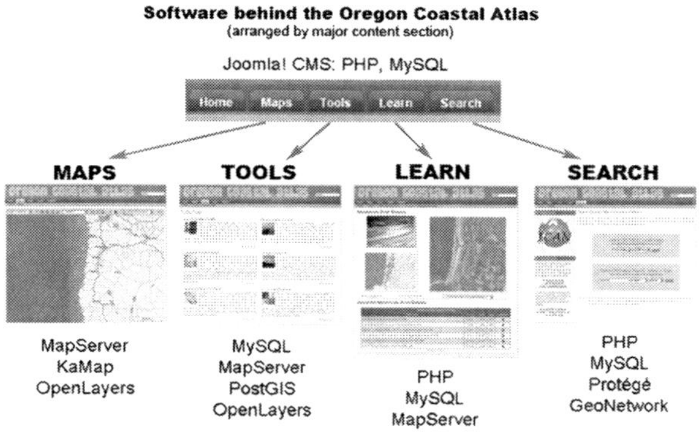

Future Directions

In the immediate future, the section of the OCA interface most likely to evolve further is the Search section. Currently, only discovery metadata are managed in the OCA extension to Joomla, as opposed to entire metadata records. The full metadata records are available when clicking on the results of a search but the results are served from flat files rather than from within the Joomla MySQL database. This is likely to be an area where the OCA is improved in the in the near future as a result of the project's involvement with the International Coastal Atlas Network (ICAN) and anticipated OCA implementation of the GeoNetwork Open Source product. GeoNetwork is specifically designed for metadata management that would enable the OCA to serve both basic discovery metadata and full XML FGDC and ISO-compliant metadata records. A full transition of the OCA data catalogue to GeoNetwork will also allow the catalogue to be searched externally by any entity capable of consuming a Catalogue Service for the Web (CSW) feed.

Looking further forward, an exciting new avenue of OCA development has been the project's involvement with the emerging ICAN. The Oregon Coastal Atlas team has participated in ICAN since its inception and through this involvement has begun work on development of a comprehensive Oregon Coastal Zone Management Ontology based on the catalogue of coastal zone management data in the existing OCA archive. Participation in the ICAN prototype has also entailed an initial installation of GeoNetwork open source to facilitate sharing of a portion of OCA catalogue. As this prototype information-sharing framework stabilizes, the OCA plans to transition the full data catalogue and to utilize more aspects of the GeoNetwork product to deliver links to data and metadata as well as OGC services and KML products to users, in order to keep pace with emerging user capabilities.

USAGE OF THE ATLAS

Server logs give a picture of the usage of the OCA and reveal some interesting information. The monthly statistics generated between March 2008 and August 2009 enable some basic metrics to be calculated. On average, the current version of the OCA receives about 13,945 visits per month resulting in about 280,475 page views per month. These visits come from an average of 3,695 unique visitors per month, of which an average of 1,127 return to the site more than once during the course of the month. About 16% (average of 176) of these returning visitors actually visit the OCA more than 10 times per month, demonstrating some utility of the OCA products to the audience of coastal management constituents.

Table 1 shows OCA log statistics and illustrates the monthly variability that exists over time. For example, March 2008 marked the launch of the newly updated OCA and the publicity surrounding the upgrade resulted in some higher than average traffic for a time. More significantly, the month of January 2009 shows a huge increase in web site traffic far outside of the typical range. The explanation for this anomaly is that severe winter storms over the 2008 Christmas holiday period resulted in high run-off in the early New Year and several instances of landslides in the Coast Range Mountains. Because the OCA features a data set of potential rapidly moving landslide risk zones published by the Department of Geology and Mineral Industries, several news outlets that reported on the January landslide events, including the state-wide *Oregonian* newspaper and at least one local television station (KATU channel 2), mentioned the OCA as a resources for landowners interested in learning more about landslide risk. Both these news outlets also put a link to the OCA on their respective websites, resulting in over 1,250 direct referrals to the website from just 3 external links. Instances such as these illustrate the power of conventional media as drivers of traffic to destinations on the web and show that projects such as the OCA can be introduced to many additional users if broadly advertised.

User Origins

Traffic to the OCA is predominantly from within the United States and only an average of 8% of the traffic per month to the OCA is from non-U.S. countries. Among U.S. states, the west coast is dominant, comprising about 75% of average monthly traffic. Other top ten high-traffic states of

Table 1. Oregon Coastal Atlas monthly server statistics from March 2008–August 2009

Date	Hits	Page Views	Visits	Percent International	Unique Visitors	Returning Visitors	Returned 10 or more times
Mar-08	331,675	257,517	13,963	9.40	5,182	1,866	135
Apr-08	259,033	184,303	15,701	5.85	4,510	1,746	250
May-08	254,994	177,971	14,073	4.96	4,039	1,560	185
Jun-08	275,579	191,283	12,925	5.61	3,224	1,122	169
Jul-08	289,162	200,471	12,282	6.17	3,467	1,113	127
Aug-08	312,398	221,026	11,240	7.01	2,977	882	105
Sep-08	248,739	171,284	8,541	7.79	2,530	645	117
Oct-08	270,225	196,615	11,640	10.85	3,124	788	107
Nov-08	228,595	151,809	13,678	12.05	3,239	1,039	129
Dec-08	233,691	156,221	11,720	14.12	3,107	891	106
Jan-09	1,732,600	1,434,650	15,752	10.54	6,056	1,256	159
Feb-09	302,291	207,412	11,047	14.09	3,041	844	115
Mar-09	394,509	265,422	12,860	13.14	3,658	993	121
Apr-09	376,172	255,107	13,607	12.03	3,601	1,203	163
May-09	356,628	229,068	17,695	4.85	3,810	1,021	261
Jun-09	366,718	260,340	15,704	4.16	3,180	870	260
Jul-09	386,797	263,741	20,257	4.42	4,084	1,199	314
Aug-09	326,992	224,313	18,322	4.22	3,689	1,244	345

Figure 5. Visits to the Oregon Coastal Atlas from outside the United States

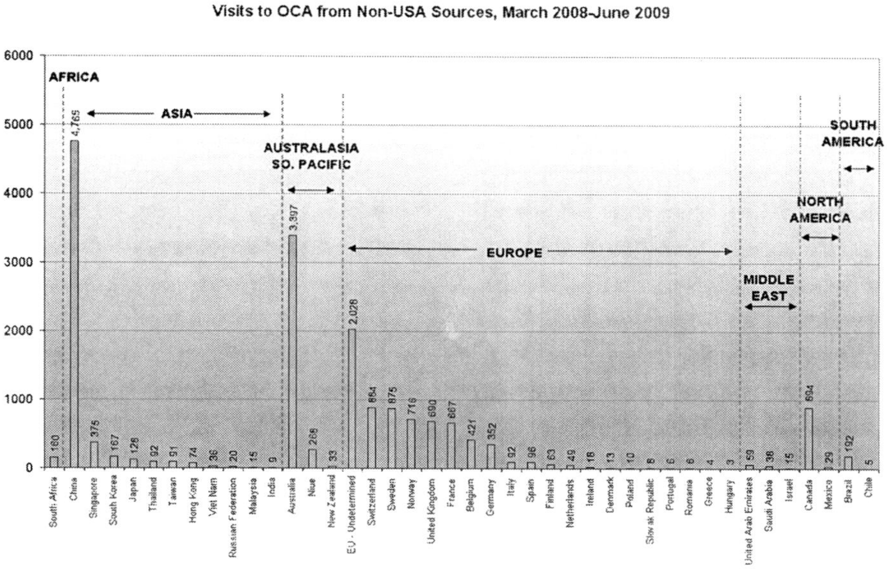

origin include Virginia, Arizona, South Carolina, the District of Columbia, Massachusetts, Texas, Colorado, New Jersey and North Carolina. A chart showing the distribution of visits from other countries over 15 months is presented in Figure 5.

Utility to Users

Beyond demonstrating that the OCA is a repeat destination for a core group of users, server logs also reveal that the project also serves as a valuable resource for casual web browsers seeking information about the Oregon coast and various coastal management-related topics. This can be documented by examining the distribution of search phrases typed into search engines that refer users to the OCA. Figure 6 illustrates an 8-month sampling of the most frequent search phrases used in the Google search engine. For clarity the three most common phrases (*Oregon Coastal Atlas*, *Coastal Atlas* and *Oregon Coast Atlas*) are excluded from the diagram.

Figure 6. Distribution of search terms used on Google that led users to the Oregon Coastal Atlas. Size of phrase in the above image is weighted by the frequency of the search phrase. Graphic created with the online tool at www.wordle.net. Images created by the Wordle web application are licensed under a Creative Commons Attribution 3.0 U.S. License.

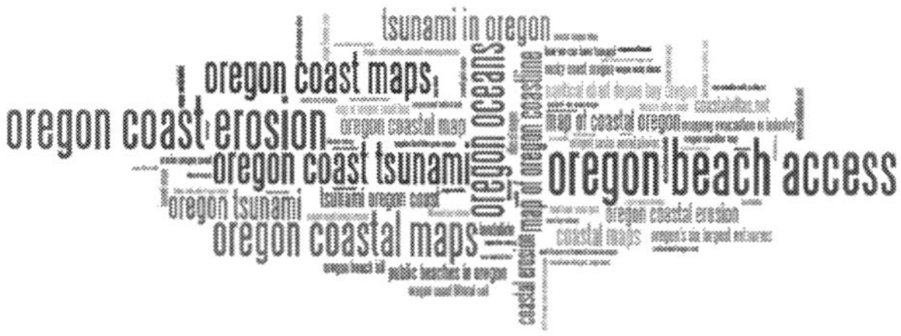

CONCLUSION

The Oregon Coastal Atlas is considered a successful project in terms of satisfying the original project goals of making it simpler for users to find, understand, and utilize data about the various resources and management issues in the Oregon coastal zone. Due to the long duration (7-8 years) of the project, many lessons were learned about the difficulty of serving such information via the web and efforts were made to incorporate some of those lessons into the current version of the OCA web product. Anecdotal feedback has implied that the revisions have resulted in an Atlas that is easier to use with a more productive CWA experience for users. At present, the project has no way to quantify this effect.

Challenges encountered over time have included both technical (e.g., low cross-browser compatibility in the early days and slow internet connection speeds at the coast) and institutional issues (e.g., initial project funding was limited in duration and when expended led to a period of transition while more permanent stewardship was established in the Oregon Coastal Management Program). Time has improved many of the original technical challenges (e.g., better browsers and faster connection speeds) while others (e.g., complex analysis tools built on proprietary software) have been avoided in the second generation OCA to simplify ongoing maintenance. Institutional stability has been achieved with a commitment from the Oregon Coastal Management Program to incorporate the OCA project into its standard ongoing portfolio of services to coastal management constituents, which assures that the project will continue to be maintained and grown over time.

The project's involvement with the emerging International Coastal Atlas Network (ICAN) has enhanced the development of new features and has strengthened the project by providing a potential direct mechanism for feeding OCA products into regional, national and international coastal information networks.

ACKNOWLEDGMENT

Over the years many people have worked on various aspects of the Oregon Coastal Atlas content, and design. Key contributors are listed here along with the organizations that they represented.

- *Oregon Coastal Management Program* (http://www.oregon.gov/lcd/ocmp/)
  - Paul Klarin, Randy Dana, Tanya Haddad, Bob Bailey, David Revell, John Marra, Steve Williams, Laurel Hillmann, Andy Lanier.
- *Oregon State University Geosciences* (Davey Jones Lab - http://dusk.geo.orst.edu/djl/)
  - Dawn Wright, Ken Crouse, Peter Bower, Chris Zanger, Danielle Pattison, Colin Cooper, Mike Tavakoli, Amythyst O'Brien, Christina Ryan, Jessica Adine, Anthea Fallen-Bailey, Jenny Allen, Ryan Field.
- *Ecotrust* (http://www.ecotrust.org)
  - Michele Dailey, Ben Donaldson, Mike Mertens.

In addition, funding for the Oregon Coastal Atlas was generously provided by the National Science Foundation (Grant No. 0113519), the NOAA Coastal Services Center, and the Federal Geographic Data Committee.

REFERENCES

DM Solutions Group. (2009). Retrieved October 15, 2009, from DM Solutions Group web site: http://www.dmsolutions.ca/.

Ecotrust. (2009). Retrieved October 15, 2009, from Ecotrust web site: http://www.ecotrust.org/.

Federal Geographic Data Committee. (2009). Retrieved October 15, 2009, from Federal Geographic Data Committee web site: http://www.fgdc.gov/.

GeoNetwork Opensource. (2009). Retrieved October 15, 2009, from the GeoNetwork Opensource web site: http://geonetwork-opensource.org/.

Google. (2009). Retrieved October 15, 2009, from Google web site: http://www.google.com/.

Haddad, T. C., Wright, D. J., Dailey, M., Klarin, P., Marra, J., Dana, R., & Revell, D. (2005). The tools of the Oregon Coastal Atlas . In Wright, D. J., & Scholz, A. (Eds.), *Place Matters: Geospatial Tools for Marine Science, Conservation, and Management in the Pacific Northwest* (pp. 133–149). Corvallis, OR: Oregon State University Press.

Joomla! Content Management System. (2009). Retrieved October 15, 2009, from Joomla! - the dynamic portal engine and content management system web site: http://www.joomla.org/.

ka-Map! (2009). Retrieved October 15, 2009, from Ka-MapMapTools.org website http://ka-map.maptools.org/

KATU. (2009). Retrieved October 15, 2009, from KATU.com - News, Weather and Sports - Portland, Oregon web site: http://www.katu.com/.

National Oceanic and Atmospheric Administration Coastal Services Center. (2009). Retrieved October 15, 2009, from NOAA CSC web site: http://www.csc.noaa.gov/.

National Oceanic and Atmospheric Administration Coastal Zone Management Program. (2009). Retrieved October 15, 2009, from Office of Ocean and Coastal Resource Management web site: http://coastalmanagement.noaa.gov/.

Ominiverdi. (2009). Retrieved October 15, 2009, from Ominiverdi web site: http://www.ominiverdi.org/.

Open Geospatial Consortium. (2009). Retrieved October 15, 2009, from Open Geospatial Consortium web site: http://www.opengeospatial.org/.

OpenLayers. (2009). Retrieved October 15, 2009, from OpenLayers: Free Maps for the Web web site: http://openlayers.org/.

Oregon Coastal Atlas. (2009). Retrieved October 15, 2009, from the Oregon Coastal Atlas web site: http://www.coastalatlas.net/.

Oregon Coastal Management Program. (2009). Retrieved October 15, 2009, from Oregon Coastal Management Program web site: http://www.oregon.gov/lcd/ocmp/

Oregon Department of Geology and Mineral Industries. (2009). Retrieved October 15, 2009, from Oregon Department of Geology and Mineral Industries web site: http://www.oregongeology.org/.

Oregon State University Department of Geosciences. (2009). Retrieved October 15, 2009, from Oregon State University Department of Geosciences web site: http://www.geo.oregonstate.edu/.

Protégé. (2009). Retrieved October 15, 2009, from The Protégé Ontology Editor and Knowledge Acquisition System web site: http://protege.stanford.edu/.

The Oregonian. (2009). Retrieved October 15, 2009, from *The Oregonian* newspater web site: http://www.oregonlive.com/oregonian/.

Wordle. (2009). Retrieved October 15, 2009, from Wordle - Beautiful Word Clouds web site: http://www.wordle.net/.

KEY TERMS AND DEFINITIONS

Coastal Web Atlas: A collection of digital maps and datasets with supplementary tables, illustrations and information that systematically illustrate the coast, oftentimes with cartographic and decision support tools, all of which are accessible via the Internet. Also known as web atlas, digital atlas, digital coastal atlas.

Atlas Interoperability: The ability to interrogate two or more unrelated atlas systems and potentially exchange metadata, images, vector and raster data

Coastal and Marine Information: Both spatial and non-spatial information regarding the coastal and marine areas of a region or country.

Integrated Coastal Zone Management: A dynamic, multi-sectoral approach to managing the coast which takes into account social, economic and environmental concerns of all parties with an interest in the coastal space.

Metadata: Information that fully describes or documents a dataset, such as its geographic coverage, quality, completeness, accuracy, etc. are a "pedigree" of sorts for a data set and helps you to judge its "fitness for use" or reliability, thereby helping one to use it more appropriately and efficiently. Metadata allow a potential user, for comparative purposes, to understand how the data were collected. They also provide the all-important details of how you can actually obtain the data in question, or who best to contact. Data that do not have accompanying metadata are often hard to find, difficult to access, troublesome to integrate, and perplexing to understand or interpret.

OCA: Oregon Coastal Atlas.

Spatial Data Visualization: The ability to view digital data with a spatial dimension in a

Chapter 7
Ireland

Edward Dwyer
University College Ireland

Kathrin Kopke
University College Cork, Ireland

Valerie Cummins
University College Cork, Ireland

Elizabeth O'Dea
Washington State, USA

Declan Dunne
University College Cork, Ireland

ABSTRACT

The Marine Irish Digital Atlas (MIDA) is an Internet resource built in a web GIS environment, where people interested in coastal and marine information for Ireland can visualize and identify pertinent geospatial datasets and determine where to acquire them. The atlas, which is being constantly maintained, currently displays more than 140 data layers from over 35 coastal and marine organizations both within Ireland and abroad. It also features an "InfoPort" which is a repository of text, imagery, links to spatial data sources and additional reference material for a wide range of coastal and marine topics. The MIDA team has been active in the creation of the International Coastal Atlas Network and the Atlas was chosen as one of the nodes for the Semantic Interoperability Demonstrator.

INTRODUCTION

Data and information regarding Ireland's coastal and marine environment are held by a broad range of organizations with both terrestrial and marine remits, including national government departments and agencies, local authorities, research and teaching institutions, industry, private consultants and non-governmental organizations. Only a limited number of these (such as the Environmental Protection Agency (EPA), the Marine Institute, the Geological Survey of Ireland (GSI) and the Ordnance Survey of Ireland (OSI)) provide ready access to their holdings (Dwyer *et al.* 2003). The time-consuming process of data sourcing and acquisition from these various data owners is

DOI: 10.4018/978-1-61520-815-9.ch007

Copyright © 2011, IGI Global. Copying or distributing in print or electronic forms without written permission of IGI Global is prohibited.

complicated by the lack of data catalogues within many organizations, nonexistent or poor-quality metadata and variations in data quality (Bartlett 1999; McCormack 2003; O'Dea *et al.* 2004).

In order to address some of these issues and to improve the visibility of and access to coastal and marine related data and information, the Coastal & Marine Resources Centre (CMRC) at University College Cork developed the Marine Irish Digital Atlas (MIDA). The overall aim of the MIDA project was to collate island-wide coastal and marine spatial data and make it freely accessible over the Internet. The Atlas was intended as a resource for a broad audience and aimed to cater for anyone interested in coastal and marine matters on the Island of Ireland, encompassing the casual user searching for general information as well as the specialist that required particular data sets. This interactive, online tool, centered on a web GIS, was created with a number of specific objectives in mind:

- to develop a web site for presentation of geo-referenced coastal and marine datasets,
- to provide greater accessibility to data and information in the form of a web-enabled, customized GIS,
- to provide flexibility of use via a range of tools that allow users to select, overlay and compare geospatial layers,
- to allow users to search for and identify sources of data, information and expertise on the marine environment, and
- to encourage a greater appreciation of Ireland's coastal regions by incorporating educational and informational materials based on multi-media technology.

Development of the Atlas was funded by a grant from the Higher Education Authority of Ireland under the Program for Research in Third Level Institutions as part of the National Development Plan (2000-2006). Additional funding was provided by the Environment and Heritage Service of the Department of the Environment, Northern Ireland.

Since the MIDA was launched in 2006, the project has had multiple impacts and outcomes in areas as diverse as coastal governance in Ireland to technology development within the European Union (EU). In terms of international significance, the MIDA, together with the Oregon Coastal Atlas, served as a catalyst in the establishment of the International Coastal Atlas Network (ICAN). The partners involved in the development of the respective atlases organized and co-chaired the initial and subsequent ICAN workshops. The current operational version of the MIDA was launched at the first ICAN workshop held in Cork in July 2006 (O' Dea *et al.* 2007). Subsequently MIDA was chosen as one of the atlases for the Semantic Interoperability Demonstrator, which is presented in chapter 4. Ongoing enhancement and improvement of the Demonstrator keep MIDA at the forefront of technical development within ICAN.

This chapter presents the many aspects involved in the development of the MIDA including an assessment of the Atlas' impact to date. The following are discussed in detail:

- approaches to and challenges involved in spatial data and information collection, including data and metadata management issues,
- Atlas design and layout from both a user and developer perspective,
- Atlas usage, statistics and publicity,
- challenges in maintaining and enhancing the Atlas,
- a partnership for creating distributed systems between key data holders in Ireland and
- current Atlas developments in the context of ICAN.

DATA AND INFORMATION IN THE MIDA

Data and Information Collection

Regardless of the intended web GIS purpose, developers and spatial data managers often face similar difficulties when sourcing and acquiring data in addition to displaying them on the web. Challenges include variability in data quality, scale, data licensing and metadata (O'Dea *et al.* 2004; Department of Environment, Food and Rural Affairs 2002). When a web GIS contains spatial data from multiple data sources as opposed to from a single organization, these issues are compounded by differences in data management practices, including the existence of or differences in data catalogues and standard specifications. For the MIDA team, addressing the underlying data issues proved to be as challenging as the technological ones.

When the MIDA project started in 2002 a decision was taken to set up a centralized repository to hold all the spatial datasets to be displayed via the Atlas. This approach was taken as few organizations provided Internet access to their data holdings and those that did were not in a position to make their data available via distributed, standards-based, interoperable, web-sharing services.

In some countries, such as the United States, base datasets are easy and free/inexpensive to acquire. In Ireland licensing and significant data costs apply. The Ordnance Survey of Ireland (OSI), the Ordnance Survey of Northern Ireland (OSNI) and the United Kingdom Hydrographic Office (UKHO) are responsible for Ireland's terrestrial and marine base data, and all charge for the licensing of their data as well as for datasets derived from their base maps. Base data such as coastline, bathymetry, digital terrain models, hydrographic chart data, roads, rivers and imagery can often be the most expensive acquisition cost in a project (O'Dea *et al.* 2004). Access may also be difficult due to strategic and commercial factors that limit their availability (Bartlett 1999). The MIDA team encountered problems in regard to data cost and licensing restrictions therefore limiting the detail and amount of base data that was included in the Atlas. However, multi-annual licensing agreements at academic rates were reached with OSI, OSNI and SeaZone Solutions Ltd (distributors of UKHO data) for the base data used in the Atlas and this helped to contain data costs. An all-island raster map at 1:450,000 and coastal tiles at 1:50,000 were included. Cost prohibited the inclusion of a more extensive range of scales or the use of vector base data.

The broad spectrum of coastal data owners across Ireland made it challenging to find out who held what. Data acquisition involved personal knowledge, word of mouth and significant time investment. There were few resources in place to make data discovery easier (O'Dea *et al.* 2004). In some cases months were spent locating and sourcing important national or island-wide datasets for the MIDA. It was not uncommon to find that some desirable datasets, expected to exist in GIS-ready format, were not available, for example data layers pertaining to commercial, fishing and ferry ports. In such cases a decision was taken to generate the spatial datasets in-house using all available information such as published reports, web-sites and direct contact with relevant organizations. Having located datasets additional time passed before they could be included in the Atlas as issues such as the owner's data preparation time, the establishment of license agreements and the determination of licensing costs had to be resolved (O'Dea *et al.* 2004). In order to ensure clarity and avoid misunderstanding each data owner was offered the possibility of completing and signing a memorandum of understanding (MOU). The MOU specified the conditions under which the data owner supplied the data to the MIDA and also those that the CMRC adhered to in their use of the data. This MOU also helped to avoid any misunderstandings with regard to data use, when the personnel, either in the owner's organization or in the MIDA team changed.

One early issue addressed by the MIDA team was that of the extent of coverage of the Atlas. Defining a coastal strip as a specific distance inland and offshore from the coast was seen as too prescriptive. However, datasets were only considered for inclusion if they included information pertinent to coastal and/or marine areas. Where a dataset also included information on inland areas (e.g., national monuments) the complete dataset was presented rather than clipped. In other cases datasets naturally fell on the coast (e.g., lighthouses). On the seaward side datasets extending to the British coast on the East and the continental shelf claim area to the west were included (e.g., sea-surface temperature). Moreover, only datasets with a national coverage (island wide, Republic or Northern Ireland) were prioritized. The inclusion of regional data was left as a future development. This means that there are few spatial gaps in coverage.

Data Organization and Sources

One of the initial project activities was the listing of all possible themes and layers that could potentially be included in the Atlas. The list was established through consultation and brainstorming efforts with experts and specialists from diverse areas of coastal and marine research to establish a comprehensive inventory. A hierarchical data classification structure evolved in the process of consultation to facilitate organization of the initial list. The data hierarchy was structured around four main categories: Management, Physical Environment, Biological Environment and Socio-Economic Activities, each emerging with several meaningful sub-categories in which similar entities were grouped to enable quick and easy navigation of data and information. Figure 1 shows the first category levels and all sub-categories for the Management category. The data structure has proven to be very robust and facilitates the addition of data layers as they become available without impacting on the overall classification so that the Atlas content can be easily expanded.

In order to focus data collection efforts a priority was assigned to each dataset listed. The data were allocated priorities of 1, 2 or 3. The priority 1 list contained a balance of datasets that were most sought after by professionals and highly desired by general users, including high and low water marks, territorial limits, protected areas, and bathymetry. Focus was placed on datasets with complete coverage of the Island of Ireland. Priorities 2 and 3 contained datasets of somewhat more specialist interest (e.g., seabed sediment types, seabird observations, shipping routes). During data collection most effort focused on sourcing priority 1 data. However, lower priority datasets were gathered and incorporated in the Atlas when readily available or easy to access.

Appendix A lists the datasets currently included in the Atlas organized under the four main data hierarchy categories. A dataset in the MIDA can contain several individual data layers, which hold different information belonging to the overall set. For example the rainfall dataset has thirteen

Figure 1. The hierarchical data classification structure for the MIDA, illustrating some subcategories and specific data layers for the Management category

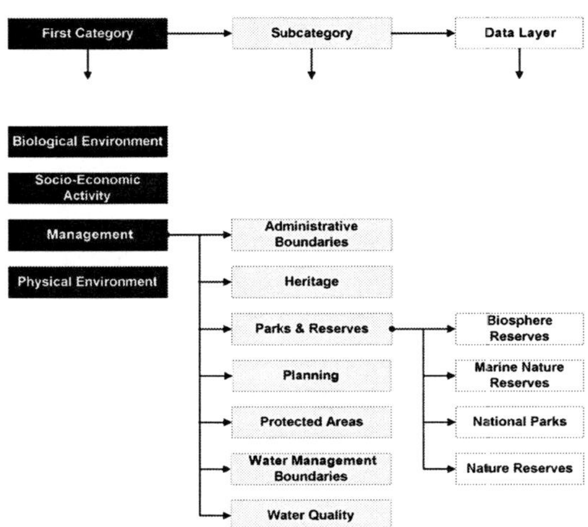

Figure 2. (a) The primary data supplier organizations (n=7) according to percentage of data layers provided to the Atlas. All those that contributed less than 3% of datasets are grouped as other (n=28). (b) Sectors to which data supply organizations (n=35) belong.

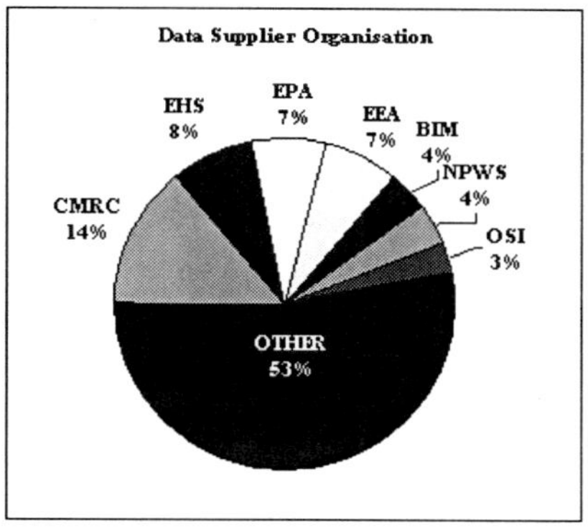

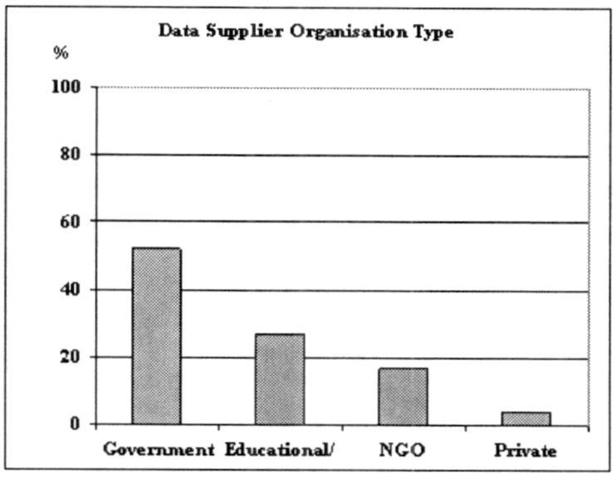

separate layers representing mean monthly and mean annual rainfall. Over 140 data layers from thirty five different organizations were gathered as of June 2009.

Figure 2 (a) shows that although 47% of the data layers came from seven organizations, the rest originated in the remaining 28 organizations. This broad distribution of data illustrates the many sectoral interests in the coastal and marine space and underlines the need for tools such as the MIDA, which can improve discovery and accessibility of spatial data.

Although data layers in the MIDA came from numerous organizations (Figure 2 (a)) many of these can be grouped as government organizations. Figure 2 (b) shows that over half of the MIDA data layers were from such organizations, a further 27% were supplied by educational institutions, followed by 17% from non-governmental organizations, while only 4% of data layers originated in the private sector. The MIDA collated and integrated many publicly available datasets from government organizations, which is encouraged through the EU Directive on the Re-use of Public Sector Information (European Union, 2003). Therefore it indirectly facilitates implementation of this Directive for public coastal and marine data in Ireland.

The seemingly large contribution of the CMRC to datasets in the MIDA is due to the fact that a significant number of datasets were constructed from other publicly available sources as directly usable GIS-ready spatial datasets did not exist (e.g., marinas, sailing clubs, saltmarshes). The private sector has a low representation among the data providers to the Atlas. This may be due to the commercial sensitivity of certain data and for less sensitive data companies market them directly. However, the MIDA offers a potential shop-window to private organizations to publicize their data holdings. The MIDA team hopes to develop a referrer capability for such organizations in the future.

THE MIDA LOOK AND FEEL

The Atlas Interface

The core of the MIDA is a web-based mapping system. The main page, illustrated in Figure 3, consists of three key areas: the map and toolbar, the layer/legend area and the information area. Users control what they see in the map by loading and displaying one or more layers. The layer/legend area lists those data layers that are currently loaded

Figure 3. The main atlas page of the Marine Irish Digital Atlas illustrates the map and toolbar, layer/ legend area and the information area

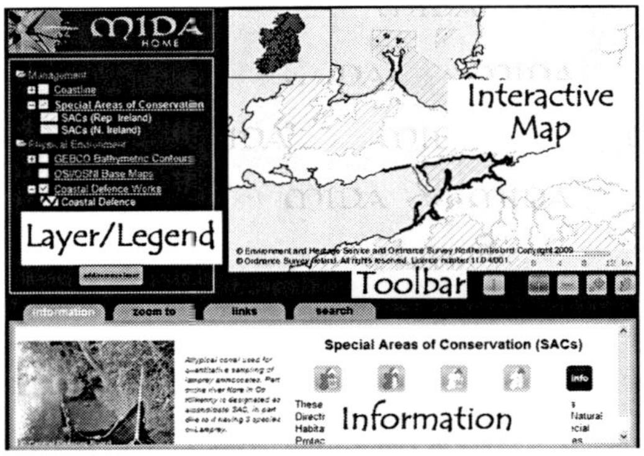

in the map, according to the four major categories mentioned previously. A tick-box is used to show or conceal the loaded layer in the map area while an expansion button allows control over the display of the individual layer's legend. Clicking on the "add/remove layer" button activates a pop-up window, which provides access to the full list of layers. One or more layers may be chosen to be added to or removed from the map. All layers loaded are automatically displayed in the map. Tools to navigate and query the layers within the map area include zoom-in, zoom-out, zoom to full extent, re-centre and feature query. These tools can be selected from the toolbar available underneath the map.

It is not appropriate to view together layers generated at very different scales without the possibility of misinterpretation by the user. Therefore, scale factor limits are associated with each layer in the Atlas display, so that when a user zooms in on an area, certain layers are turned off if the data are being viewed at a scale inappropriate to the level of detail in the layer (von Meyer *et al.*, 1999). The data owner who supplies each dataset specifies which of its attributes can be viewed. The attribute table opens in an external window when the user clicks on a specific feature after selecting the feature query tool from the toolbar. The attributes for the features in all layers directly below the point clicked are displayed. This was deemed to be more user- friendly than having to activate a specific layer in order to view its attributes. In some cases, the feature attributes contain a link to live data feeds (e.g., marine data buoys, webcams).

One of the innovative areas of the MIDA is the information area (see Figure 3) that provides access to additional functionality. From the "search" tab a basic and advanced search facility allows users to search for specific layers based on combinations of title, keyword, theme, region, downloadable dataset and a defined geographical area. The "links" tab provides a list of links to specific areas of the site. The "zoom to" tab allows users focus on a specific marine or terrestrial region or a specific subject theme (e.g., marine safety) by loading only those layers having a keyword associated with that region or theme. The "information" tab provides access to thematic information, metadata and the digital spatial data itself in cases where the data is freely downloadable.

Thematic Information

A brief description of the thematic content and a relevant photograph or image is provided for each data layer via the information tab. Currently 62 of the spatial layers are linked to specific additional thematic information in the MIDA InfoPort, which can be accessed directly from the information tab or via the Information link on the main Atlas page. The InfoPort is a repository of text, imagery, links to spatial data sources and additional reference material for a wide range of topics under the four main sections of the data hierarchy. In this regard the MIDA follows the approach of traditional, printed atlases, which provide descriptive information in addition to thematic maps. Each section has a brief description of the thematic areas addressed in the specific topic information pages within that sub-section.

As shown in Figure 4 each information page has five tabs providing access to:

- Overview – This is a short introduction to the specific thematic topic.
- Details – This section presents the key aspects of the topic, organized under a number of headings to which the user can navigate. Images, tables and links to map views of relevant layers within the MIDA web GIS are provided.
- Data Sources – These are links to spatial data resources that are pertinent to the thematic topic. The spatial data may be in the MIDA itself or available via external websites.
- Links – These provide access to additional information and allow users explore a topic in more depth.
- References – This is a list of the documents and sources consulted in compiling the information page.

Each page is reviewed by an expert in the thematic area to ensure that the key information is complete and correct. The material is written so as to be accessible to and understandable by a wide audience whilst also being of relevance to specialists and professionals.

Metadata Display

In many web GIS applications metadata presentation is less than ideal. Long and extensive records, using the implemented standard's hierarchy, element names, technical terminology and structure are displayed. For users with limited familiarity with metadata, this can be incomprehensible and off-putting. Indeed, it can be difficult for users to find the basic information required on the dataset. For the purpose of the MIDA it was decided that metadata should be presented in a user-friendly and easy to follow manner. XSLT style sheets were used to display the XML encoded discovery metadata as HTML. Elements which are obliga-

Figure 4. Via the InfoPort users can access thematic information, images, links to spatial data sources and additional reference material for a wide range of coastal and marine topics

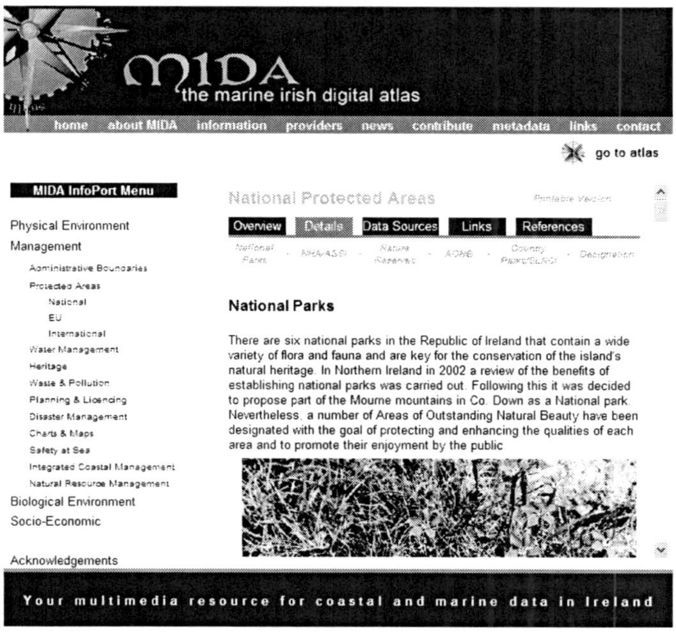

tory in the implementation of the ISO metadata standard but deemed to be of little interest to the general user (e.g., metadata language, character encoding) are not displayed although they are available in the XML. Moreover a tiered approach to viewing the metadata was adopted in order to enhance usability. Three tiers or levels of metadata may thus be viewed in the MIDA:

- **Abstract Metadata:** This is a succinct description of each dataset. It is the first level that users see when they choose to view the metadata for a particular layer. Five elements (title, abstract, geographical extent, owner, publication date) are simply extracted from the Discovery Metadata database and displayed. There is a link to the complete Discovery Metadata if users wish to view more information.
- **Discovery Metadata:** This contains all the metadata elements defined in the customised ISO profile for each dataset. It forms the main catalogue of the Atlas and may be queried. Appendix B lists and defines the elements that form the Discovery Metadata. If the user wishes to see more information, a link to Full Metadata is available.
- **Full Metadata:** This may be provided by the data owner who contributes the dataset. The format, detail and quality of the metadata are the data owner's responsibility and it is displayed as supplied. Full metadata is available for 82 of the over 140 data layers.

FACILITATING DIFFERENT USER-GROUPS

One of the MIDA developers' main intentions was to cater for a broad audience, which would enable professionals, students and casual visitors to find coastal and marine data and information of interest. In order to attract and facilitate such a wide range of user groups, the team developed a number of options to access and search the Atlas. A tutorial, which can be opened from the Atlas start page, was developed to help fist time visitors understand the various ways of using the atlas and it demonstrates the different search options to access specific data layers and information.

A list of all the spatial data layers in the Atlas, which can be accessed from the home page, provides an immediate but extensive overview for first time and casual MIDA visitors. A more controlled exploration of data is possible by using the "add/remove layer" button in the Layer/Legend area (see Figure 3) of the Atlas interface. Here the datasets are organized in pre-grouped themes.

The MIDA search facility was developed with the professional user in mind. The tool provides search options on dataset title, keyword, theme, region or by downloadable datasets. An advanced search tool (Figure 5) permits selection of a combination of these options in addition to selection of geographical area. The search results are presented in a table (Figure 5) providing a summary paragraph on the data. The table provides direct links to the data, metadata, and additional thematic information (if available).

To support research on coastal and marine topics, the MIDA InfoPort (see Figure 4) provides a good starting point for a number of thematic areas. This could be of particular value to students when researching associated subjects. In addition the student can use the InfoPort to view relevant spatial datasets available in the MIDA via in-built links in the information page of interest. It also provides links to external web sites holding related spatial data via the data sources tab. In addition the links tab of the InfoPort lists and links to web sites that can support further research of the thematic information in question.

Figure 5. The advanced search tool within the MIDA and the results of a search presented in an informative scroll down table

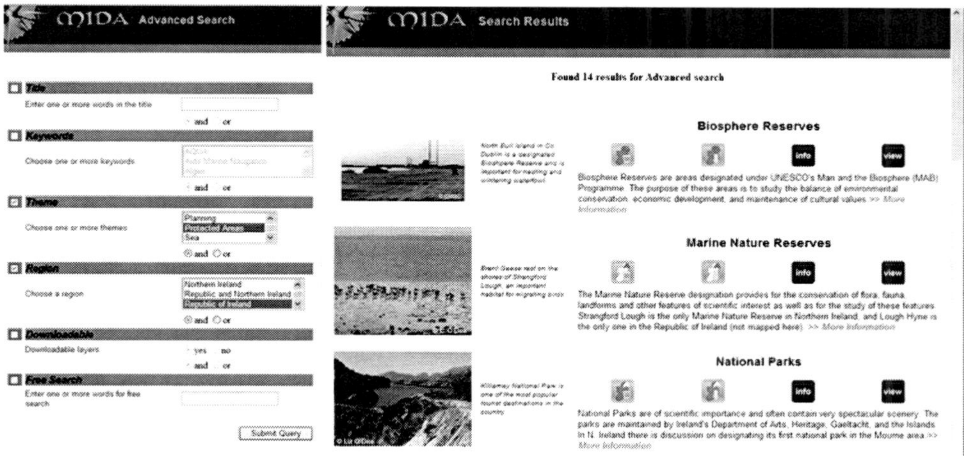

PUBLICITY AND ATLAS USAGE

Workshops

A workshop which targeted some key user groups and potential data contributors was held in Dublin in December 2003 to demonstrate the MIDA prototype. The feedback was used in design changes for the version of the MIDA released in May 2005. Following the release three workshops were held in different locations around the country. The main aims of these were to raise awareness and publicize the Atlas as well as get additional feedback to aid ongoing development of the MIDA. Each workshop was attended by approximately twenty people representing a typical cross-section of users. The attendees had the opportunity to work with the Atlas and had to carry out a number of tasks specified by the workshop facilitators. Both written and oral feedback was gathered. The "one-stop-shop" nature of the Atlas was identified as a key benefit by many participants. Feedback proved to be very useful in tailoring certain aspects of the Atlas. For example a number of participants expressed interest in spatial data for specific additional themes; as of June 2009 over 85% of those themes suggested contain spatial information.

Publicity

The Atlas was promoted at both national and international specialist conferences. Initially articles were written for a number of popular publications in the marine domain which are distributed in Ireland and Britain. To enable quick and easy dissemination of MIDA news and to raise the profile of the resource an email list of those with an interest in the coastal and marine environment and web GIS developments was compiled. Furthermore an Atlas brochure was published and distributed widely. A specific email address (mida@ucc.ie) was set up when the Atlas went online so that users could send queries and comments to the MIDA team. This has been less used than expected with just over 50 emails received since June 2005. Most have been requests for data or information not included in the Atlas whilst others have been questions on specific data layers or technical aspects of the Atlas. In addition there have been many positive comments on the Atlas.

The Atlas was officially launched by the President of University College Cork during the Coastal Mapping and Informatics Trans-Atlantic Workshop held in Cork in July 2006. This generated significant media interest and led to articles

Figure 6. Number of unique visitors to the MIDA and visits per months since June 2006

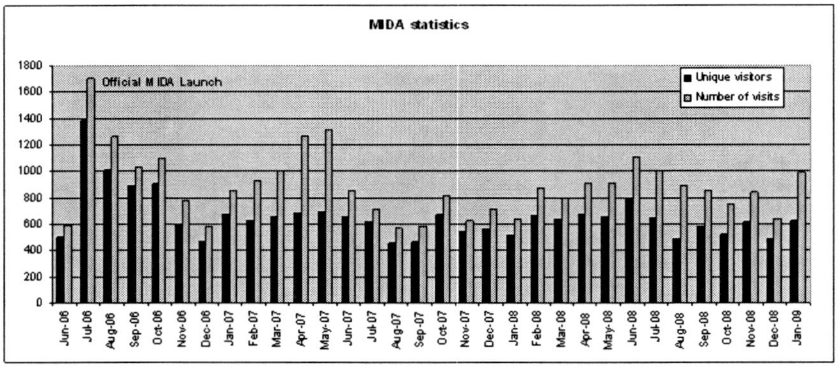

being published in the national and regional press. Furthermore, through all the ICAN Workshops and the Network's outreach work, the MIDA has become known as a good example of Coastal Web mapping to an international audience. MIDA statistics show external links to the MIDA are incorporated in at least 40 web sites, while there are dozens of references to the Atlas in various online documents and publications (e.g., González Del Campo, 2009; IWEA, 2008; CSA Group/Circa Group/Swiftsure, 2005).

Usage

Figure 6 shows usage statistics for the Atlas since June 2006. The publicity associated with the official launch in July 2006 led to an increase in visits to the Atlas which was sustained for about three months, visits thereafter reducing somewhat. No additional significant publicity events have been carried out since July 2006 but with approximately six hundred unique visitors per month, the MIDA is proving to be a relevant and popular resource.

A further analysis of the 2008 MIDA user statistics confirmed that most Atlas visitors were from the Republic of Ireland, followed by visitors from the UK and the USA. High UK interest was expected as the MIDA was a cross border collaboration featuring island-wide datasets as well as some specific Northern Irish data layers.

The 2005 workshop participants stated that the most appealing aspects of the MIDA included the amount of data in one site as well as the downloadable data layers and metadata records. The 2008 MIDA statistics showed that of the 53 data layers that were directly downloadable from the Atlas, the most popular were from the socio-economic activity section, specifically water-based recreation. Three of the five most popular downloads in 2008 belonged to the sub-section of water based recreation:

1. The Blue Flag Beaches in the Republic of Ireland,
2. The Blue Flag beaches in Northern Ireland,
3. The Location and relevant information of Marinas and Pontoons on the Island of Ireland,
4. A MODIS Satellite Image of Ireland,
5. The Territorial and Fisheries Limits of the Republic of Ireland.

The keen interest in spatial data regarding water based recreation, as shown by the MIDA visitor statistics, indicate that this type of information was not readily available elsewhere for the Island of Ireland, therefore supporting the decision of the MIDA team to create such layers in-house.

The MIDA is used in teaching on a number of courses organized by the department of Geography within UCC. Technology and web GIS aspects are presented in GIS courses and many students use the Atlas as a resource for their research in coastal management related courses. Furthermore, since 2004, the CMRC facilitates work experience for trainees through the Atlas developing skills in web mapping and GIS and data and metadata management. Trainees improve their writing skills through work on the information pages, which also raises their awareness of coastal and marine concerns. To date nine trainees from five different countries have joined the MIDA team on placements that last from three to twelve months. Most of these placements were funded through the European Union's Leonardo da Vinci vocational training program. Local students and secondary school transition year students on shorter term placements with the CMRC have also worked on the MIDA. Feedback given by the students indicates that MIDA is utilized as a tool by teachers in some secondary schools.

The Marine and Coastal Heritage Directory (http://www.coastalheritage.ie/), developed by the Heritage Council features text and imagery related to Ireland's coastal heritage. Links from the directory to specific spatial data layers within the MIDA are used to illustrate the text within this Directory. Such collaborations help to expand the user community for the Atlas.

THE ATLAS SYSTEM: A DEVELOPER PERSPECTIVE

Selection of Web GIS Software

Visualization and overlay of the many data layers on a map within a web browser was one of the key objectives of the MIDA. Therefore, a web mapping software package had to be selected. Two were evaluated: ESRI ArcIMS and University of Minnesota MapServer. Table 1 presents the main advantages and disadvantages of each system. MapServer was chosen as the most appropriate solution as display and navigation of layers rather than elaborate processing were sufficient for the Atlas; significant programming experience was available with the CMRC and the absence of a license fee made it a more low-cost, long-term solution.

Data and Metadata Preparation

As the number of data layers to be included in the MIDA was in the hundreds, a geospatial database

Table 1. Advantages and disadvantages of the ESRI ArcIMS and the Minnesota MapServer

ESRI ArcIMS	Minnesota MapServer
Advantages	**Advantages**
"out of the box" web GIS	Non-proprietary technology
High level of built-in GIS functionality	No licensing fees
Large user base	Large community of users
Good online support	Simple configuration structure
Integrated metadata server	
Disadvantages	**Disadvantages**
Significant cost	Not "out of the box" web GIS
Proprietary technology	Limited GIS functionality
Complex configuration structure	Require programming skills to build and customise web GIS
Require programming skills for customization	Lacks integrated metadata system

was considered unnecessary. Instead, standard file formats are used. Shapefiles store vector data and GeoTIFF store imagery/raster data. Data preparation principally involved conversion, cleaning, reprojection, tiling and attribute table editing. Most data preparation tasks were carried out in ESRI ArcGIS. However, for certain operations that enhanced MapServer performance other tools were employed. For example, the open source utilities *gdaltindex* was used for tiling large raster layers and *shptree* for optimizing vector display.

Metadata forms an integral part of the MIDA by consistently documenting each dataset held. Although no national metadata standard existed in Ireland at the project's outset, many key organizations that handle spatial data have been adopting ISO 19115 in view of its inclusion as the standard of choice in the implementation of the European Union's INSPIRE Directive (European Union 2007). After a review of a number of metadata standards, and taking into account the limited availability of metadata associated with spatial data in Ireland, it was decided to devise a customized profile of the ISO 19115 standard for the MIDA. A total of 55 elements were selected as the Discovery Metadata for each layer in the Atlas (see Appendix B). These included all elements defined as obligatory in the ISO standard and a number of optional ones believed to be necessary in order to have a minimum acceptable amount of information for each dataset. This metadata was saved in Extensible Markup Language (XML), using the Short Names defined in the ISO 19115 as tags (ISO 2003). ISO 19139 tags were not used as this standard had not yet been released when this phase of the project was undertaken.

A number of metadata entry tools available at the project's outset, such as the ESRI ArcCatalog ISO wizard, M^{3}Cat™ (http://www.intelec.ca/html/en/technologies/m3cat.html), Enraemed (http://geoinfo.uneca.org/geoinfo/ethiopia/enraemed.html) and the ANZLIC Metadata Collector (http://www.walis.wa.gov.au/anzlic_met_project) were reviewed but were not deemed suitable for MIDA metadata collection. Therefore a customized, HTML web-enabled, data entry wizard was designed and implemented in order to ease the task of Discovery metadata collection and entry. This wizard was developed using HTML, JavaScript and JavaServer Pages (JSP) technologies. Metadata were entered in a sequence of HTML forms. Once submitted they were saved on the web server in a flat XML document. The MIDA administrator collected this metadata, reviewed it, and registered it, in the MIDA system. An XSLT (Extensible Stylesheet Language Transformation) stylesheet was created in order to visualize the XML metadata in HTML.

ESRI's ArcCatalog was used to help the MIDA team with management of the datasets included in the Atlas. This system was separate from the Discovery Metadata described above. ArcCatalog provided a way of recording details vital for data management and update, such as when and how data from external organizations were acquired and what modifications were made in order to display those datasets in the Atlas. When the project started only a limited profile of ISO elements was available within ArcCatalog therefore the more complete and extensive FGDC (Federal Geographic Data Committee) standard was chosen for the management of metadata.

The MIDA Prototype

The MIDA prototype consisted of the main Atlas page (including the map and toolbar, the layer/legend area and the information area), and the InfoPort page. The technologies used were HTML, JavaScript and MapServer, running in CGI (Common Gateway Interface) mode. It was deployed within the Apache web server in a Linux Operating System. This prototype was inherently static, which led to update and maintenance problems. In order to incorporate a new layer, the MIDA administrator needed to edit the HTML code manually, and register the layer both in the layer/legend area, and the information area. Another problem was the system's inability to perform dynamic searching for datasets using metadata elements. In order to overcome these problems the 'MIDA Engine' was conceived.

The MIDA Engine

The "MIDA Engine" is a customized mini content management system (CMS) designed to handle the main Atlas page. At the time of the development, no operational framework for implementing MapServer web GIS applications existed. Although generic frameworks such as CartoWeb (http://www.cartoweb.org/), Mapbender (http://www.mapbender.org/) and Openlayers (http://openlayers.org/) now exist, they still require customizations to meet specific web GIS development needs. The "MIDA Engine" consisted of three main components: the map and toolbar, the layer/legend area and the information area. The content for these components was entered into two database tables which interact with MapServer's main configuration file (known as the mapfile). One table dynamically controls the layer/legend area. It contains layer information such as the filename, keywords and location of downloadable data among others. This structure facilitates user data search. The full XML metadata for each layer is also stored in this table. Currently each new metadata record must be manually uploaded to the database by the MIDA administrator, but there are plans to automate and streamline this process. The other table contains: information theme title; a brief description of the thematic content of the layer; a photo/image and a caption.

The "MIDA Engine" is principally implemented using PHP (hypertext preprocessor) and MapServer's MapScript/PHP module. MapScript provides a scripting API (application programming interface) to MapServer which enables a programmer to extend/customize MapServer's default functionality contained in the CGI version. MapScript supports programming languages/ environments including PHP, Python, Perl, Ruby, Java, and .NET. PHP was chosen because it is the best supported MapScript module and it is also used in frameworks such as CartoWeb and Mapbender. The PHP code controls the Atlas functions including layer display and overlay, the zoom extent, watermarking and attribute display.

When a user queries a vector feature in the map, its attributes are displayed. For this the CGI version of MapServer requires that HTML templates are created for each dataset. As this is a cumbersome task, the "MIDA Engine" uses PHP code to register a layer's attributes in the mapfile and control their display in a pop-up window.

The MIDA search facility is implemented via SQL queries on the two database tables described above. The "MIDA Engine" can connect to any RDBMS (relational database management system) database supported by PHP; in this case the open source database PostGreSQL is used.

The MIDA Engine Legacy

The "MIDA Engine" has proven to be a robust platform which can be deployed on both UNIX and Windows platforms. It has provided excellent added value as it has been used to support other initiatives requiring web GIS portals. For example, the European Commission, Framework Program 6, funded InterRisk project which provided tools for management of marine pollution events (Hamre *et al.,* 2009) and the Interreg III A funded IMAGIN project which studied sustainable aggregate exploitation in the Irish Sea (O' Mahony *et al.*, 2008) have both used the "MIDA Engine." As neither project required a dynamic database search facility, the "MIDA Engine" was deployed without an RDBMS and used dynamic layer management via the mapfile directly. Moreover, as the "MIDA Engine" uses MapServer, it automatically supports OGC (Open Geospatial Consortium) technologies including WMS (Web Map Service), WFS (Web Feature Service) and WCS (Web Coverage Service), as demonstrated in InterRisk. In early 2009 an agreement was reached with the National Institute of Marine Sciences and Technologies (INSTM), Carthage, Tunisia to use the "MIDA Engine" in the development of a Tunisian Marine Atlas (Y. Lassoued, pers. comm., 2009).

There is a need for constant innovation in web GIS data presentation and management in order to meet the expectation of users. New open source technologies such as OpenLayers are being investigated in order to enhance the user interface. With regard to metadata management, as part of the ICAN semantic interoperability initiative, selected MIDA metadata records have been transposed into the open source GeoNetwork metadata management system, which supports the ISO 19115/19139 metadata standards and the CSW (Catalogue Services for the Web) standard (described elsewhere in this book). This will improve metadata management and interoperability with data sharing initiatives.

DISCUSSION

An Assessment of the MIDA

The MIDA has met its primary goal as a key access point for spatial data and information on Ireland's coastal and marine areas for a broad audience. The user statistics confirm that the Atlas is constantly visited attesting to its relevance and usefulness. Recognizing that the Atlas is an extremely valuable resource for the coastal and marine community, the CMRC maintains the Atlas from its own resources, since the initial development concluded in 2007, when the project funding period ended. Overcoming difficulties and finding solutions to problems that emerged early in Atlas development have equipped the MIDA team with experience and skills that are invaluable in maintaining and enhancing the Atlas. The use of the 'MIDA Engine' in other national and international initiatives requiring web GIS portals illustrates the added value of the project as it highlights the technical skills within the Centre which in turn has led to additional project work.

The MIDA team was able to overcome initial obstacles concerning data licensing agreements and cost but will have to continuously address these issues, adopting new solutions that will benefit the data owners and the Atlas simultaneously. The lack of data catalogues and the non-existence of metadata for most layers, except for a few institutions was a major barrier. Almost all the Discovery Metadata had to be compiled by the development team. This often involved phone calls with the data providers in order to collect a minimum of information on the datasets. New developments such as the European INSPIRE Directive should lead to an improvement in coastal and marine data documentation and discovery ability. The requirement to provide metadata with all spatial data generated by public bodies should not only benefit existing CWAs such as the MIDA but aid new CWA developments in Europe. Initial MIDA data classification and organization has proven to be extremely useful and has developed into a strong data classification system. However, a shortcoming of the Atlas from a developer's perspective is the lack of a streamlined and automated data management system. MIDA metadata entry and update is currently a time consuming process but the emergence of GeoNetwork and other data management tools now offer the possibility for improved and more streamlined metadata management and can also enhance interoperability within distributed systems.

The MIDA has proven to have numerous uses from providing data and information to different end-user groups to being a valuable training resource for skills in web mapping, data and metadata management. Throughout the design and development phase MIDA developers consulted a representative cross-section of end-users. Such consultation is vital in CWA development as end-user input can ensure that the atlas meets their needs. Face-to-face meetings are more effective than anonymous web-based surveys as a means of gathering feedback. In addition to keeping the atlas up to date, sustained promotion and publicity is vital for any CWA in order to raise awareness and attract new users. Analysis of the Atlas usage statistics has highlighted the most popular aspects of the Atlas and is informing ongoing development work. Furthermore, the MIDA is useful with regard to the implementation of Integrated Coastal Zone Management and the emergence of Marine Spatial Planning initiatives as supported by the proposed Integrated Maritime Policy for Europe (Commission of the European Communities 2007). As the Atlas allows the visualization of multiple data layers from a wide range of sectors it can help inform marine and coastal planners and managers. Although it is not a decision support tool, access to the data and information that can form the basis for such tools is provided. It also complements the European Atlas of the Seas which is being developed by the European Commission (Commission of the European Communities 2007)

Challenges

A major challenge for any atlas developer and the current MIDA team is the on-going financing of the atlas. Research funds are often available for the development of new and innovative solutions. However, it is more difficult to get funding for the maintenance of existing tools. There is a need for constant innovation to keep the Atlas relevant. Advances in the display of spatial data in environments such as Google Earth (http://www.openioos.org/real_time_data/gm_sos.html) and Google Maps (e.g., http://marinemap.org/marinemap/) raise expectations of web GIS users concerning the look and feel of such applications. The MIDA team needs to address not only maintenance but also updates to develop a new technology interface as well as a data management system. Furthermore these updates have to take into consideration people with limited knowledge of GIS functionality, who have found it somewhat difficult to operate the current interface. However these challenges also offer opportunity for technical innovation and therefore the ability to target appropriate funding programs.

It is important to keep the Atlas in the public eye and to engage in ongoing publicity and promotion. This will be an ongoing challenge but also creates opportunities to address other issues such as funding. It is imperative that the data and information in the Atlas are kept up to date and that new data are constantly added. Many such initiatives fail because web-sites are not updated once the initial projects which brought them into being are concluded. The CMRC has assured the MIDA's future by dedicating resources to maintain it. Moreover active participation in initiatives such as ICAN provides constant exposure for the Atlas and a stimulus to undertake additional development work.

Future Developments

With advances in technology and CMRC expertise gained through the MIDA and related projects the potential exists to develop a new technology interface and data management system. New open source technologies such as OpenLayers for data visualization and GeoNetwork for metadata management will be investigated in order to enhance the system. This could provide the requisite funding to update the technological aspects of the Atlas as well as resources to address ongoing issues with regards to data acquisition and maintenance. Efforts will be made to complete the InfoPort therefore strengthening the thematic aspects of the Atlas. Ideas for further development of the Atlas include a tiered approach to address the needs of different audiences. For example, the MIDA team has evidence that the Atlas is sporadically used in some schools to aid teaching. Creating a tailored version to compliment the secondary school geography curriculum for use by both teachers and students would not only fulfill the MIDA potential as a teaching tool but also raise awareness and profile of the resource in general.

Other ideas include a second tier targeting tourism and recreation and a third tier for professionals, designed with advanced mapping tools for use in coastal management and spatial planning.

Since the MIDA project began, there has been significant progress in interoperable solutions and the development of OGC standards to facilitate the sharing of geospatial data and metadata. Indeed the CMRC together with the Marine Institute, the Environmental Protection Agency (EPA) and the Department of Communications, Energy and Natural Resources (DCENR) have set up the Irish Spatial Data Exchange (http://www.isde.ie). This Exchange enables the sharing of metadata from these institutions' online catalogues across a distributed network. The MIDA is using open source solutions whilst the other partners are using proprietary tools from ESRI, however with the use of CSW and ISO standards, it is possible to share the metadata seamlessly. The next step will be to expand the efforts to include the seamless display of spatial data within each portal's web GIS. This will allow access to the data source instead of potentially outdated versions on individual servers. As part of the ISDE network, the ability to implement WMS and potentially WFS/WCS will be investigated further.

On an international level, as presented in Chapter 4, the MIDA is one of the nodes of the ICAN global atlas prototype which is testing semantic interoperability between two coastal web atlases. This work is currently incorporating additional nodes and extending OGC services to include CSW and WMS. This project is relevant in regard to the emerging EU Integrated Maritime Policy, in which the need for data and infrastructure is highlighted as well as awareness raising tools such as the European Atlas of Seas. Interoperability approaches and coastal web mapping in general are innovations that can facilitate those visions on a very practical level.

CONCLUSION

The MIDA is facilitating better sourcing of and improved access to spatial data in Ireland and will continue to do so with ongoing maintenance assured by CMRC. Usage statistics illustrate that the site is in constant use and that many of the accessible data layers are being downloaded. The Atlas has also been extremely valuable as a technology demonstrator as acknowledged by its use in other web GIS initiatives including the ICAN semantic interoperability prototype. However, ongoing development is crucial to the MIDA's long term survival. Funding, technology and data management issues all need to be addressed. The MIDA's key driving role in regard to developing catalogue and atlas interoperability on a national and international level offers new opportunities for advanced deployment and future development of the Atlas.

ACKNOWLEDGMENT

This work was funded for the period September 2002 – August 2005 by the Higher Education Authority (HEA) of Ireland under the third Program for Research in Third Level Institutions (PRTLI 3) Program as part of the National Development Plan 2002 – 2006. Additional funding was received from the Environment and Heritage Service, Northern Ireland. The authors would like to thank Ciara Herron, Didac Perales, Juan Arévalo, Carlo Brondi, Iban Ameztoy, Kristel Coutel, Natasha Faucher, Diego del Villar, Paula Domingo, and Yvette Harrington for their contributions to the MIDA development. Furthermore we are grateful to all those who provided feedback on the MIDA at the various workshops and other events organized around the project.

REFERENCES

Bartlett, D. J. (1999). Working on the Frontiers of Science: Applying GIS to the Coastal Zone . In Wright, D. J., & Bartlett, D. (Eds.), *Marine and Coastal Geographical Systems* (pp. 11–24). London: Taylor & Francis.

Commission of the European Communities. (2007). *Communication from the commission to the European parliament, the Council, the European economic and social committee and the committee of the regions- An Integrated Maritime Policy for the European Union, COM*(2007) 575 final.

CSA Group/Circa Group/Swiftsure. (2005). *An Assessment of the optimal use and application in the immediate to long term future of the Irish National Seabed Survey Deliverables to Date (Phase 1) and to contribute towards the development of phase 3*. Retrieved June 29, 2009, from Infomar website: http://www.infomar.ie/documents/Assessment_of_the_INSS.pdf

Department of Environment, Food, and Rural Affairs. (2002). *Delivering Integrated Marine Mapping for the UK*: Report of DEFRA-funded workshop held at Church House, London, 11 September 2002, compiled by F.L. Franklin, DEFRA, Burnham Laboratory.

Dwyer, N., O'Dea, L., & Cummins, V. (2003). *The Marine Irish Digital Atlas: a Web Portal to Coastal and Marine Data in Ireland*. Paper presented at CoastGIS 2003, Genova, Italy. Retrieved June 24, 2009, from MIDA website: http://mida.ucc.ie/assets/documents/ MIDA_CoastGIS2003.pdf

European Union. (2003). Directive 2003/98/EC of the European Parliament and of the Council on the re-use of public sector information. *Official Journal of the European Union. L&C, L345*(90), 46.

European Union. (2007). Directive 2007/2/EC of the European Parliament and of the Council establishing an Infrastructure for Spatial Information in the European Community (INSPIRE). *Official Journal of the European Union. L&C, L108*(1), 50.

González Del Campo, A. (2009). *Current Practice and Potential on the Application of Geographic Information Systems as a Support Tool in Strategic Environmental Assessment of Irish Land Use Plans. GISEA Manual – Consultation Draft (1st April 2009)*. Ireland: Environmental Protection Agency.

Hamre, T., Krasemann, H., Groom, S., Dunne, D., Breitbach, G., & Hackett, B. (2009). Interoperable web GIS services for marine pollution monitoring and forecasting. *Journal of Coastal Conservation, 13*, 1–13. doi:10.1007/s11852-009-0046-y

International Organization for Standards (ISO). (2003). *International Standard ISO 19115, geographic Information*. Metadata.

IWEA. (2008). *Best Practice Guidelines for the Irish Wind Energy Industry*. The Irish Wind Energy Association and Sustainable Energy Ireland. Retrieved from http://www.iwea.ie

Mahony, O. C., Sutton, G., McMahon, T., O'Cinneide, M. & Nixon, E. (2008). *Issues and Recommendations for the Development and Regulation of Marine Aggregate Extraction in the Irish Sea*. Galway, Ireland: Marine Institute.

McCormack, B. (2003). *Irish Spatial Data Infrastructure*, paper presented at Irish Organization for Geographic Information (IRLOGI) 2003 Conference, Dublin, Ireland. Retrieved June 24, 2009, from IRLOGI website: http://www.irlogi.ie/pdf/ISDI_IRLOGI_2003.pdf

O'Dea, L., Cummins, V., & Dwyer, N. (2004). *Developing an Informational Web Portal for Coastal Data in Ireland: Data Issues in the Marine Irish Digital Atlas*. Paper presented at Oceanology International London 2004, 18 March, London, UK. Retrieved June 24, 2009, from MIDA website: http://mida.ucc.ie/assets/documents/oceanology2004_mida.pdf

O'Dea, L., Cummins, V., Wright, D., Dwyer, N., & Ameztoy, I. (2007). *Report on Coastal mapping and Informatics Trans-Atlantic Workshop 1: Potentials and Limitations of Coastal Web Atlases*. University College Cork, Ireland. Retrieved June 24, 2009, from ICAN website: http://workshop1.science.oregonstate.edu/final_rpt

Von Meyer, N., Foote, K. E., & Huebner, D. J. (1999). In Information Quality Considerations for Coastal Data. In D.J. Wright & D. Bartlett (Ed.), *Marine and Coastal Geographical Systems* (pp. 295-308). London: Taylor & Francis.

KEY TERMS AND DEFINITIONS

Coastal Web Atlas: A collection of digital maps and datasets with supplementary tables, illustrations and information that systematically illustrate the coast, oftentimes with cartographic and decision support tools, all of which are accessible via the Internet.

Atlas Interoperability: The ability to interrogate two or more unrelated atlas systems and potentially exchange metadata, images, vector and raster data.

Coastal and Marine Information: Both spatial and non-spatial information regarding the coastal and marine areas of a region or country.

Integrated Coastal Zone Management: A dynamic, multi-sectoral approach to managing the coast which takes into account social, economic and environmental concerns of all parties with an interest in the coastal space.

Ireland: European Island in the Northeast Atlantic.

Metadata: Metadata in the geographical domain is structured information on a dataset, which helps the data owner to document and catalogue the data, whilst helping a data user to understand the content and fitness for use of a dataset.

MIDA: Marine Irish Digital Atlas.

Spatial Data Visualization: The ability to view digital data with a spatial dimension in a computer environment containing a coordinate reference system.

Web GIS: A geographical information system which can be accessed over the Internet and allow visualization and interaction with spatial data via a map as well as providing analysis functionality such as spatial analysis, querying and buffering.

APPENDIX A

Data layers currently contained in the Marine Irish Digital Atlas organized by the four main categories

Table 2. Management

First Category: Management		
Name	**Description**	**Owner**
AONB	Areas of Outstanding Natural Beauty	Environment and Heritage Service Northern Ireland (EHSNI)
ASSI	Areas of Special Scientific Interest	Environment and Heritage Service Northern Ireland (EHSNI)
Biosphere Reserves	UNESCO sites to promote solutions to reconciling conservation of biodiversity with sustainable use.	Coastal and Marine Resources Centre (CMRC)
Border	Border of the Rep. of Ireland	Environmental Protection Agency (EPA), Ordnance Survey Ireland (OSI)
Celtic Sea Cod Spawning Box	Location of Cod restriction area	Coastal and Marine Resources Centre (CMRC), European Commission
Coastline	Coastline of Ireland	Ordnance Survey Ireland (OSI); Global Self-consistent, Hierarchical, High-resolution Shoreline Database (GSHHS)
Country Parks	Country Parks of Northern Ireland	Environment and Heritage Service Northern Ireland
Clár Areas	Rural areas targeted for investment	National University of Ireland Maynooth (NUIM)
Current PAD Authorizations	The current exploration authorizations given by the Petroleum Affairs Division (PAD)	Department of Communications, Energy and Natural Resources (DCENR)
Hydro-chemical properties	Station sites. North of Ireland Joint Agency Coastal Monitoring Programming (NI-JACMP).	Department of Agriculture and Rural Development (NI) / Agri-Food and Biosciences Institute (AFBI)
High & Low Watermarks	High & Low Watermarks	Ordnance Survey Ireland (OSI)
ICES areas	ICES fishing blocks	International Council for Exploration of the Sea (ICES)
Irish Conservation Box	Irish Conservation Box	Coastal and Marine Resources Centre (CMRC), European Commission
Gaeltacht Boundaries	Gaeltacht Boundaries	GAMMA company
MNR	Marine Nature Reserves	Environment and Heritage Service Northern Ireland (EHSNI)
National Monuments	National Monuments in state care	National Parks and Wildlife Service (NPWS)
National Parks	National Parks	National Parks and Wildlife Service (NPWS)
NHAs	Natural Heritage Areas	National Parks and Wildlife Service (NPWS)
Nature Reserves	Statutory Nature Reserves	National Parks and Wildlife Service (NPWS), Environment and Heritage Service Northern Ireland (EHSNI)
Outline NI	Coastline and border of Northern Ireland	Ordnance Survey of Northern Ireland (OSNI)

continued on following page

Table 2. continued

First Category: Management		
Name	**Description**	**Owner**
Petroleum Licensed Blocks	License Blocks designated by the Petroleum Affairs Division (PAD)	Department of Communications, Energy and Natural Resources (DCENR)
Ports	Commercial and Ferry Ports	Coastal and Marine Resources Centre (CMRC)
Ramsar Sites	Wetlands of international importance	National Parks and Wildlife Service (NPWS); Environment and Heritage Service Northern Ireland (EHSNI)
Regions	Local Authority Regions of the Republic of Ireland	Environmental Protection Agency (EPA), Ordnance Survey Ireland (OSI)
River Basin Districts	River Basin Districts for the Water Framework Directive	Environmental Protection Agency (EPA)
Rockall Haddock Box	Location of Fishery Protected Area Rockall Haddock Box	Coastal and Marine Resources Centre (CMRC), European Commission
SACs	Special Areas of Conservation (NATURA 2000)	National Parks and Wildlife Service (NPWS); Environment and Heritage Service Northern Ireland (EHSNI)
SLNCI	Sites of Local Nature Conservation Importance	Environment and Heritage Service Northern Ireland (EHSNI)
SPAs	Special Protection Areas (NATURA 2000)	National Parks and Wildlife Service (NPWS); Environment and Heritage Service Northern Ireland (EHSNI)
Straight Baselines	Irish Straight baselines	Irish Naval Service (INS)
Territorial limits	Location of 6,12 and 200 nautical mile limits	Irish Naval Service (INS)
Towns	Towns over 2,000 people (1995)	Environmental Protection Agency (EPA); Ordnance Survey Ireland (OSI)
Water Quality	Bathing Water Quality	Environmental Protection Agency (EPA); The Northern Ireland Environment Agency (NIEA)
Whitefish Restriction Area	Location of Whitefish Restriction Area	Coastal and Marine Resources Centre (CMRC), European Commission
World Heritage Sites	World Heritage Sites on the island of Ireland	Coastal and Marine Resources Centre (CMRC)

Table 3. Physical environment

First Category: Physical Environment		
Name	**Description**	**Owner**
450,000 base map	1:450,000 base map image	Ordnance Survey Ireland (OSI)
50,000 Webmap	1:50,000 OSI Webmap images (coastal tiles only)	Ordnance Survey Ireland (OSI)
Bathymetry	GEBCO vector and gridded bathymetry	Natural Environment Research Council (NERC)
Bedrock Geology	1:500,000 bedrock geology map	Geological Survey Ireland (GSI)

continued on following page

Table 3. continued

First Category: Physical Environment		
Name	**Description**	**Owner**
Coastal Geology	Coastal protection classified in the European EUrosion Project	European Environmental Agency (EEA)
Coastal Geomorphology	Coastal protection classified in the European EUrosion Project	European Environmental Agency (EEA)
Coastal Defense Works	Coastal protection classified in the European EUrosion Project	European Environmental Agency (EEA)
Coastal Lagoons	Distribution of lagoons	Dr. Marinus Otte; Coastal and Marine Resources Centre (CMRC)
Coastal Waters	Coastal Waters	Environmental Protection Agency (EPA)
CORINE Landcover	Satellite interpreted landcover for 1990 and 2000	Environmental Protection Agency (EPA); European Environmental Agency (EEA)
CORINE Landcover Change	Satellite interpreted landcover change between 1990 and 2000	Environmental Protection Agency (EPA); European Environmental Agency (EEA)
Erosion Trends	Coastal erosion trends classified in the European EUrosion Project	European Environmental Agency (EEA)
General Soil Map	General Soil Classification Map	Teagasc
General Soil Map of Northern Ireland	Soil Classification Map of Northern Ireland at a scale of 1:250,000	Department of Agriculture and Rural Development (DARD); Agri-Food and Biosciences Institute (AFBI)
Landsat Image	LANDSAT Satellite Image of Ireland	European Environmental Agency (EEA); Joint Research Centre (JRC)
LaCoast	Coastal Land Cover Change between 1975 and 1990	ERA Maptec Ltd; European Environmental Agency (EEA)
LCA	Landscape Character Areas	Environment and Heritage Service Northern Ireland (EHSNI)
Lifeboat Stations	Lifeboat stations	Royal National Lifeboat Institution (RNLI)
Lighthouses	Lighthouses and other navigational aids	Commissioners of Irish Lights
Main Lakes	Main Lakes	Environmental Protection Agency (EPA)
Main Rivers	Main Rivers	Environmental Protection Agency (EPA)
Mean Tidal Amplitude	Mean Tidal Amplitude classified in the European EUrosion Project	European Environmental Agency (EEA)
MODIS Image	MODIS Satellite Image	National Aeronautics and Space Administration (NASA)
Rainfall	Average monthly rainfall grid	Met Éireann
River Basins	River Basins	Environmental Protection Agency (EPA)
Salt marshes	Significant salt marshes on the island of Ireland	Coastal and Marine Resources Centre (CMRC); T. Curtis and M. Sheehy Skeffington
Seabed Survey	Irish National Seabed Survey - areas surveyed	Geological Survey Ireland (GSI)
Sea Level Rise	Sea level rise classified in the European EUrosion Project	European Environmental Agency (EEA)
Sea Waves	Sea wave height classified in the European EUrosion Project	European Environmental Agency (EEA)

continued on following page

Table 3. continued

First Category: Physical Environment		
Name	**Description**	**Owner**
SPOT Image	SPOT (Satellite Pour l'Observation de la Terre) -Satellite Image of Ireland	LANDMAP (a service to provide spatial data to academic users in the UK and Ireland)
SST	Average Monthly Sea Surface Temperature (SST) grid	National Aeronautics and Space Administration (NASA)
Stable lights	Image of Irish urban areas taken from space	National Oceanic and Atmospheric Administration (NOAA)
Submarine Cables	Location of underwater cables	Coastal and Marine Resources Centre (CMRC); Kingfisher Information Service
Swell Waves	Swell wave height classified in the European EUrosion Project	European Environmental Agency (EEA)
Tide Gauges	Location of tide gauges	Coastal and Marine Resources Centre (CMRC)
Topography	Shaded Topographic Relief at 50m spatial resolution	ERA Maptec Ltd; Shuttle Radar Topography Mission (SRTM); Coastal and Marine Resources Centre (CMRC)
Marine Data Buoys	Location of marine data buoys around the island of Ireland	Coastal and Marine Resources Centre (CMRC)
Waves	Wave height average classified in the European EUrosion Project	European Environmental Agency (EEA)
Weather Stations	Weather Data Collection Stations	Met Eireann; Coastal and Marine Resources Centre (CMRC)

Table 4. Biological environment

First Category: Biological Environment		
Name	**Description**	**Owner**
Brown Crab	Brown Crab fishing areas in Northwest and Southwest	Bord Iascaigh Mhara – Irish Sea Fisheries Board (BIM)
Cetacean sightings	Cetacean (whales, dolphins, porpoises) sighted in Irish waters	Coastal and Marine Resources Centre (CMRC)
Cockle	Cockle fishing areas around Dundalk bay	Bord Iascaigh Mhara – Irish Sea Fisheries Board (BIM)
Crayfish	Crayfish fishing areas	Bord Iascaigh Mhara – Irish Sea Fisheries Board (BIM)
Harbour Seals 1963-2002 (N. Ireland)	Harbour Seals sighted in NI	CEDaR, Ulster Museum
IBAs	Important Bird Areas	Birdlife International; Bird Watch Ireland (BWI); Royal Society for the Protection of Birds (RSPB)
Lobster	Lobster fishing areas	Bord Iascaigh Mhara – Irish Sea Fisheries Board (BIM)
Scallop	Scallop fishing areas	Bord Iascaigh Mhara – Irish Sea Fisheries Board (BIM)
Shrimp	Shrimp fishing areas	Bord Iascaigh Mhara – Irish Sea Fisheries Board (BIM)

continued on following page

Table 4. continued

First Category: Biological Environment		
Name	**Description**	**Owner**
Seaweeds	Seaweeds	NUI Galway
Seals - Harbor	Harbor Seal Distribution	Coastal and Marine Resources Centre (CMRC); National Parks and Wildlife Service (NPWS); Environment and Heritage Service Northern Ireland (EHSNI)
Seals - Grey	Grey Seal Distribution	Coastal and Marine Resources Centre (CMRC); NPWS
Periwinkles	Distribution of significant periwinkle sites	Coastal and Marine Resources Centre (CMRC), MI
Whelk	Whelk fishing areas	Bord Iascaigh Mhara – Irish Sea Fisheries Board (BIM)

Table 5. Socio-Economic activity

First Category: Socio-Economic Activity		
Name	**Description**	**Owner**
Blue Flag Beaches	Blue Flag Beaches	An Taisce; Tidy Northern Ireland; Coastal and Marine Resources Centre (CMRC)
Fishing Ports	Fishing Port Locations	Coastal and Marine Resources Centre (CMRC)
Irish Surfing Association Clubs	Members of ISA	Irish Sailing Association; Coastal and Marine Resources Centre (CMRC)
Marinas	Location of marinas and pontoons around the island of Ireland	Coastal and Marine Resources Centre (CMRC)
Moorings	Location of tourist moorings	Coastal and Marine Resources Centre (CMRC)
Sailing Clubs	Location of ISA affiliated sailing clubs	Coastal and Marine Resources Centre (CMRC), Irish Sailing Association (ISA)
Surf Spots	Location of Surfing Spots	Coastal and Marine Resources Centre (CMRC)
Webcam	Location of coastal web cameras	Coastal and Marine Resources Centre (CMRC)

APPENDIX B

Table 6. Description of the 55 elements from ISO 19115 included in the Discovery Metadata

Plain English Description	Short Name	Definition
Title of the dataset	ResTitle	Name by which the cited resource is known
Filename of dataset to which this metadata corresponds	ResAltTitle	Short name or other language name by which the cited information is known
Brief description, or abstract for the dataset	IdAbs	Brief narrative summary of the content of the resource
When the dataset was created or published	ResRefDate	Reference date for the cited resource
See table	*RefDateType*	*Event used for reference date*
Main themes of dataset (see table)	TpCat	Main theme(s) of the dataset
A list of keywords	Keyword	Commonly used word ....used to describe the subject
	GraphOver	A graphic that illustrates the resource
	DataLang	*Language used within the dataset*
	DataChar	*Full name of the character coding standard used for the dataset*
Use a scale value (e.g., 1:50000) for vector data	EquScale	Level of detail expressed as the scale of a comparable map or chart
Use pixel spacing (e.g., 30 m) for raster data	ScaleDist	Ground sample distance
Is it vector, grid, data etc	SpatRpType	Method used to spatially represent geographic information
Describe which area the dataset covers	GeoDesc	Description of the geographic area within which data is available
What area does the data set cover	ExDesc	Spatial and temporal extent for the referring object
Name of the projection system used	RefSysId	Name of reference system used
Western most coord in lon	WestBL	Western-most coordinate of the limit of the dataset extent, expressed in longitude in decimal degrees (positive east)
Eastern most coord in lon	EastBL	Eastern-most coordinate of the limit of the dataset extent, expressed in longitude in decimal degrees (positive east)
Southern most coord in lat	southBL	Southern-most coordinate of the limit of the dataset extent, expressed in latitude in decimal degrees (positive north)
Northern most coord in lat	northBL	Northern-most coordinate of the limit of the dataset extent, expressed in latitude in decimal degrees (positive north)
Local projection system (west)	*WestEx	Western-most coordinate of the limit of the dataset extent
Local projection system (east)	*EastEx	Eastern-most coordinate of the limit of the dataset extent
Local projection system (south)	*SouthEx	Southern-most coordinate of the limit of the dataset extent

continued on following page

Table 6. continued

Plain English Description	Short Name	Definition
Local projection system (north)	*NorthEx	Northern-most coordinate of the limit of the dataset extent
When was the data collected or created	TM_CalDate	Date for the content of the dataset
	DqScope	*The specific data to which the data quality information applies*
How was the dataset generated	Lineage Statement	General explanation of the data producer's knowledge about the lineage of a dataset
Is dataset, complete, being updated, etc	IdStatus	Status of the resource
How often the dataset is updated	MaintFreq	Frequency with which changes and additions are made to the resource after the initial resource is completed
In what format is the data distributed	FormatName	Name of the data transfer format
Is there a version number	FormatVer	Version of the format (date, number, etc)
What media is the data distributed on	MedName	Name of the medium on which the resource can be received
Is the data free, commercial, research price etc	ResFes	Fees and terms for retrieving the resource, including monetary units
What constraints exist on accessing and using the data	OthConsts	Other restrictions and legal prerequisites for accessing and using the resource
Name of distributor	RpIndName	Name of the responsible person – surname, given name, title separated by a delimiter
Their organization	RpOrgName	Name of the responsible organization
Their role (see table)	Role	Function performed by the responsible party
Address	DelPoint	Address line for the location
	City	City of the location
County. For NI or GB insert postcode here as well	AdminArea	County
	Country	Country of the physical address
	VoiceNum	Telephone number
	FaxNum	Fax number
	EMailAdd	Email
	Linkage/ orName/ orDesc	Location for on-line access to additional info.
This filename	MdFileID	Unique identifier for this metadata file
Language	*MdLang*	*Language used for documenting metadata*
Character coding	*MdChar*	*Full name of the character coding standard used for the metadata set*
When was this MD record created	MdDateSt	Date that the metadata was created
Metadata standard	*MdStanName*	*Name of the metadata standard (including profile name used)*
Standard version	*mdStanVer*	*Version (profile) of the metadata standard used*
Who is responsible for this metadata	RpOrgName	Name of the responsible organization for metadata

continued on following page

Table 6. continued

Plain English Description	Short Name	Definition
	Role	Function performed by the responsible party
Availability of extended metadata	SuppInfo	Supplementary information regarding the dataset
Online link to extended metadata, if available	CI_Contact – Online Resource	Online linkage to further information

Those elements shown in italics are required for interoperability of the XML, but are hardcoded in the file and not displayed in the atlas. Elements marked with * are not part of the ISO standard and are defined in the implementation.

Chapter 8
Virginia and Maryland, USA

Marcia Berman
College of William and Mary, USA

Catherine McCall
Maryland Chesapeake and Coastal Program, Maryland Department of Natural Resources, USA

ABSTRACT

The Chesapeake Bay Watershed, the largest estuary in the United States, spans 62,000 square miles and includes six states and the District of Columbia. A stewardship agreement exists among the three primary states; Maryland, Virginia, and Pennsylvania that calls for a commitment to implement regulation and uphold practices that maintain or improve the Bay's ecosystem as a whole. To meet these and other coastal challenges Virginia and Maryland have independently developed Internet based products through which data, maps, and information are served. This chapter will summarize some of the highlights of each state's coastal web atlas. The type and format of resources available through each site will be reviewed. The user community will be defined. And a brief description of the site management structure will be presented. Both efforts have been spear- headed and supported by the states' Coastal Zone Management Program, a program within the National Oceanic and Atmospheric Administration (NOAA).

INTRODUCTION

A digital coastal atlas is a product where data are displayed, analyzed and/or distributed. A cursory review of selected sites across the country reveals dramatic differences in objectives and capabilities, despite some common themes. This chapter will describe products coming out of the Chesapeake Bay region, and focus specifically on the states of Virginia and Maryland. The following discussion highlights the history, goals and attributes of the Maryland and Virginia products and how they seek to resolve management challenges within the region.

Overview of Chesapeake Bay Regional Needs

The Chesapeake Bay is the largest estuary in the United States with more than 62,000 square miles

DOI: 10.4018/978-1-61520-815-9.ch008

Copyright © 2011, IGI Global. Copying or distributing in print or electronic forms without written permission of IGI Global is prohibited.

of surface area. The watershed includes six states: Maryland, Virginia, Pennsylvania, Delaware, New York, and West Virginia, as well as the District of Columbia. There are more than 100,000 streams in the Chesapeake Bay Watershed (Chesapeake Bay Foundation, 2003) and the population of 16.6 million, reported in 2005, all lives within minutes of one of them.

Virginia, Maryland and Pennsylvania share the majority of the Chesapeake Bay watershed. The governors of these three states have entered into multiple stewardship agreements that explicitly call for the adoption of policies and regulations necessary to meet standards for improvement in water quality and ecosystem health. While many of these policies have been in place for several years, goals have not been met and the governing bodies have been criticized for the short falls. Most recently the designation of the Chesapeake Bay as a "national treasure" by current president Barack Obama redirects attention to the persisting issues of water quality and bay restoration.

The urgency to prepare for and manage for climate change has also been raised in the region. The low-lying coastal landscape enhances the potential for land loss and increased inundation associated with storms and sea level rise. Model simulation using actual storm events compare well with measured observations and indicate that inundation of the Bay shoreline is influenced by sea level rise (Shen *et al.*, 2005). This presents obstacles and challenges for sustaining economic growth, public and private land holdings, cultural and natural heritage resources, and essential ecosystem habitat.

Solutions to the challenges faced within the Chesapeake Bay region begin with education, research, and tools. Whether you are a coastal manager trying to identify hot spots for non-point source pollution, an outreach coordinator developing a workshop on hurricane impacts, or a citizen deciding how to best manage shoreline erosion on his property, the ability to reach your desired outcome is driven by how much you know about the issues, the ease with which you can access current information and the availability of resources that support your objectives.

In a proactive move, the Coastal Zone Management Programs within Virginia and Maryland have independently directed funds toward the development of digital atlas products that offer a venue within which data are presented, information is disseminated, and tools are developed that encourage spatial planning within the coastal zone. These programs, funded through the National Oceanic and Atmospheric Administration's (NOAA) Office of Ocean and Coastal Resource Management, work within their designated states to develop partnerships with state agencies, municipalities, and non-governmental organizations (NGOs). Both the Virginia and the Maryland programs have a long history of collaborating with regional groups to develop strategies for enhanced coastal zone management.

In the sections below, the Virginia and Maryland coastal web atlas products will be described separately. A summary will attempt to bring the information together to reflect on common problems, positive aspects, and finding as they may pertain to the objectives of the International Coastal Atlas Network (ICAN).

VIRGINIA COASTAL GEOSPATIAL AND EDUCATIONAL MAPPING SYSTEM

In 2001, the Virginia Zone Management Coastal Program (VA CZMP) began spare-heading an effort to identify and map the best remaining terrestrial and aquatic resources within the coastal lands of Virginia. Projects were funded among partners to collect and disseminate the inventory of available coastal data sets within the state. These initial activities resulted in three basic findings. First, there was a considerable amount of digital natural resource data available, although they were archived in more than twelve different agencies.

Second, there was no central data repository in the state consistently used by data developers. Finally, software availability, skill levels, and a basic connection to data sources limited data access.

The Virginia Coastal Geospatial and Educational Mapping System (GEMS) (2007) was developed as a collaborative partnership between VA CZMP, Virginia Commonwealth University (VCU), and a private consultant to address these shortfalls. The initiative is funded entirely by VA CZMP. Under the current structure of the program, renewable grants are in place to enable maintenance and enhancements for Coastal GEMS to continue. With GEMS, VA CZMP has improved the decision-making capacity among agency partners in Virginia by providing access to spatial data within a medium that is both informative and easy to use.

Information access to coastal data resources is the principal goal of Coastal GEMS. It accomplishes this through the display of data in a digital map environment, and through fact sheets that summarize information displayed in the digital atlas. GEMS has also simplified the output of complex spatial models in an effort to elevate the level of information contained in the map atlas without requiring users to have advanced GIS or spatial modeling skills. In doing this GEMS can makes modeled data easy to understand and therefore available to a much wider audience

Virginia Coastal GEMS: Data Collection

With a focus on coastal planning and conservation, data collection to support Coastal GEMS is broad based and flexible. The program accepts most data sets that support both terrestrial and aquatic planning. The Coastal Policy Team, a steering committee composed of partners within the coastal zone, provides input on data gaps, and direction for future projects that will result in data collection.

To date, data providers span regional planning organizations, state agencies, academicians, NGOs and the federal government. Individual members of the Coastal Policy Team voluntarily provide data, where appropriate, from their respective agencies. Many contributions are through grantees funded under VA CZMP's grant program. The responsibility for data currency rests with the VA CZMP GIS data manager. However, updates and data uploads to the interface occur at the server site managed under a continuing contract with VCU.

No user restrictions or licensing requirements are needed to use Coastal GEMS. The program clearly and specifically defines use limitations for the data within a Disclaimer; however, the user can bypass this page and go directly to the map interface (Figure 1). Currently more than 50 data layers are available within Coastal GEMS.

Virginia Coastal GEMS: System Design

From the User's View

Coastal GEMS is presented within a simple interactive mapping interface that applies navigation tools very similar to Google Maps. In the custom interface, layers are grouped into categories: features, tools, boundary layers, and reference data. These categories can be expanded with a mouse click to reveal the full data list under each heading.

"Features" are subdivided into "water," "shoreline," "land," "recreation," and "wildlife" features. "Conservation planning tools" represent the output of different models that range from aquaculture vulnerability, to land growth, or tidal flushing rates. "Boundary" layers include examples of conservation strategies that have been implemented in the coastal zone. An example would be the "Sea Side Heritage Program" boundary. "Reference" data include baseline boundary or framework layers such as watershed boundaries and roads. Scale restrictions set by the developer control when lay-

Figure 1. Coastal GEMS homepage. The user goes directly to the application when the "launch" button in the upper right is selected

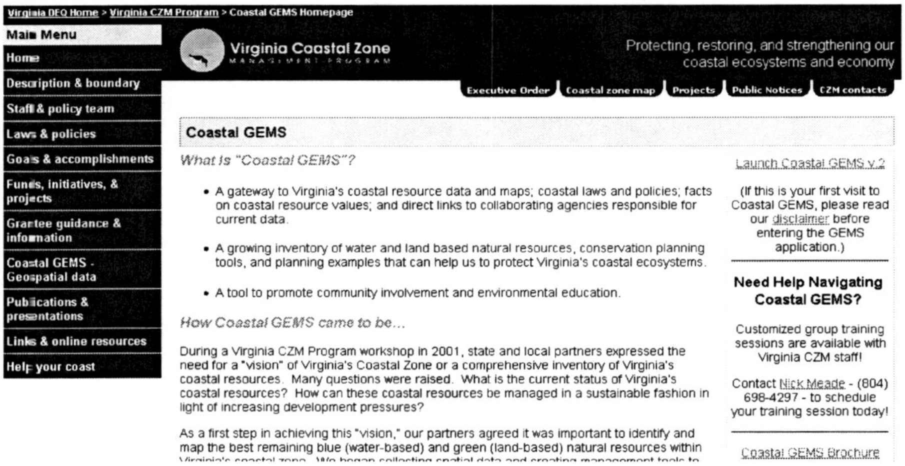

Figure 2. A partial data layer list of Wildlife Features, Recreational Features and Conservation Tools are shown. At the drawn map scale, all features are highlighted and can be viewed.

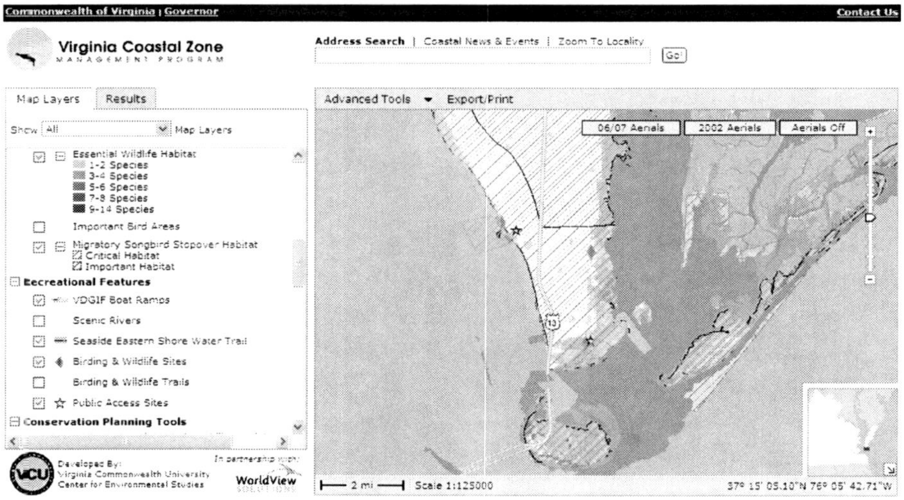

ers can be viewed in the map window. When the scale requirements are met or exceeded the layer is highlighted in the data layer list (Figure 2). This speeds the rate that large data sets are uploaded to the map window. High-resolution imagery from two different years can also be selected in the map window and displayed when the user zooms to an acceptable map extent.

Zooming, distance measuring, draw shape/area, identify, and print tools are part of the interactive interface. The "identify" tool allows the user access to a selective amount of attribute data retrieved from the database. Address search, zoom to locality, and zoom to latitude/longitude features are nice features when a user has a specific area of interest.

A unique feature that appears in Coastal GEMS is an indicator tab that defines the number of independent locations a particular data layer appears in the map extent the user has drawn. For example, in one specific field of view, the map may illustrate nine wetland polygons and one beach. This selection would be displayed in the results window. With the identify tool these become individually listed with associated attribute information and highlighted in the map window.

Aside from the map interface, fact sheets provide succinct, but detailed information about each layer (Figure 3). Along with source, data currency, and data content, fact sheets assess resource management details pertaining to the data. A summary explains why the attribute is important for coastal conservation. Links to related and more detailed information are provided. This resource includes links where a user may find access to downloadable data and metadata. No direct links to data or metadata are available within Coastal GEMS.

Behind the Scenes

Coastal GEMS uses a variety of software packages for data storage, access, and viewing. Spatial and tabular data are stored in a Microsoft SQL database. ESRI's ArcSDE manages the spatial data and provides access. ESRI's ArcIMS® (v.9.2) powers the map interface and generates the map image. The interface is a customized .NET and AJAX application designed to allow users on slower Internet connections to still retrieve maps and information. Two map caches are also used in Coastal GEMS to display a base layer or background and aerial imagery.

The GIS Coordinator within VA CZMP serves as the liaison between data providers and the offsite application manager. This individual is responsible for creating or acquiring the data files and transferring them to the application manager for uploading, which includes both shape files and fact sheet files.

Figure 3. A fact sheet on conservation lands is shown as a pop up window when the feature is selected

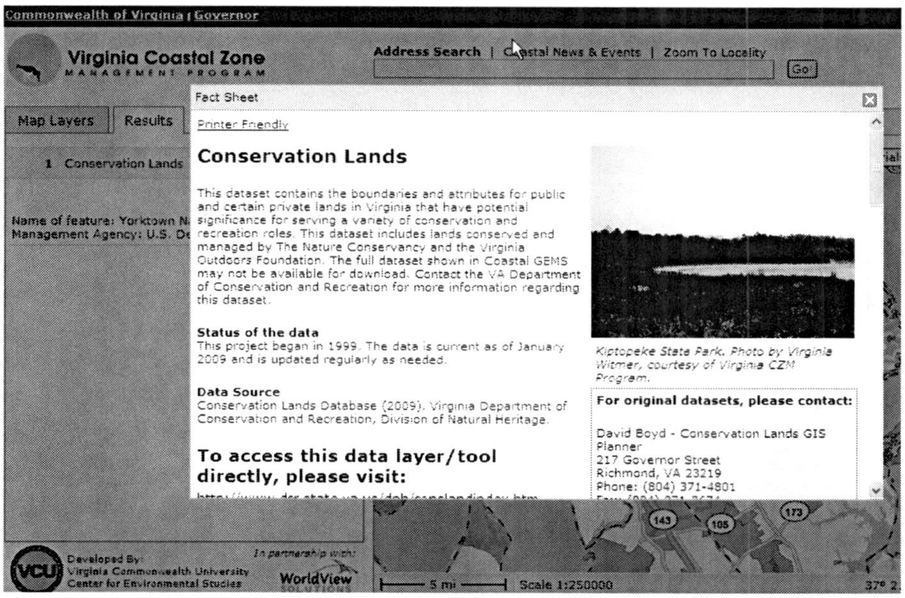

Virginia Coastal GEMS: Atlas Usage

In 2008, the Coastal GEMS webpage received over 5,000 hits. VA CZMP believes this is an underestimate of the actual number of visitors. Repeat visitors are bypassing the GEMS homepage and linking directly to the map interface through a URL bookmark. This year, tracking has been installed to confirm this; however, no numbers were available to report.

VA CZMP has used a number of public venues including newsletters, conferences, and the press to build awareness regarding the availability of Coastal GEMS. Training sessions for potential users can be arranged through the program. Feedback to date has been positive, and the developers consider Coastal GEMS to be a successful venture that serves an important role in facilitating coastal management among Virginia stakeholders.

Discussion of Coastal GEMS

In a state with more than 10,000 miles of shoreline, multi-jurisdictional oversight, and an aggressive agenda for environmental improvements, a demonstrated need for a coastal atlas exists. As in other coastal states, Virginia's NOAA funded CZMP has taken the lead in the development of Coastal GEMS, and has succeeded in filling a critical data gap.

GEMS has many positive features that include a growing data catalogue, a simple and well-organized interface, and fact sheets that summarize information in easy to understand language. The close connection Virginia's CZMP maintains with its partners ensures GEMS evolves to meet the growing demands on coastal managers and stakeholders within the coastal community.

A few problems with Coastal GEMS are worth mentioning. Perhaps the most obvious to the user is the speed in which data loads. Individuals familiar with this type of interface have come to expect these delays. Speed declines markedly when imagery is selected. There is also a decline in functionality when a user tries to access GEMS in a browser other than Explorer. Alternative software could decrease loading times and system crashes. A number of atlases, including the Oregon Coastal Atlas (OCA), use open source technology to drive the application. It is unclear if this is responsible for the rapid loading of data to the OCA map interface.

A random selection of layers revealed a significant number had little or no retrievable attribute information using the "identify" tool. In some cases, access to information may be intentionally restricted. However, the end user can be left feeling in error.

The structure of GEMS' administration is also somewhat problematic. Two outside contractors administer the maintenance and updates to the interface. VA CZMP managers acknowledge that using an offsite administrator can be inefficient. Delays in updates and revisions when necessary can be slow to occur. Ideally GEMS should be managed in-house, but a number of internal obstacles currently make this not possible. However, the addition of a GIS manager to the VA CZMP has significantly improved a number of aspects in the development and maintenance of GEMS.

VA CZMP has plans to expand Coastal GEMS to include data layers that extend offshore to the twelve-mile territorial sea boundary. Currently, GEMS is largely focused on the coastal zone within the Chesapeake Bay Estuary and seaside lagoons on Virginia's Eastern Shore. The expansion along the open ocean coast will be critical to support upcoming marine spatial planning initiatives.

The interface will also begin incorporating output from storm surge models that have been run for various regions of the coastal zone. This modification will have great value for local governments currently planning for climate change impacts. Finally, the Virginia portion of the Chesapeake Bay Shoreline Inventory, funded in part by VA CZMP, will be nearing completion, and that data will also become an integral component of GEMS.

With these features, along with an easy to navigate interface, GEMS is reaching a broad audience ranging from coastal planners and policy makers, to marine contractors and private citizens. Coastal GEMS remains a work in progress and the evolution of the database that supports GEMS continues to grow and become more comprehensive. The program expects to continue expanding GEMS through future grant and contract activities funded through the program.

MARYLAND'S SHORELINES ONLINE

In 1999, the Governor of Maryland established a task force to address shoreline erosion. This action set the stage for a series of statewide initiatives. Among them was the development of a Maryland coastal atlas that would focus on coastal hazards. The Maryland Chesapeake and Coastal Program (MD-CCP, 2005), administered within the state's Department of Natural Resources (DNR) took the lead in the development of the coastal atlas, currently known as "Maryland Shorelines Online" or for this review "MSO." The project has evolved over several years and continues to expand, both in data content and information resources. Like Virginia's product, MSO is funded through Coastal Zone Management grants from NOAA.

MD-CCP developed MSO to address some specific shortfalls identified by the governor's task force. First, a lack of information available to citizens regarding shoreline erosion was apparent, and second, no data or rationale existed to enable state coastal managers to prioritize shoreline protection and restoration. Accordingly, a focus on coastal hazards and resource protection has been driving MSO since its inception.

The MSO portal combines outreach and educational information, guidance on coastal regulation, map information and links to data in a structured web interface (Figure 4). Lesson plans on coastal issues, how to apply for a permit, as well as technical and financial support are all available. The developers have also taken the initiative to identify specific areas of interest for a defined user audience. Direct links to topics of interest are highlighted for coastal managers, property owners, marine contractors, and educators. While these suggested users on the home page clearly outline the intended audience, the site is public and there are no access restrictions.

General information and tools are accessible using rollover menu links on the MSO home page. Seven links divide the information resources available through MSO into seven categories. One category is Interactive Mapping. The remainder of this discussion will focus on interactive mapping and specifically, the Maryland Shoreline Changes Online (MD-SCO) digital coastal atlas.

Maryland's Shorelines Changes Online: Data Collection

MD-CCP has partnered with regional experts in the fields of geology, ecology, coastal mapping/GIS, and resource management to assemble a vast array of data and products to support coastal hazard preparedness and aquatic resource protection in Maryland. MSO serves as the vehicle to disseminate information to the user audience. These partnerships insure a coordinated and comprehensive assessment of resources and data needs.

Among the various components of MSO is the digital coastal atlas referred to as "*Shoreline Changes Online*" (MD-SCO). Access to this interactive mapping interface is through a rollover menu link on the home page. While the digital atlas has many noteworthy elements only a few are discussed in detail.

To address the issue of shoreline erosion identified early on in the Shoreline Erosion Task Force an assessment of shoreline response to natural and anthropogenic influences would be vital. The Maryland Geological Survey (MGS) completed a historical shoreline-mapping project that developed a set of shoreline change maps for coastal Maryland. A shoreline change analysis

Figure 4. The Maryland Shorelines Online homepage

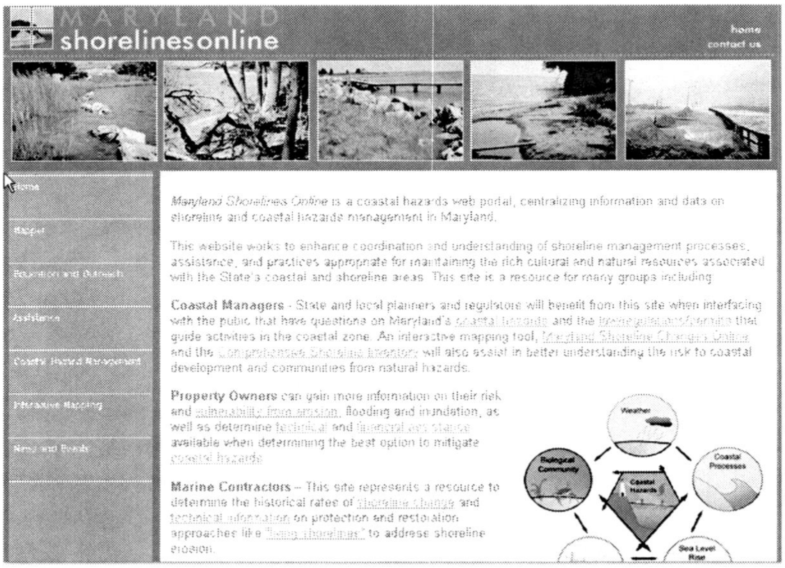

was computed by comparing the digital shoreline records and calculating rates of erosion and accretion alongshore. These data can be viewed along with the historical shoreline records through the online atlas and custom shoreline change rate calculations can be made for any length of shorelines depending on a user's interest or the scope of a project (Figure 5).

MD-CCP partnered with the Virginia Institute of Marine Science to develop a comprehensive shoreline inventory that mapped current condition along the tidal shoreline of Maryland. The surveys mapped land use, bank stability, erosion control structures, and public access. The data provides important information for local and state planners charged with regulation of tidal shoreline, expansion of waterfront development and public access, and managing for climate change impacts. The survey results have been added to the MD-SCO digital atlas (Figure 6).

Growing interest in aquatic spatial planning led MD-CCP to complete an aquatic resource inventory and assessment known as Blue Infrastructure. It is the most recent addition to the digital atlas. This product was a joint collaboration between MD-CCP and Towson State University, who currently administers MD-SCO. The inventory includes more than 30 layers of aquatic resources and conditions found in Maryland waters including fish spawning and nursery areas, native oyster bars, submerged aquatic vegetation, salinity distributions and bottom sediment characteristics to name a few (Figure 7).

Other noteworthy data layers in the digital atlas include the output from a storm surge inundation model that maps surface area flooding resulting from hurricanes of various forces. Another layer illustrates elevation contours and areas at risk to various levels of sea level rise. In addition, other layers are available associated with hydrologic boundaries, municipal boundaries, and transportation networks.

Prior to entering the digital atlas a user is prompted to read and acknowledge a product disclaimer. No mechanism exists to bypass this disclaimer. There are no access restrictions to the site and there are no GIS files downloadable from MSO or MD-SCO. The Maryland Department of Natural Resources has a Geospatial Download Service, and a link to that service is accessible from

Figure 5. Shoreline Changes Online illustrates erosion rates and historic shoreline position

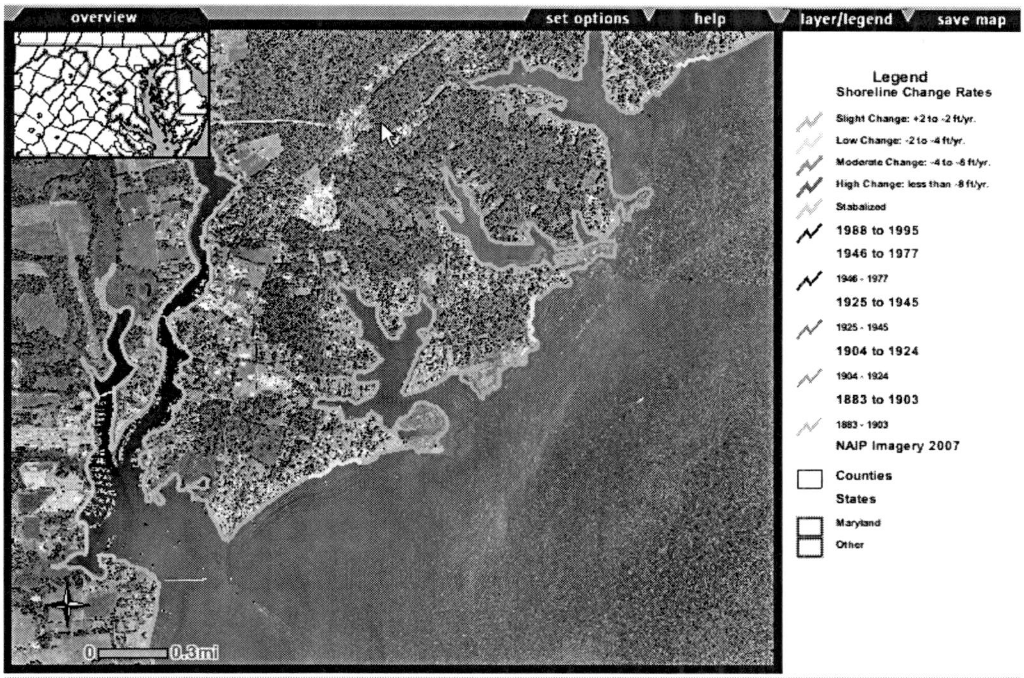

Figure 6. The Shoreline Inventory is incorporated into MD-SCO

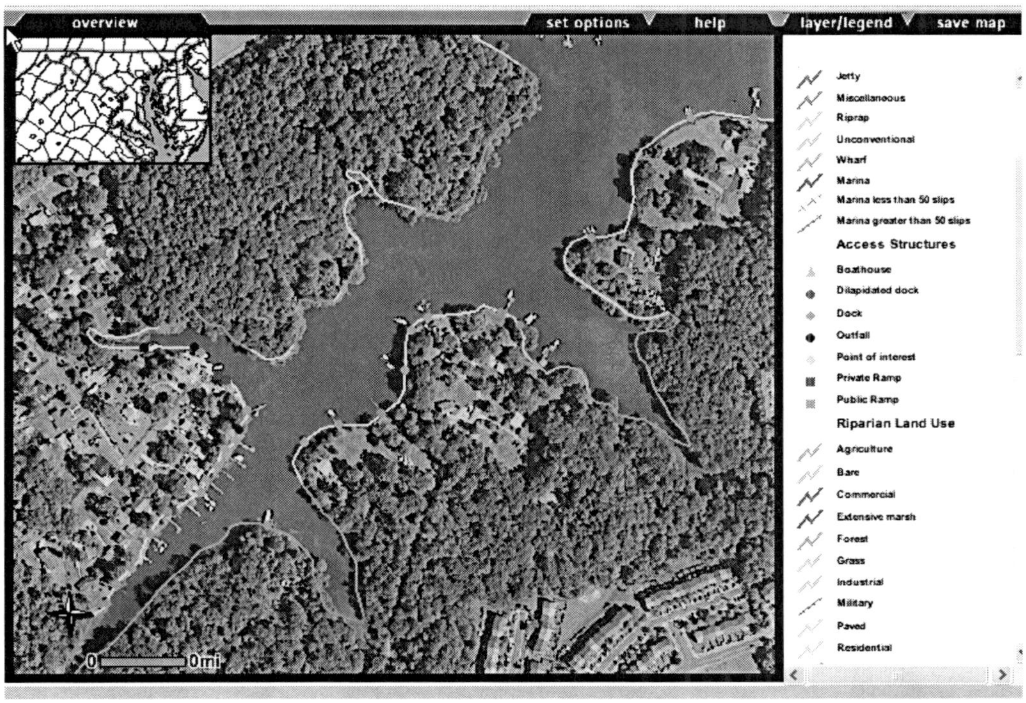

Figure 7. Blue Infrastructure contains more than 30 layers. Shown here is bottom sediment type and location of oyster bars. A pull down menu in the bottom left allows the user to select an active layer. Data for a specific polygon is displayed below the map using the identify tool.

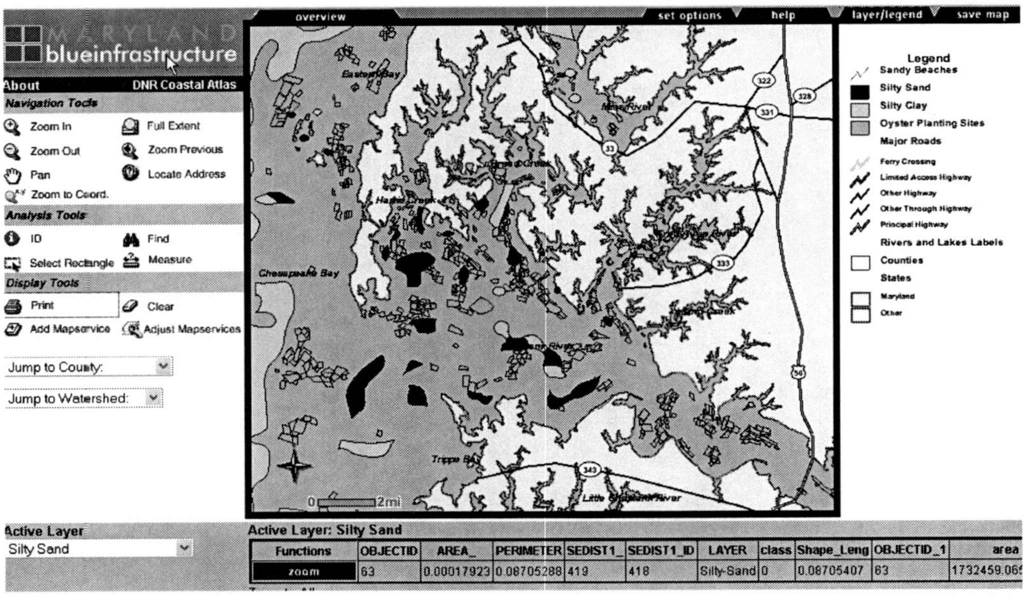

a menu link on the MSO homepage. Licensing agreements may be necessary and links to other geospatial data displayed in the digital atlas are also available, but are stored elsewhere.

Maryland's Shoreline Changes Online: System Design

From the User's View

The digital atlas is a clean and simple customized interactive map interface with standard features including, zoom, pan, and identify. Both metadata and attribute information is displayed in a window below the map display. Only general information is initially revealed in the metadata window, but links to more detailed files are given. Data developed within the agency includes metadata, but some layers reveal no available metadata.

The layers, which include but not limited to, Blue Infrastructure, Historic Shorelines, and the Shoreline Inventory, collapse and expand in a list depicted to the right of the map window. Metadata is revealed when clicking on a layer in the expanded list view. This feature is easy and clean to decipher. The user can toggle between the list of data layers and the map legend. This allows for a relatively large viewing window.

Imagery is automatically loaded when the user zooms to a predetermined scale requirement. There is a noticeable decline in speed when the user zooms to this extent, selects new layers and tries to refresh the map. With no apparent way to turn the imagery off, there is no way to overcome the limitations brought on by reduced loading speeds.

Help is readily available. The interface contains help on various needs. A link provides directions from the disclaimer page where the application is launched. These directions provide an expanded explanation of some major features in the atlas, what the user might expect, and where to go for additional help.

Behind the Scenes

The user interface was developed by Towson State University's Center for Center for Geographic Information Sciences (CGIS). It uses the ArcIMS® platform and database and web servers run SQL server. The digital atlas is maintained by CGIS. New data to the atlas is delivered electronically to CGIS through the MD-CCP staff. MD-CCP must provide all files and documentation.

Maryland Shoreline Changes Online: Atlas Usage

Like Virginia's Coastal GEMS, the Maryland digital coastal atlas does not track or report usage statistics. With a broad user audience, the MD-CCP relies on partners for feedback, as well as public forums, workshops and training sessions. This forum provides the guidance they need to redirect data development and application enhancement. The program believes property owners are particularly interested in their erosion control options and reviewing strategies used by their neighbors. Non-government organizations (NGOs) use the atlas to help identify locations for community-based projects. Local governments use the atlas to review parcel conditions prior to issuing permits. The growing addition of tools is expected to increase and enhance their usage. MD-CCP believes that many local governments use MSO to support daily tasks.

Discussion of Maryland Shorelines Online

Maryland has more than 7,500 miles shoreline. The shoreline is variable and complex. MD-CCP has made a significant investment in the development of a coastal toolbox to increase the capacity of local governments to make informed shoreline management decision. MSO is the portal to many of these tools. Among them, the MD-SCO digital atlas expands access to digital GIS data to the non-GIS user community.

The digital coastal atlas includes a focused set of data layers that support the underlying theme of coastal hazards and aquatic resource restoration. The program has received positive response from the user community, particularly local governments and NGOs. Since MD-CCP collaborates regularly with local governments, it is no surprise that MSO and the MD-SCO atlas are well integrated into local programs.

The simplicity of the application design makes MD-SCO easy to use and understand. The program sponsors and partners have provided training opportunities across the state as well as help guides throughout the portal. The MSO portal, as well as the digital atlas, has been expanded and there are plans to expand again. MSO will add tools that provide direct guidance on the subject of erosion control and how to best protect your shoreline. An independent Blue Infrastructure application is planned as well. This will likely include the layers of MD-SCO currently categorized under the heading "*Blue Infrastructure.*" Once complete, a near-shore assessment layer will be added to the site that combines many of the aquatic resource and habitat data, shoreline inventory information, and watershed characteristics.

The challenges MD-CCP currently faces with MSO and the atlas will most likely carry forward as tools and data layers are added to the current interface. MD-CCP specifically identifies metadata as a leading problem to overcome. The absence of metadata and the completeness of metadata have been noted previously in this review.

Hurdles exist because MD-SCO is maintained offsite by CGIS. The frequency at which new data are loaded to the site can be slow. How new layers integrate cartographically with existing layers in the application are not always correct. In this review different layers used common or similar symbols and colors, which masked the presence of data in the map window. On the server side, problems with downtime are not always reported or resolved quickly.

MD-CCP plans to launch a new MSO application through the Maryland iMap initiative. At this time there is some uncertainty if Maryland iMap will integrate with MSO and what changes to the functionality will occur. Portions of the new mapping application are being developed in-house which presents an opportunity for a new interface to be served within MD-DNR, the host agency.

This may overcome some of the off-site maintenance problems MSO and MD-SCO currently incur.

SUMMARY

This chapter reviewed two digital coastal atlases developed by the states of Virginia and Maryland. These Mid-Atlantic States have suffered extreme losses due to hurricanes and storm events in the last decade. They have both entered into agreements that commit them to improve ecological conditions in the Chesapeake Bay Watershed. In addition, they are both actively engaged in activities to manage for climate change impacts. The need, therefore, to develop shoreline management tools that provide guidance for state and local governments is a priority.

The Virginia and Maryland NOAA funded Coastal Zone Management Programs have recognized and backed this effort by allocating resources toward the development of a digital coastal web atlas. In Maryland and Virginia these successful projects have relied heavily on partnerships between agencies and organizations within the state to prioritize data development and acquisition, contribute data directly, organize training, and provide technical support. Presently, there appears to be little contribution from local governments as a data provider, but both developers have sought input from this user audience at various stages of development. In both states, the programs have turned to academic institutions for technical support in the development and maintenance of the web interface.

While both web atlases have similarities, their differences from the user end are far greater. Virginia's Coastal GEMS is designed exclusively as an interactive map interface. The fact sheets, which provide valuable information for each data layer and guidance on its ecological importance and relevance to the Commonwealth's coastal management goals, are tied to the map interface. The fact sheets contain the only links to more detailed information, data downloads, and metadata.

Maryland developed Maryland Shorelines Online (MSO) as a stand-alone web portal serving a plethora of information and resources supporting a common theme – coastal hazards. The portal organizes resources customized by the agency as well as links to external resources through other providers. Among the resources available is the Shoreline Changes Online (MD-SCO) interactive map interface. MD-SCO parallels Coastal GEMS most closely in application and information content.

In this review, the authors found the lead agencies for each coastal web atlas shared common problems. Both atlases performed noticeably better in Explorer versus a browser like Mozilla. Issues like metadata completeness and cartographic correctness were concerns. Since the technical managers were disconnected from the data, new additions to the interface could be ordered improperly, or displayed with duplicative or incorrect symbology. Speed and loss of functionality within the interface while in use was observed during this review. Since both sites are maintained by an external agency, both VA CZMP and MD-CCP stated response times to correct problems were often slow and unpredictable.

Since one goal of ICAN is to provide guidance to programs building a coastal web atlas interface, this review has identified some positive features and common problems that will likely be found in other coastal web atlases around the globe. Feedback from the user community will be an important source of information to improve the web atlas interface in support of user needs. In both Maryland and Virginia mechanisms to track use or acquire structured feedback are not in place. ICAN is encouraging these and other interfaces to provide links to new surveys that will collect and synthesize feedback for future recommendations.

As ICAN expands its efforts to encourage interoperability between coastal atlases worldwide, the states of Maryland and Virginia have begun discussions that set the stage for experimentation in this important area. With several common datasets there are a number of focal areas that will provide excellent use cases.

ACKNOWLEDGMENT

The authors would like to thank the reviewers for their suggestions and editorial corrections. There are many individuals who have contributed to the development of the Maryland and Virginia atlases. The number of data contributors alone is far too long to list. In Maryland, the authors would like to acknowledge Audra Luscher, former coastal planner with the Maryland Department of Natural Resources and now with the National Oceanic and Atmospheric Administration, who conceptualized Maryland Shorelines Online and oversaw the development of the atlas. Lamere Hennessee of the Maryland Geological Survey developed the historic shoreline assessment and VIMS' Comprehensive Coastal Inventory Program generated the Maryland Shoreline Inventory. Today, Maryland Shorelines Online is maintained by Towson State University's Center for Geographic Information Sciences. Staff of the Virginia Coastal Zone Management Program, particularly program director Laura McKay, must be recognized for their contribution to the development and support of Virginia Coastal GEMS. Today Coastal GEMS is technically supported and maintained by Nick Meade and Will Shuart of Virginia Commonwealth University. In both states, the NOAA support the products.

REFERENCES

Chesapeake Bay Foundation. (2003). *Chesapeake Bay Fact Sheet*. Retrieved May 15, 2009 from the Chesapeake Bay Foundation web site: http://www.cbf.org/site/PageServer?pagename=resources_facts_general.

Coastal Geospatial and Education Mapping System. (2007). *Virginia Coastal Zone Management Program, Virginia Department of Environmental Quality*. Retrieved June 3, 2009 from the Virginia Department of Environmental Quality web site: http://www.deq.state.va.us/coastal/coastalgems.html.

Maryland Chesapeake and Coastal Program. (2005). *Maryland Shorelines Online, Maryland Department of Natural Resources*. Retrieved June 15, 2009 from the Maryland Department of Natural Resources web site: http://shorelines.dnr.state.md.us.

Shen, J., Gong, W., & Wang, H. (2005). Simulation of Hurricane Isabel using the Advanced Circulation Model (ADCIRC). In K.G. Sellner (Ed.), *Hurricane Isabel in Perspective: Proceedings of a Conference* (pp. 107-116). Baltimore, MD: Chesapeake Bay Consortium.

Wang, H. V., Cho, J., Shen, J., & Wang, Y. P. (2005). What has been learned about storm surge dynamics from Hurricane Isabel model simulations? In K.G. Sellner, (Ed.), *Hurricane Isabel in Perspective: Proceedings of a Conference* (pp. 117-112). Baltimore, MD: Chesapeake Bay Consortium.

KEY TERMS AND DEFINITIONS

Coastal Web Atlas: A collection of digital maps and datasets with supplementary tables, illustrations and information that systematically illustrate the coast, oftentimes with cartographic and decision support tools, all of which are accessible via the Internet. Also known as web atlas, digital atlas, digital coastal atlas.

Atlas Interoperability: The ability to interrogate two or more unrelated atlas systems and potentially exchange metadata, images, vector and raster data.

Coastal and Marine Information: Both spatial and non-spatial information regarding the coastal and marine areas of a region or country.

Integrated Coastal Zone Management: A dynamic, multi-sectoral approach to managing the coast, which takes, into account social, economic and environmental concerns of all parties with an interest in the coastal space.

GEMS: Geospatial and Educational Mapping System.

Metadata: Information that fully describes or documents a dataset, such as its geographic coverage, quality, completeness, accuracy, etc. are a "pedigree" of sorts for a data set and helps you to judge its "fitness for use" or reliability, thereby helping one to use it more appropriately and efficiently. Metadata allow a potential user, for comparative purposes, to understand how the data were collected. They also provide the all-important details of how you can actually obtain the data in question, or who best to contact. Data that do not have accompanying metadata are often hard to find, difficult to access, troublesome to integrate, and perplexing to understand or interpret.

Web GIS: A geographical information system which can be accessed over the Internet and allow visualization and interaction with spatial data via a map as well as providing analysis functionality such as spatial analysis, querying and buffering.

Chapter 9
Wisconsin, USA

David Hart
University of Wisconsin Sea Grant Institute, USA

ABSTRACT

Simply stated, a coastal web atlas (CWA) is a means of organizing, presenting, and sharing spatial data for the coast. Once in place, a CWA can function as a coastal spatial data infrastructure and a platform for developing coastal management decision support tools. While Wisconsin has been actively applying geospatial technologies to coastal issues since 1994, development of a CWA is in its infancy. Wisconsin Sea Grant has learned much about key components of a CWA during the past decade through its role leading four coastal spatial data integration projects. Several technical and institutional issues surfaced as the projects moved from discovery, acquisition, and integration of spatial data from multiple sources to analyze regional coastal issues to the development of interoperable web mapping services and spatial data catalogs. These issues are associated with the following research topics: web portal design and evaluation, choosing appropriate web mapping technologies, GIS cartography, domain spatial data infrastructures, geospatial data archives, and spatial ontologies. Building the Wisconsin Coastal Atlas will provide insight on these important research topics.

INTRODUCTION

While Wisconsin has long been applying geospatial technologies to coastal issues, the process of implementing a coastal web atlas (CWA) is only beginning. The past 15 years has seen the development of many coastal GIS applications and experimentation with multiple platforms for web mapping and spatial data catalogs. The Wisconsin Coastal GIS Applications Project (http://coastal.lic.wisc.edu/) has yielded many benefits, but it has not taken full advantage of current "Web 2.0" technologies that promote information sharing and collaboration. The emergence of an international effort to promote interoperable CWAs provides a good reason to re-examine coastal GIS in Wiscon-

DOI: 10.4018/978-1-61520-815-9.ch009

Copyright © 2011, IGI Global. Copying or distributing in print or electronic forms without written permission of IGI Global is prohibited.

sin. It is now a good time to extend coastal GIS beyond a campus research and outreach project to a broader platform for discovering, accessing, integrating, and utilizing geospatial data about the Great Lakes coasts. This chapter will explore the evolution of coastal GIS in the state, examine the benefits and drawbacks of the different web mapping/spatial data catalog-portal approaches that have been developed, and share the strategy that will be used to build a Wisconsin Coastal Atlas.

WISCONSIN COASTAL GIS APPLICATIONS PROJECT (1994-2009)

Since 1994, the University of Wisconsin Sea Grant Institute has actively supported the application of geospatial technologies to better understand coastal issues facing the Great Lakes. The Wisconsin Coastal GIS Applications Project, a collaborative project between University of Wisconsin-Madison (UW) Sea Grant and the Land Information and Computer Graphics Facility at UW, was the vision of Allen H. Miller, then Assistant Director for Outreach at UW Sea Grant and a pioneer in the modernization of land records in Wisconsin. An important mission of the project was to leverage the sizeable investment in local government land information systems in Wisconsin (over $160 million since 1991) to support decision-making about Great Lakes coastal management.

The project has passed through several phases since its inception (see Figure 1). The first phase focused on the use of GIS "teaching models" to address specific coastal issues in specific places, such as shoreland zoning, coastal erosion, floodplain management, and water quality (Rink *et al.,* 1998). The models were used in numerous workshops and classes, translated into on-line GIS tutorials, and disseminated to thousands worldwide. The second phase examined the use of GIS for integrated coastal management – moving from specific local issues to complex regional issues. This phase included an early demonstration of how digital parcel mapping, tax assessment databases, and other spatial data could be acquired from local governments and aggregated to support regional-scale analysis and mapping (Hart, 2000). The third phase explored the use of GIS for "dynamic and distributed" coastal management. The result was a system where custodians, whether they be local, regional, state, federal, academic, or non-profit entities, maintain and provide access to the most current spatial data and multiple remote users can access and integrate data in real-time from multiple sources. The fourth phase showed the value of visualization and virtual globe tools to promote a more intuitive understanding of coastal issues (Stone *et al.*, in press).

LESSONS LEARNED FROM FOUR COASTAL DATA INTEGRATION PROJECTS

The following section explores four data integration efforts undertaken as components of the broader Wisconsin Coastal GIS Applications Project and examines the issues and obstacles faced as one moves from discovery, acquisition, integration, and analysis of spatial data from multiple sources to the development of interoperable web mapping services and spatial data catalogs.

Lake Michigan Potential Damages Study

The centerpiece of the second phase of the Wisconsin Coastal GIS Applications Project (use of GIS for integrated coastal management) was an extensive effort to discover, acquire, and integrate local government spatial data to analyze coastal erosion hazards for the Lake Michigan coast of Wisconsin. This work was a component of the Lake Michigan Potential Damages Study, a multi-year project funded by the U.S. Army Corps of Engineers to examine the impacts of extreme variations

Figure 1. The four phases of the Wisconsin coastal GIS applications project

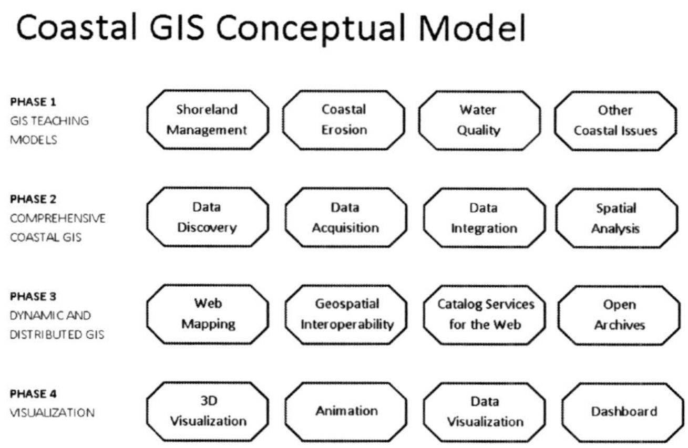

in Great Lakes water levels on coastal property. Multiple spatial data sets were acquired from land information offices in eleven coastal counties in 1998. These included parcels, tax assessment data, planimetric and topographic mapping, land use, and digital orthophotos. Several institutional issues surfaced during the process of acquiring the spatial data. These included the actual cost of acquiring the data, the time required to obtain the data from the moment the initial requests were made, and restrictions placed on sharing and use of the data. Several technical issues arose during the process of integrating the data. These included lack of metadata provided by data custodians, as well as dealing with multiple and confusing data formats and map projections (Hart, 2000).

With an integrated spatial database for the Lake Michigan coast in place, it was then possible to attempt to conduct the coastal erosion analyses that were formulated at the beginning of the study. These involved specific spatial queries such as determining the assessed value of land and improvements within the jurisdiction of shoreland zoning regulations. While at the time of the study, it was not possible to fully complete all the analyses due to gaps in the availability of local data, it was recognized that a valuable, integrated spatial database had been compiled that could be used to analyze a variety of regional-scale coastal issues. A resulting problem was that many of the data sets (parcels, topography, planimetric features, and imagery) needed to be updated on a regular basis to reflect the changing nature of the coast. Also, it was difficult to make the data available to others. Several organizations involved in state and regional coastal issues on the Great Lakes were interested in access to the integrated coastal data, but, as researchers and outreach specialists, it was difficult to fulfill the role of a spatial data librarian.

Lake Superior Coastal Mapping Portal

As the Lake Michigan Potential Damages Study wound down in 2000, the Open Geospatial Consortium (OGC) was implementing standards to promote geospatial interoperability and software emerged that allowed integration of spatial data from distributed sources. An opportunity emerged to move the focus of the Wisconsin Coastal GIS Applications Project to the Lake Superior coast of Wisconsin and test whether it was possible to build a "dynamic and distributed" coastal GIS. The NOAA Coastal Services Center provided funding to collaborate with local governments and other coastal data custodians to author interoperable web mapping services and build a web-mapping portal that integrated these services to support coastal management decision-making along the Lake Superior coast. The project team was influenced by the Gulf of Maine Ocean Observing System (GoMOOS), who had developed the cutting edge Gulf of Maine Mapping Portal (http://www.gommap.org/). GoMMaP was constructed using Minnesota MapServer (http://www.mapserver.org/), an open-source web map server software initiated at the Department of Forest Resources at the University of Minnesota, and Chameleon (http://chameleon.maptools.org/) – an open source environment for developing web mapping applications launched by DM Solutions Group, a geospatial consulting group based in Ottawa, Canada.

The Lake Superior Coastal Mapping Portal (http://maps.aqua.wisc.edu/lscmp/) includes a rudimentary catalog of interoperable web mapping services and allows users to select a "context file" in the mapping interface that groups services by theme and/or geography. For example, portal users can select a combination of services (point observations, NEXRAD radar, etc.) that display weather conditions for the entire Great Lakes (see Figure 2) or a different combination of services that integrate local spatial data for the four Wisconsin counties that border Lake Superior. While the LSCMP served as an early testbed showing the potential of integrating web mapping services in Wisconsin, it suffered from slow performance and maintenance issues. The Chameleon interface that was developed proved to be poorly suited to maintain an operational catalog of interoperable web mapping services.

Figure 2. Lake Superior coastal mapping portal map interface

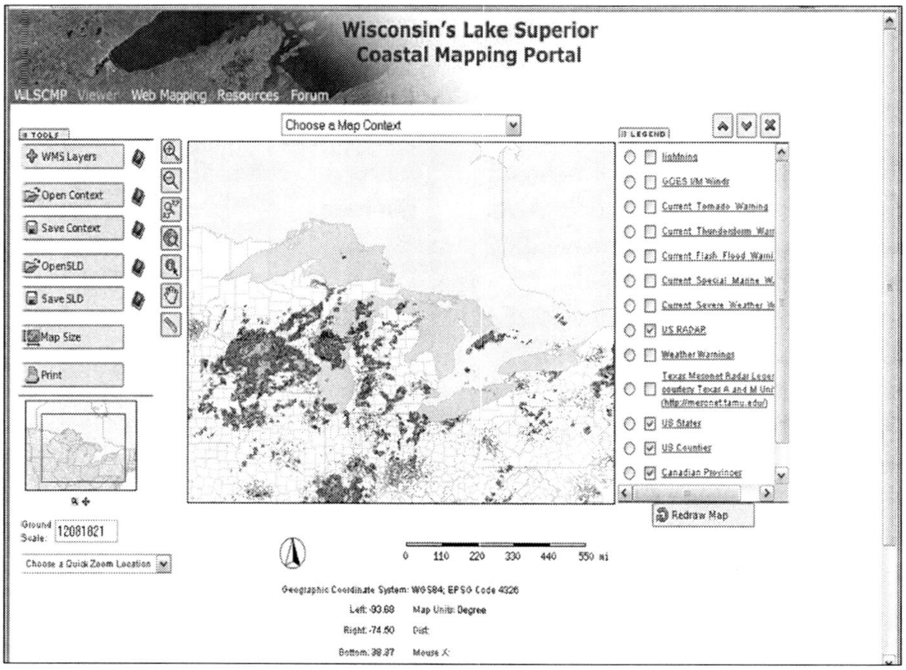

Wisconsin Coastal Guide

A third project initiated in 2005 allowed UW Sea Grant to test new approaches to web mapping. The Wisconsin Coastal Guide (http://www.wisconsincoastalguide.org/) is an interactive web-mapping site that supports coastal heritage tourism along the Great Lakes Circle Tour route in Wisconsin and was funded by the Wisconsin Coastal Management Program. The Great Lakes Circle Tour is a scenic driving route circumnavigating all of the Great Lakes. The route, which is marked by distinctive signs, tends to follow busy state and federal highways and sometimes veers away from the lakes. The Wisconsin Coastal Guide shows the designated driving route, but also includes local roads and coastal attractions such as lighthouses, shipwrecks, boating access, beaches, and parks (see Figure 3). Panorama photos have been taken for nearly all of the public access sites to the Great Lakes. The primary mapping interface utilizes the Google Maps Application Programming Interface (API) (http://code.google.com/apis/maps/). When users click on the icon for a specific coastal attraction, a dialog box appears with more information, including links to external web sites with rich content about the attraction. A user can, for example, click on a lighthouse and gain access to any of the independently maintained web sites that provide detailed historical information and photos about Great Lakes lighthouses.

As an alternative to the Google Maps interface, links are provided to Keyhole Markup Language (KML) files for each of the coastal features, as well as other reference layers such as the Great Lakes Circle Tour route and oblique aerial photos of the coast. In addition, a KML file was created that combines all the coastal features and reference

Figure 3. Wisconsin coastal guide Google maps interface

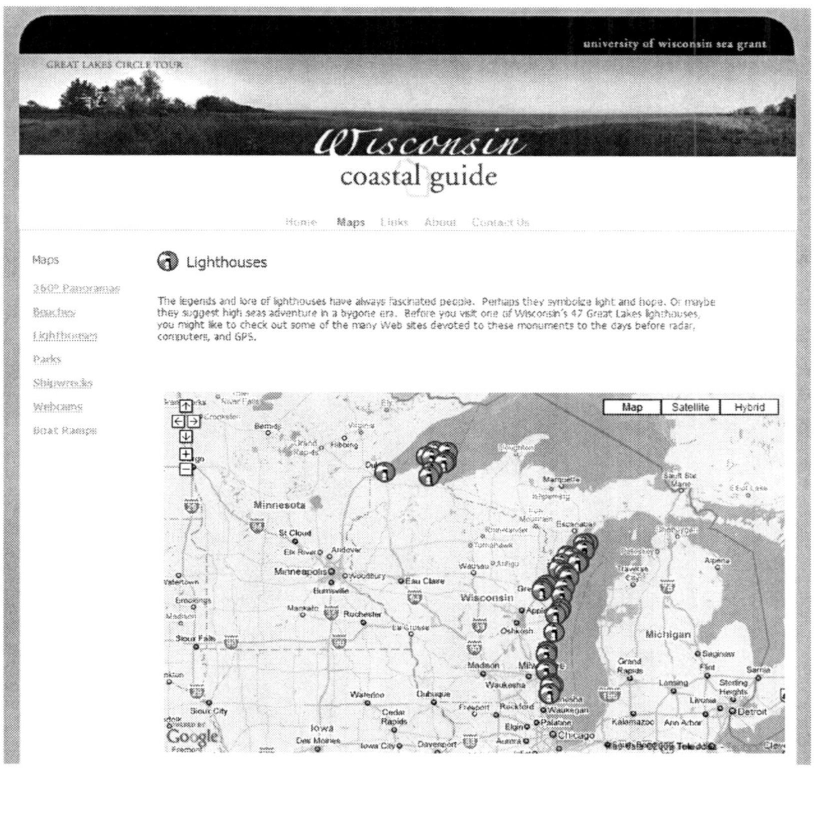

layers (see Figure 4). Users who have downloaded virtual globe applications such as Google Earth (http://earth.google.com/) or NASA World Wind (http://worldwind.arc.nasa.gov/) find they have much improved ability to navigate the Circle Tour and explore coastal attractions. The main drawback is that users need to download very large files to install these "thick software clients" on their computer. This may not be a problem for those with high bandwidth and administrative privileges on their computer, but there are computer users without these advantages. Another benefit to this approach is that KML is now an OGC standard that promotes interoperability. Providing a stable URL to the KML files promotes their use in other web mapping applications. As an example, the Wisconsin Coastal Guide KML files have been used in the HarborView web-mapping application (http://glos.us/hview/) developed by the Great Lakes Observing System (GLOS). HarborView is an interactive web-mapping site that integrates weather and current Great Lakes observations with coastal attractions to promote safe and pleasurable recreational boating (see Figure 5). This initial foray into the world of mapping mashups has been beneficial in promoting coastal heritage tourism, but showed how little guidance exists on deciding among the various approaches to developing web-mapping interfaces.

Wisconsin Coastal Data Catalog

Influenced by a presentation on the International Coastal Atlas Network interoperability prototype and coastal erosion use case at the Coastal Zone '07 conference in Portland, Oregon, UW Sea Grant decided to explore adoption of some of the supporting technologies for a CWA in Wisconsin. Most of this effort involved installing and customizing GeoNetwork (http://geonetwork-opensource.org/), an open source geospatial catalog software developed by the United Nations.

The Wisconsin Coastal Data Catalog (http://speedy.ersc.wisc.edu:8080/geonetwork/) was populated with spatial data sets developed for the Lake Michigan Potential Damages Study in Ozaukee County, Wisconsin. These included mapping of coastal bluff tops and toes in 1956 and 1999, and coastal orthophotos for the same time periods. Metadata was included for each of the data sets. The GeoNetwork interface allows a user to search for data using a variety of techniques; view an abstract, key words, and a display graphic of the data set; access the detailed metadata; view layers in an interactive web mapping application;

Figure 4. Wisconsin coastal guide combined KML file in Google Earth

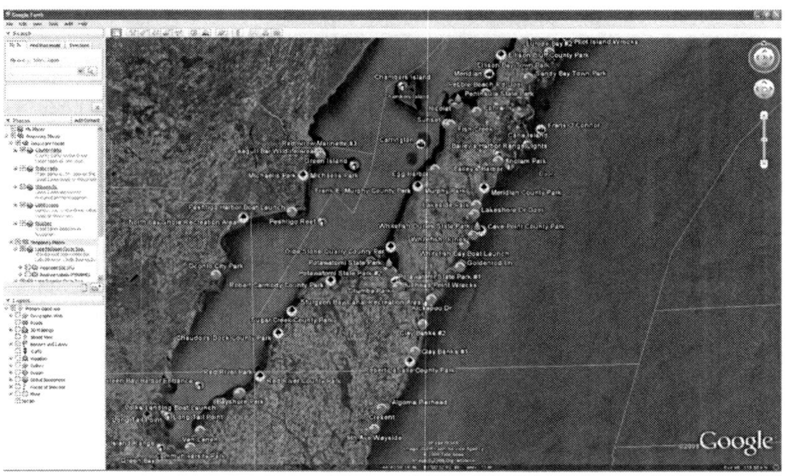

Figure 5. GLOS HarborView web mapping application

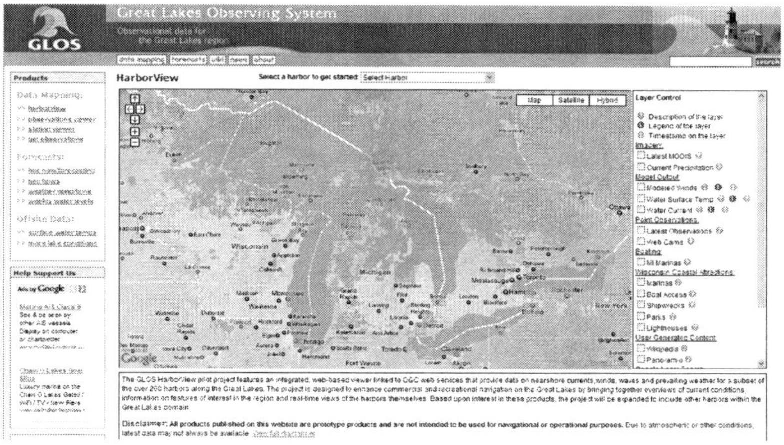

and download data in a variety of formats (see Figure 6).

GeoNetwork has many powerful capabilities. It is one of the few open-source geospatial data catalog software packages available and supports a variety of OGC interoperability protocols. Perhaps its greatest strength is the collaborative nature of the software. Because it is a web application, a remote user with credentials can login and enter information for the data sets they maintain. If some data custodians want to set up their own GeoNetwork site, communication can be established between these sites.

Some problems were experienced during the process of installing and utilizing GeoNetwork. First, it proved difficult to customize the GeoNet-

Figure 6. Screenshot of Wisconsin coastal data catalog interface

work interface. It was difficult to modify the code that governed the appearance of the interface in the database format used by GeoNetwork. Also, importing metadata in the FGDC format did not work. GeoNetwork is better suited for importing metadata in the ISO format, while many spatial data custodians in Wisconsin maintain their metadata in FGDC format using ESRI's ArcCatalog. This issue regarding metadata uptake may affect later participation in the Wisconsin Coastal Data Catalog by local governments. Next, some of the data distribution formats supported by GeoNetwork did not work. In particular, it was not possible to view interactive maps through the GeoNetwork interface and KML files generated by the catalog did not display properly. Finally, it appears that work was still needed to successfully harvest other OGC Catalog Services for the Web (CSW) using version 2.2 of GeoNetwork. It is important to note that many of these issues could likely be solved with more programming experience at UW Sea Grant.

Despite these drawbacks, GeoNetwork still provides the best open source option for implementing a coastal spatial data catalog. It is worth watching to see if anyone "geo-enables" one of the existing open access institutional repository software environments (ePrints, DSpace, Fedora), but that may be a long way off. Over the next several months, there are plans to: (1) include outside coastal data custodians with robust interoperable web mapping services in the GeoNetwork testbed; (2) complete an ontology of Wisconsin coastal hazards terms using the Protégé software developed by Stanford University (http://protege.stanford.edu/); and (3) document other coastal data sets and reports in the Wisconsin Coastal Data Catalog. This is possible because GeoNetwork is set up to catalog more that just geospatial data. The first data sources to be cataloged include oblique aerial photos of the coast taken in 2007 and 2008 by Dr. David Mickelson, a geologist at the University of Wisconsin-Madison, and the many coastal hazards technical studies completed for the Great Lakes coast in Wisconsin during the past 20 years.

BUILDING THE CASE FOR A WISCONSIN COASTAL ATLAS

In August 2003, a retreat on the future of the Great Lakes Information Network (GLIN) was held at Wingspread near the Lake Michigan shore outside Racine, Wisconsin. Wingspread is a unique conference facility designed by Frank Lloyd Wright and operated by the Johnson Foundation. Many deliberations on important policy issues ranging from local to international have been held at Wingspread over the years. Photos on the wall show dignitaries such as Henry Kissinger engaged in serious debate. In this auspicious location, experts with an interest in the intersection of Great Lakes policy and information technology met on the tenth anniversary of GLIN to chart a course for its future. GLIN (http://glin.net/) is an innovative information portal and discussion forum for the Great Lakes that was initiated in 1991 using some of the earliest web protocols such as Gopher (Manninen, 1999). One important topic was raised multiple times during the retreat – the need for an integrated spatial database for the Great Lakes to allow exploration of what-if alternatives for Great Lakes management. The metaphor of having all the information readily available to conduct "SimGreatLakes" simulations was advanced. This is the spirit that will guide development of the Wisconsin Coastal Atlas.

A coastal web atlas is defined as "a collection of digital maps and datasets with supplementary tables, illustrations, and information that systematically illustrate the coast, oftentimes with cartographic and decision-support tools, and all of which are accessible via the Internet" (O'Dea *et al.*, 2007). The Oregon Coastal Atlas (http://www.coastalatlas.net/) and the Marine Irish Digital Atlas (http://mida.ucc.ie/) are early examples of robust CWAs. In particular, the Oregon Coastal Atlas provides a good model for a CWA in Wisconsin. It was developed by the Oregon Coastal Management Program in cooperation with Oregon State University and other partners beginning in 2000. It consists of four components (maps, search, tools, learn) that enable "users to search and find data, but also to understand its original context, and put it to use via online tools in order to solve a spatial problem" (Haddad *et al.*, 2005). The atlas has been a resounding success. It now serves more than 3,500 data sets and received about 2.5 million hits in 2008 (Oregon Department of Land Conservation and Development, 2009). Moreover, the map interface has become a common framework for discussing coastal management issues (National Oceanic and Atmospheric Administration, Coastal Services Center, 2003).

As the Wisconsin Coastal Atlas is built, there will be an effort to balance the practical needs of developing a CWA with the work that is needed to advance the most significant research issues concerning implementation and interoperability of CWAs. This includes development of a template for design and evaluation of CWA projects; promotion of sound cartographic design principles for web mapping interfaces; identifying and reducing the technical and institutional barriers to the development of domain spatial data infrastructures that draw heavily on local government data; development of an effective archive for coastal geospatial data, and; application of ontology tools to promote semantic mediation of local government spatial data sets to conduct spatial analyses of coastal issues at a regional scale.

UW Sea Grant is actively working with potential partners to communicate the vision for a Wisconsin Coastal Atlas. Development of the atlas will start in February 2010 with an initial focus on the mapping interfaces and spatial data needed to address coastal hazards on the Great Lakes. This will be synchronized with a two-year NOAA Coastal Management Fellow who will help build a Great Lakes spatial decision-support toolbox to address coastal hazards resilience and more effective implementation of comprehensive plans in coastal communities.

CONCLUSION

Wisconsin has a long history of applying geospatial technologies to Great Lakes issues. Coastal GIS projects have touched upon many of the components of a CWA, but Wisconsin currently lacks a consolidated web portal for maps, data, and decision-support tools for the Great Lakes. Funding for a two-year period starting in February 2010 has been secured to build the Wisconsin Coastal Atlas. Key project partners anticipate the benefits of the atlas. The Wisconsin Department of Natural Resources expects that an atlas will enable the analyses needed to implement the Wisconsin Great Lakes Strategy and the Great Lakes Compact. The State Geographic Information Officer sees the atlas as a template for state GIS initiatives both on and away from the coast. Completing an interoperable Wisconsin Coastal Atlas could promote development of other regional, state, and provincial CWAs. Sharing among interoperable CWAs could help realize a long-time goal of scientists and coastal resource managers – management of the Great Lakes using a whole ecosystem approach (Caldwell, 1988; MacKenzie, 1996; Sproule-Jones, 2002).

REFERENCES

Caldwell, K. (1988). *Perspectives on ecosystem management for the Great Lakes: A reader*. Albany, NY: State University of New York Press.

Chameleon. (2009). Retrieved October 31, 2009, from the Chameleon web site: http://chameleon.maptools.org.

GeoNetwork Opensource. (2009). Retrieved October 15, 2009, from the GeoNetwork Opensource web site: http://geonetwork-opensource.org/.

Google Earth. (2009). Retrieved October 31, 2009, from the Google Earth web site: http://earth.google.com.

Google Maps Application Programming Interface. (2009). Retrieved October 31, 2009, from the Google Maps Application Programming Interface web site: http://code.google.com/apis/maps.

Great Lakes Information Network. (2009). Retrieved October 31, 2009, from the Great Lakes Information Network web site: http://glin.net.

Gulf of Maine Mapping Portal. (2009). Retrieved October 31, 2009, from the Gulf of Maine Ocean Observing System web site: http://www.gommap.org.

Haddad, T. C., Wright, D. J., Dailey, M., Klarin, P., Marra, J., Dana, R., & Revell, D. (2005). The tools of the Oregon Coastal Atlas . In Wright, D. J., & Scholz, A. (Eds.), *Place Matters: Geospatial Tools for Marine Science, Conservation, and Management in the Pacific Northwest* (pp. 133–149). Corvallis, OR: Oregon State University Press.

HarborView Web Mapping Application. (2009). Retrieved October 31, 2009, from the HarborView web site: http://glos.us/hview.

Hart, D. (2000). *Building a Horizontally and Vertically Integrated Coastal GIS using Local Government Spatial Data: The Case of Coastal Erosion Hazards on the Lake Michigan Coast of Wisconsin*. Doctoral dissertation. Madison, WI: University of Wisconsin-Madison.

Lake Superior Coastal Mapping Portal. (2009). Retrieved October 31, 2009, from the University of Wisconsin web site: http://maps.aqua.wisc.edu/lscmp

MacKenzie, S. (1996). *Integrated Resource Planning and Management: The Ecosystem Approach in the Great Lakes*. Washington, DC: Island Press.

Manninen, C. (1999, December 5-9). *The Great Lakes Information Network: Trials and Triumphs of an Integrated Approach to Web Design*. Paper presented at the Symposium on Water Resources and the World Wide Web, American Water Resources Association, Seattle, WA. Retrieved on October 25, 2009 from http://www.awra.org/proceedings/www99/w10/index.htm

Marine Irish Digital Atlas. (2009). Retrieved October 31, 2009, from the Marine Irish Digital Atlas web site: http://mida.ucc.ie

Minnesota MapServer. (2009). Retrieved October 31, 2009, from the Minnesota MapServer web site: http://www.mapserver.org

National Oceanic and Atmospheric Administration, Coastal Services Center. (2003). Opening a gateway to Oregon's coastal information. *Coastal Services*, *6*(6), 2–3.

O'Dea, L., Cummins, V., Wright, D., Dwyer, N., & Ameztoy, I. (2007). *Report on Coastal Mapping and Informatics Trans-Atlantic Workshop 1: Potentials and Limitations of Coastal Web Atlases*. Cork, Ireland: Coastal and Marine Resources Centre, University College Cork.

Oregon Coastal Atlas. (2009). Retrieved October 31, 2009, from the Oregon Coastal Atlas web site: http://www.coastalatlas.net/

Oregon Department of Land Conservation and Development. (2009). *Protecting and managing Oregon's coastal and ocean resources, 2007-2009 biennial report, Oregon Coastal Management Program*. Salem, OR: Oregon Coastal Management Program.

Protégé. (2009). Retrieved October 15, 2009, from The Protégé Ontology Editor and Knowledge Acquisition System website http://protege.stanford.edu/

Rink, M., Hart, D., & Miller, A. (1998). *Customized geographic information systems training for coastal and marine resource managers*. Paper presented at the 1998 ESRI User Conference, San Diego, CA. Retrieved on October 25, 2009 from the ESRI User Conference website http://proceedings.esri.com/library/userconf/proc98/PROCEED/TO650/PAP604/P604.HTM

Sproule-Jones, M. (2002). *Restoration of the Great Lakes: promises, practices, performances*. Vancouver, BC: University of British Columbia Press.

Stone, J., Harrower, M., Hart, D., & Koch, T. (in press). Coastal erosion on the Great Lakes: Using remote sensing and geovisualization for public education . In Madden, M. (Ed.), *Landscape Analysis Using Geospatial Tools*. New York: Springer-Verlag.

Wisconsin Coastal Data Catalog. (2009). Retrieved October 31, 2009, from the Wisconsin Coastal Data Catalog web site: http://speedy.ersc.wisc.edu:8080/geonetwork.

Wisconsin Coastal Guide. (2009). Retrieved October 31, 2009, from the Wisconsin Coastal Guide web site: http://www.wisconsincoastalguide.org

World Wind, N. A. S. A. (2009). Retrieved October 31, 2009, from the NASA World Wind web site: http://worldwind.arc.nasa.gov

KEY TERMS AND DEFINITIONS

Coastal Web Atlas: A collection of digital maps and datasets with supplementary tables, illustrations and information that systematically illustrate the coast, oftentimes with cartographic and decision support tools, all of which are accessible via the Internet. Also known as web atlas, digital atlas, digital coastal atlas.

Atlas Interoperability: The ability to interrogate two or more unrelated atlas systems and potentially exchange metadata, images, vector and raster data.

Coastal and Marine Information: Both spatial and non-spatial information regarding the coastal and marine areas of a region or country.

Integrated Coastal Zone Management: A dynamic, multi-sectoral approach to managing the coast which takes into account social, economic and environmental concerns of all parties with an interest in the coastal space.

Metadata: Information that fully describes or documents a dataset, such as its geographic coverage, quality, completeness, accuracy, etc. are a "pedigree" of sorts for a data set and helps you to judge its "fitness for use" or reliability, thereby helping one to use it more appropriately and efficiently. Metadata allow a potential user, for comparative purposes, to understand how the data were collected. They also provide the all-important details of how you can actually obtain the data in question, or who best to contact. Data that do not have accompanying metadata are often hard to find, difficult to access, troublesome to integrate, and perplexing to understand or interpret.

Spatial Data Visualization: The ability to view digital data with a spatial dimension in a computer environment containing a coordinate reference system.

Web GIS: A geographical information system which can be accessed over the Internet and allow visualization and interaction with spatial data via a map as well as providing analysis functionality such as spatial analysis, querying and buffering.

Virtual Globe: A three-dimensional visual rendering of the Earth or of other planets, usually available as online software application, or as an application for the desktop. A virtual globe provides the user with the ability to navigate throughout the virtual world freely in all directions, by changing viewing angle, azimuth, and heading. A virtual globe is primarily a visualization tool (oftentimes open source) and is not completely synonymous with a geographic information system in terms of spatial analysis capabilities. Examples include Google Earth, NASA World Wind, and Microsoft Virtual Earth.

Chapter 10
Belgium

Kathy Belpaeme
Coordination Center on Integrated Coastal Zone Management, Belgium

Hannelore Maelfait
Coordination Center on Integrated Coastal Zone Management, Belgium

ABSTRACT

The Belgian Coastal Atlas was published as a book in 2004, triggered by reporting obligations regarding Integrated Coastal Zone Management (ICZM) for the European Commission on the state of the Belgian coast. Initially there was no intention to move towards a web version. The many enquiries for digital information and interactive maps led to the development of a web based atlas in 2005. This chapter describes the content and set-up of the Belgian coastal atlas, which is quite different from other online atlases. The usage statistics and the success of the atlas are analyzed. Future directions for the atlas are discussed and the possibility to link the atlas to coastal sustainability indicators is considered. The Belgian atlas should develop towards a policy support tool, supporting the ICZM process for a wide range of coastal actors, planners and managers.

INTRODUCTION AND OVERVIEW

The Belgian coast forms a unique region with a variety of wildlife areas, coastal developments and economically important areas. Tourism is the main economic activity and generates €1.9 billion while the two main ports had a turnover of € 2.2 billion in 2006 (statistics from the Provincial Government of West-Flanders, 2007).

DOI: 10.4018/978-1-61520-815-9.ch010

The Belgian coastal municipalities are home to 210,712 people (2008). With an average age of 45.6 (2008), people there tend to be 5 years older than in the rest of Belgium. Year on year the number of elderly people with respect to younger people is increasing. In 2007 the coast had 157 inhabitants over 60 for every 100 younger than 20. Almost 50% of residential properties located at the coast are estimated to be second and/or holiday homes (Kahane *et al.*, 2009).

The coast also fulfils a key function for nature conservation. Just over 20% or 50,720 ha of the

Copyright © 2011, IGI Global. Copying or distributing in print or electronic forms without written permission of IGI Global is prohibited.

total surface of the coastal area benefits from a legal form of nature conservation protection (Maelfait & Belpaeme, 2007).

The scientific community has been instrumental in gathering data and gaining knowledge on the coast. In 2008 over 1000 scientists were active in marine and coastal science, working in 98 different research groups at scientific institutes and universities (Mees, 2008; IMIS, Integrated Marine Information System). They study the coastal and marine environment of the North Sea but also that of other parts of the world, thus generating a significant amount of coastal information in many disciplines: not only the classical science domains like biology, chemistry, engineering, geology, geography and sedimentology, but also disciplines like history, sociology, linguistics, archaeology and law. Management and policy related to the coast is the responsibility of several administrations and directorates, all developing their own visions and policies. This also means that all these institutes and administrations hold coastal information in their databases. However, it is highly fragmented, often sector or discipline-specific and not recorded in a uniform or consistent way.

INITIATIVES AIMED AT SUPPORTING COASTAL MANAGEMENT

Several initiatives in Belgium have been instrumental in the management of coastal data and information.

Coastal Indicators

To pursue effective policies for the coast, a wide variety of Information and data are needed. A set of sustainability indicators can provide an answer in an easy reference format. The Coordination Centre on Integrated Coastal Zone Management (ICZM) in Belgium provides a regular update via a set of twenty-one indicators for the Belgian coast. Forty-three yearly updated and maintained datasets form the basis for these indicators. The underlying datasets capture data on socio-economic and ecological themes at the lowest statistic level possible. These data are stored in a management system, which contains the raw data, but also the calculations needed for determining the value of each indicator. The indicators can have added value when integrated in an atlas, as they make the link between data and policies.

Capturing Marine Data

The Marine Data and Information Centre of Flanders Marine Institute has a mandate to capture, document, integrate and redistribute coastal and marine datasets. These data are stored in the Institute's Integrated Marine Information System (IMIS). The information is openly available and can be used to support initiatives such as the coastal atlas.

Coastal Atlas

The Belgian Coastal Atlas is an initiative of the Coordination Centre on Integrated Coastal Zone Management. It was published in 2004 as a book, triggered by the reporting obligations to the European Commission about the state of the Belgian coast related to the ICZM recommendation (Recommendation of the European Parliament and of the Council of 20 May 2002 concerning the implementation of Integrated Coastal Zone Management in Belgium, 2002/413/EC). From the beginning it was not the intention to give detailed information and data, but mainly to present a general and synthesized overview of the main characteristics of the Belgian coast. The target audience was defined rather broadly: scientists, civil servants, politicians, coastal players and the wider public. The book contains about forty informative maps and eighty photos of the Belgian coast. One thousand three hundred Dutch copies and five hundred English copies of the atlas were produced. The Dutch version of the atlas went on sale at bookshops along the Belgian Coast.

The book proved to be very popular and sold out quickly. The Coordination Centre on ICZM also received many requests from planners, scientists, lecturers and other interested parties to use the maps in a digital format. Consequently, the management board of the Coordination Centre decided to produce an online version of the atlas. This web version was launched in November 2005.

The book and the online version of the Belgian Coastal Atlas were financed by the project partners of the Coordination Centre: the province of West-Flanders and the two main Flemish coastal administrations, and by European structural funds within the objective 2 program for the Belgian coast.

MOVING FROM A TRADITIONAL ATLAS BOOK TO AN ONLINE TOOL

The aim of the online atlas was to produce an easily accessible, simple, user-friendly version of the hardcopy atlas, with additional features, like printing and downloading of maps.

In developing the online version, the following choices were made:

1. to keep the same structure, look and feel as the successful printed version of the atlas;
2. to make the atlas available in four languages: Dutch, English, French and German;
3. to include hyperlinks with the sustainability indicators for the Belgian coast. A subset of indicators was added to each thematic chapter; and
4. to provide limited options for interactivity: clicking on and off data layers on online maps, PDF, map and data-download options.

Data Collection

As the process for developing a coastal atlas stemmed from an idea to publish a book rather than an online tool, the choice of what data to include was possibly quite different from other online atlas initiatives. Such initiatives often intend to give an extensive overview of all the data available, provide frequent updates, and they tend to work in a dynamic and interactive environment. For example, the Marine Irish Digital Atlas (MIDA) contains over 130 datasets and the Oregon Coastal Atlas over 3,300 datasets. Both atlases provide several basic mapping tools, such as layer list control, feature identification, zoom and re-centering functions.

The Belgian atlas has the following thematic chapters:

- **Spatial setting:** a general description of the Belgian coast, its geographical position in Europe and the North Sea, and the coastal and marine boundaries.
- **Physical environment:** a description of the dynamic and fragile features of the coast, the coastal landscape and its historic perspective, characteristics of the coastal dunes and the seabed, and more details on Belgian weather, climate, tides and currents.
- **Spatial structure:** an overview of the spatial planning system and the policy priorities for the coast.
- **Integrated Coastal Zone Management:** highlighting the importance of an integrated approach to coastal issues, the European perspective and the Belgian approach.
- **Use of the sea:** a description of the multifunctional use of the sea, and related monitoring and research efforts.
- **Environment and nature:** a focus on the natural assets of the Belgian coast on land and at sea, including examples of the most valuable coastal habitats: beaches, dunes,

estuaries, and polders.

- **Tourism and recreation:** an illustration of the different tourist activities on the coast and an introduction to the 10 coastal municipalities and the 13 seaside resorts.
- **Industry and business:** introducing the main economic activities and giving details on the ports of Zeebrugge and Oostende.
- **Fishery and agriculture:** general facts about fisheries in Belgium and agriculture in the coastal plains, including a brief historical perspective.
- **Culture and heritage:** giving the remaining historical features, and the main events and activities links to coastal heritage.
- **Living near the sea:** giving a general overview about the settlement types and some notable facts about coastal architecture.
- **Coastal defense:** focus on the erosion and sedimentation processes and coastal protection measures along the Belgian coast.

From this list, it is clear that the option was taken to organize the data in a conventional, sector-oriented way. This is a site structure that readers are used to and they can easily navigate their way to the relevant information. An identical structure was followed when designing the online atlas.

All data used in the atlas are publicly available, mostly provided by public authorities and their administrations. These authorities gather the information themselves, as part of their objectives and policies, or the data are generated by universities, consultants or other institutions commissioned by the authorities. None of the data were purchased and no licensing or data release issues apply. As explained earlier, the atlas contains a limited set of aggregated data, but by linking the atlas to the databases of the sustainability indicators for the coast, a rich set of data was added. The indicators are, in themselves, often aggregated data but a carefully chosen combination of relevant datasets can explain more about the state of the coast and whether it is approaching sustainability or not. In other words, all the data are interpreted within a sustainability framework. The data needed for calculating the indicators are also provided by various administrations. Wherever possible, this information is gathered at the lowest possible administrative level.

The atlas was developed in collaboration with the partners of the Coordination Centre on ICZM. Consultants carried out the cartography, web design and implementation of the website.

For the development and updating of the coastal indicators, the Coordination Centre has established a network of about 80 scientists, coastal experts and civil servants. Flanders Marine Institute (VLIZ) is responsible for developing and hosting the website and the database.

System Design

The website was set up according to web standards current in 2005 and is easy to consult using popular web browsers that support industry standards. It is hosted by the Flanders Marine Institute. It works on an Apache webserver and runs on a MSSQL relational database. The website is set up in PHP (hypertext pre-processor)/HTML (hyper text mark-up language) and, for the interactive maps, flash applications and Actionscript were used. All maps are downloadable in PDF format; the data are stored in Microsoft Excel format. For some maps vector shape files are provided. The maps were designed with graphics software and uploaded to the site as vector graphics.

As far as IT-tools are concerned, most of the maps allow clicking on and off of data layers. Zooming possibilities are provided on all maps, and a X-Y reference system is applied. For the search function, an internal search tool of the Flanders Marine Institute was integrated into the website. The maps in the original web atlas were not designed within a GIS environment, because of the limited layout and graphical possibilities of GIS. During a follow-up test phase, some maps were added in GIS format or linked to Google

Table 1. Average number of unique visitors per month and per day for www.kustatlas.be

Period	Average number of unique visitors per month	Average number of unique visitors per day
Mid Nov 05- end Jan 06	2349	76
2007	2155	71
2008	1853	61
2009 (Jan-June)	1727	57

Earth. These features will be further developed in the future.

The visualization of the sustainability indicators provides a policy support tool for coastal managers and decision-makers.

For the digital Belgian Coastal Atlas itself, no databases were needed since the amount of data included was rather limited. For the sustainability indicators, a database designed by Flanders Marine Institute is used. The database used is Microsoft Access and can be queried via an SQL interface. It has a search field for parameter (indicator) values, years, and geographical units (coordinates or municipalities). For the mapping, the in-house developed S3 (SVG Sniplet Server) system was used to build a zoom- and pan enabled map in SVG format. For graphs, the PHPChardir module was used. A specific database was developed to hold all metadata.

USAGE OF THE ATLAS

Usage Statistics

The website was launched in November 2005, and has had an average number of more than 2000 unique visitors per month ever since.

The average number of unique visitors during the first three months after the launch was 76 people per day. The launch received a lot of media coverage - all the local and regional newspapers and the regional television company featured it. In the first two weeks after the launch, the site had 4449 unique visitors, of which one third were the day after the launch. This suggests that linking the launch to a press release and bigger event is worthwhile in promoting an online atlas. In order to promote the website, a set of postcards with aerial photos of the coast was produced. These were sold in bookshops at the coast and were also used as promotional gifts.

In 2007 there were, on average, 2155 unique visitors per month or 71 per day. For 2008 and 2009 the number was 1853 and 1727 unique visitors per month respectively or 61 and 57 per day (see Table 1). In the period 2005-2008 about 17% of the visitors accessed the website more than once a day.

Analysis of the visits to the website shows that in 2008, on average, 3.23 pages were loaded per visit. In 2005-2006 this was 10.67 pages per visit. The time spent on the site was 2min46 sec in 2008 and 5min 11 sec in 2005-2006.

The site is mostly visited during the working week, with peak periods around 10:00 am to12:00 pm and 3:00 pm to 4:00 pm, and in the early evening hours. From this it can be assumed that the site is mostly visited during working hours and hence visits are work related.

In 2005-2006 over 70% of the visitors used the direct web address, 16.2% reached the site through a search engine, and 12.1% through external links. In 2008 the corresponding percentages were 41.8% directly, 53.3% via a search engine and 4.7% via external links.

For the whole period analyzed, about 53% of the visitors added the site to their favorites.

Since the website is available in four languages, it reaches an international audience. The top three visiting countries in 2005-2006 were Belgium, the Netherlands and Germany. In 2008 France took the third place.

Analysis of the top 10 pages visited and the top 10 keywords did not provide valuable information. The pages visited and the keywords used are quite diverse. 'Coast' was however often mentioned in the keywords. Some other examples are: coastal tourism, coastal municipalities, coastal atlas and coastline.

Feedback

Feedback reaches the Coordination Centre during or in relation to workshops and conferences, but also spontaneous reactions are received regularly. All reactions and questions are registered and monitored. Requests for which the atlas could provide an answer concern mostly specific or general information on coastal management, (aerial) photographs and the use of maps from the atlas. The integrated map presenting the multifunctional use of the sea has been requested many times. This shows that there is a clear need and big market for integrated information. Requests come from scientific users (e.g. universities, research institutes) as well as from the public, administrations, supporting services and schools. The feedback has always been very positive.

DISCUSSION

Has the Atlas Been A Success?

Development of a web atlas was originally NOT the intention of the Coordination Centre on ICZM. The aim was to publish a tabletop atlas as part of the reporting system to the European Commission. The challenge at the time of the web development was mainly to achieve the best possible link between the book and the online version. The Web visitors should feel like they are stepping inside the book, and should be able to use the maps and data in their own way.

The development of the online atlas has surprised the initiators with its resounding success. Although there was no active promotion for the atlas apart from the launch, the website has had on average more than 2 000 unique visitors per month, which is a good result compared to similar websites and considering Belgium has a small coastline where people are expected to be less interested than in countries with more coastline.

The feedback is unanimously positive. In particular, the layout and the ease of navigation through the website is pleasing to visitors. The simplicity of the design and the fact that visitors are not overwhelmed by information to choose from is also appreciated.

The sustainability indicators have much potential to add extra information to the atlas. These indicators will allow visitors to interpret information and provide useful input about coastal management and trends towards or away from sustainability.

A special characteristic of the Belgian coastal atlas is that it contains information on the land and the seaside of the coast, whereas some atlases only consider the sea *or* landside. By pursuing this path, the Coordination Centre wishes to stress the importance of the land-sea interface. It is precisely this interface that deserves special attention within the process of Integrated Coastal Zone Management. Atlases can help blur the low-water mark as a physical division between land and sea, and in the thinking about connecting the wet and dry sides of the coast.

The Belgian Coastal Atlas tried to visualize the complexity of the users of the sea and the coastal zone by providing integrated maps. Linking the atlas to indicators in the future should build on this and further demonstrate the integration of data between sectors and users of the coast.

Figure 1. Estimated average number of unique visitors per month to www.kustbeheer.be/indicatoren

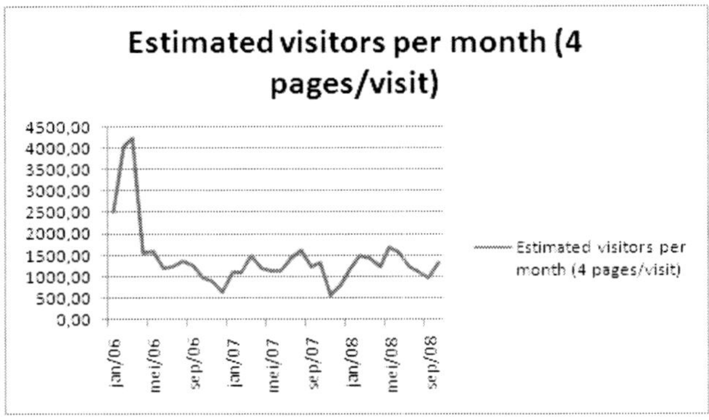

What is the Future for the Atlas?

While the atlas attracts a significant number of visitors, the website specifically developed for the sustainability indicators, known as 'the coastal barometer', reaches a much smaller audience. This website was not developed for the general public, but for policymakers, managers, scientists and coastal stakeholders. The coastal barometer contains data on 21 coastal sustainability indicators. The data are presented as graphs and tables, but not with geographically explicit spatial referencing. Some test cases for using GIS have been developed, since it is felt that trends and messages on effects of management options are sometimes more easily understood when presented as maps.

The coastal barometer website was launched in February 2004. An overview page and a short description of each indicator gives a summary of the trend. Detailed data and graphs are available for download.

In order to bring the trends on the coast even more to the attention of policymakers and coastal managers, the Coordination Centre decided to publish the results of the indicators in a publication called "The Coastal Compass". Over 50 scientists and coastal experts contributed to this publication.

There are no unique visitor data available for the indicator website, but if one considers that 4 pages are consulted per visit, then the average number of visitors per month was 1781, 1166 and 1306 for 2006, 2007 and 2008 respectively. This equals 58, 38 and 43 visitors per day as illustrated in Table 10.2 and Figure 1. From the total hits, we may safely conclude that the site had a much higher number of visitors during the month that the first "Coastal Compass" was presented. This peak in visitor numbers was not so prominent with the publication of the second Compass.

Table 2. Estimated average number of unique visitors per month and per day to www.kustbeheer.be/indicatoren

Year/period	Estimated average number of unique visitors per month (based on 4 pages/visit)	Estimated average number of unique visitors per day (based on 4 pages/visit)
2006	1781	59
2007	1166	38
2008 (Jan-Oct)	1306	43

As the indicator website contains a wealth of information about the coast, all this data could be linked to the atlas website. This would provide big advantages to both initiatives: the atlas would be fed with extra, carefully selected and relevant information to coastal management, and the indicators would be presented as interactive maps. By doing so, the Belgian coastal atlas could evolve towards an interactive, data driven ICZM tool, which is part of the coastal zone management process. This upgrade has to be considered carefully. The starting-point for this will be a thorough analysis of the current usage and wishes of the end-users.

How Does it Relate to ICAN?

The Belgian Coastal Atlas is quite different from other web atlases. It contains only a limited number of data layers, and it originated from quite different underlying motives than most atlases, which often aim to provide a comprehensive overview of data. The Belgian atlas could, therefore, be considered the odd one out when presented at the first international coastal atlas workshop in Cork in 2006. However, the other atlas initiatives were all very positive about the Belgian approach, especially in regard to its success in reaching a consistent number of visitors. Joining a network like the International Coastal Atlas Network (ICAN) will certainly help our atlas developers learn from the approaches taken by our peers.

CONCLUSION

First published as a book, the Belgian Coastal Atlas was translated into a web-application only in a later phase. Designing web-atlases requires careful consideration before starting. Many different approaches can be taken and the final product can have many different guises. The degree of interactivity and the use of GIS (or other mapping applications) will depend on the aim and target audience of the atlas. A close collaboration between managers and scientists is needed in order to produce a useful but scientifically sound management support tool.

Atlases have become more and more part of the process of coastal management, being much more than just a database. Just like the coastal sustainability indicators, the atlas has evolved towards being a required tool for ICZM. A participatory process should support its development, with a strong input from end users. The atlases should become part of a policy framework, supporting coastal scientists, managers and practitioners in their activities.

REFERENCES

Belpaeme, K., & Konings, Ph. (2004). *De kustatlas Vlaanderen-België*. Oostende, Belgium: Coordination Centre on Integrated Coastal Zone Management.

De Kustatlas. (2009). Retrieved October 7, 2009, from De Kustatlas Online Web site: http://www.kustatlas.be/

European Commission. (2002). 413. Recommendation of the European Parliament and of the Council of 20 May 2002 concerning the implementation of Integrated Coastal Zone Management in Belgium, 2002/413/EC.

European Commission. (2007). 456 FINAL. COMMUNICATION FROM THE COMMISSION TO THE EUROPEAN PARLIAMENT pursuant to the second subparagraph of Article 251 (2) of the EC Treaty concerning the Common position of the Council on the adoption of a European Parliament and Council Directive establishing a Framework for Community Action in the field of Marine Environmental Policy (Marine Strategy Framework Directive)-2005/0211 (COD).

Flanders Marine Institute. (2009). Retrieved October 7, 2009, from the VLIZ Web site: http://www.vliz.be

IMIS, Integrated Marine Information System. (2009). Retrieved October 7, 2009, from the VLIZ Web site: http://www.vliz.be/imis/imis.php?module=institute.

Kahane, L., de Laat, W., Myncke, R., & Staelens, P. (2009). *Second sojourn at the coast: applied research on the littoral profile, the effects and toolkit for sound coastal policy (original title: Tweede verblijven aan de Kust: onderzoek naar het profiel, de effecten en een toolkit voor het beleid)*. Brugge, Belgium: Idea Consult.

Maelfait, H., & Belpaeme, K. (Eds.). (2007). *The Coastal Compass, Key Indicators as Guidelines for ICZM*. Oostende, Belgium: Coordination Centre for Integrated Coastal Zone Management.

Marine Irish Digital Atlas. (2009). Retrieved October 7, 2009, from the MIDA Web site: http://mida.ucc.ie/.

Mees, J. (2008). *Het regionaal, nationaal en internationaal belang van de Universiteit Gent in de mariene wetenschappen*. Oostende, Belgium: Vlaams Instituut voor de Zee (VLIZ).

Oregon Coastal Atlas. (2009). Retrieved October 7, 2009, from the Oregon Coastal Atlas website http://www.coastalatlas.net/

Provincie West-Vlaanderen. (2009). *West-Flanders: Social and Economic Facts and Figures 2009 (original title: West-Vlaanderen sociaal economisch-feiten en cijfers editie 2009)*. Brugge, Belgium: Provinciebestuur West-Vlaanderen.

KEY TERMS AND DEFINITIONS

Coastal Web Atlas: A collection of digital maps and datasets with supplementary tables, illustrations and information that systematically illustrate the coast, oftentimes with cartographic and decision support tools, all of which are accessible via the Internet.

Coast: The strip of land and sea of varying width comprising the part of the land influenced by its proximity to the sea, and the part of the sea influenced by its proximity to the land, as well as the natural coastal systems and the areas where human activities imply the use of coastal resources.

Coastal Management: The management of coastal areas is defined as a continuous process with the general objective of realising sustainable development in the coastal zone by means of optimum management of human activities in this zone, with a view to improving the situation of the coastal environment and to maintain its diversity.

Data: Pieces of information that represent the qualitative or quantitative attributes of a variable or set of variables.

Indicator: A data measure (fact/figure/value) describing an important phenomenon and allowing the representation of the state of affairs in an application area.

Spatial Data Visualization: The ability to view digital data with a spatial dimension in a computer environment containing a coordinate reference system.

Visitors: The uniquely identified client generating requests on the web server (log analysis) or viewing pages (page tagging) within a defined time period (i.e. day, week or month). A Unique Visitor counts once within the timescale. Identification is made of the visitor's computer, not the person, usually via cookie and/or IP+User Agent.

Chapter 11
Africa

Lucy E.P. Scott
United Nations Development Programme (UNDP) Global Environment Facility (GEF) Agulhas and Somali Currents Large Marine Ecosystem (ASCLME) Project, Grahamstown, South Africa

Greg Reed
Australian Ocean Data Center Joint Facility, New South Wales, Australia

ABSTRACT

The African Marine Atlas, launched in 2007, is an online resource that provides maps, images, data and information that can be used by scientists, students, coastal resource managers, planners and decision makers from institutions and specialized agencies across Africa. The ODINAFRICA Project, funded by the Flanders Government and the IOC/UNESCO, initiated and supported the African Marine Atlas Project from the IOC Project Office for IODE in Oostende, Belgium. Regional partners were the African Coelacanth Ecosystem Program (ACEP) and the United Nations Environment Program (UNEP). A team of 16 marine scientists and Geographic Information Systems (GIS) specialists from National Oceanographic Data Centers (NODCs) in Benin, Ghana, Kenya, Mauritania, Mauritius, Mozambique, Namibia, Senegal, Seychelles, South Africa, and Tanzania participated in the design and development of the atlas. The African Marine Atlas, which is now a member of the International Coastal Atlas Network (ICAN), may be found online at http://www.africanmarineatlas.net.

INTRODUCTION

Africa has rich and varied coastal and marine resources that provide the basis for significant economic and social activities including tourism, oil and gas extraction, fishing, and shipping (UNEP, 2008). The state of existence and availability of data about these resources varies considerably from country to country, with economic, environmental, cultural and political factors all playing an influencing role. At a continental scale, a number of web sites provide access to spatial information about the coastal and marine environment, among those being the United Nations Environment Program's Geo Data Portal (UNEP, 2009) and the European Commission's Joint Research Center African Marine Information System (EC, 2009). Prior to 2007 no integrated, multidisciplinary por-

DOI: 10.4018/978-1-61520-815-9.ch011

Copyright © 2011, IGI Global. Copying or distributing in print or electronic forms without written permission of IGI Global is prohibited.

tal of information relevant to marine and coastal management existed for Africa.

The Ocean Data and Information Network for Africa (ODINAFRICA) project was initiated in 2000 and brings together forty (40) marine related institutions from twenty five (25) African Member States of the Intergovernmental Oceanographic Commission (IOC) of UNESCO. ODINAFRICA has assisted African coastal states to establish and operate National Oceanographic Data and Information centers, to develop skills for manipulation of data and preparation of data and information products, and to develop infrastructure for their archivaling, analysis and dissemination.

The national representatives of African oceanographic data and information management institutions recognized that increased access to marine and coastal data was essential for the effective management of the marine and coastal environment in the region. Considerable national-level consultation took place for representatives of the ODINAFRICA Project to identify the most important priorities for their countries to increase access to data and information (IOC, 2003). The promotion of access to spatial information was a common thread of importance around the continent and the need to develop a Pan-African marine atlas was identified as one of the objectives of the third phase of ODINAFRICA,which commenced in 2004.

The Atlas project was initiated in 2006 to synthesize geospatial data products for the African continent (emphasizing especially the marine and coastal environment). The Atlas incorporates existing geo-referenced datasets available in the public domain and also data products created from national and international marine data collections by scientists participating in the ODINAFRICA Project.

The prototype Atlas was launched in February 2007 after nine months of intensive work by a team of 16 marine scientists and GIS experts from National Oceanographic Data Centers (NODCs) in Benin, Ghana, Kenya, Mauritania, Mauritius, Mozambique, Namibia, Senegal, Seychelles, South Africa, and Tanzania. International partners in this project were the African Coelacanth Ecosystem Program (ACEP) and the United Nations Environment Program (UNEP). International ocean data management experts from Belgium, the USA and Australia provided training and advice for data synthesis and analysis.

The Atlas provides maps, images, data and information that can be used by scientists, students, coastal resource managers, planners and decision makers from institutions and specialized agencies across Africa (Scott & Brown, 2009). The African Marine Atlas has been released in two formats. A "clearinghouse" website containing a repository of over 800 downloadable data products derived from the fields of marine geosphere, hydrosphere, atmosphere, biosphere, geopolitical and the human socio-economic dimensions (http://omap.africanmarineatlas.net) A demonstration MapServer site (click on "Go to the Atlas" on http://www.africanmarineatlas.net) has also been developed as a data visualization and dissemination tool where a number of data layers can be viewed.

The Atlas project was designed to identify, collect, organize and make available datasets in an atlas of environmental themes for the 25 coastal countries participating in the ODINAFRICA project. A second aim was to provide training to promote the use of Geographic Information Systems (GIS) and derived data atlas products for the dissemination of information. The African Marine Atlas was developed to provide a solution for the identified gap in access to information, by gathering public domain data together and making it available through one portal (Scott *et al.*, 2009).

Data Collection

In June 2006 a team of scientists from the ODINAFRICA project's ocean data management group commenced the task of planning for the Atlas. An inventory of over 200 datasets was identified from a survey of the needs of the participating countries.

The Atlas planning team prepared a list of tasks, which included:

- Identifying a geographic scope for the Atlas. It was decided to include the entire African continent (including island states) and an area of interest was defined by 40° North to 50° degrees South; and from 30° West to 80° East. This ensured that all Exclusive Economic Zones (EEZs) of the ODINAFRICA countries were covered.
- Identifying relevant environmental themes as a subject taxonomy.
- Data mining: this consisted of gathering all available data from online sources and published CD-ROMs and identifying unpublished datasets that could be used in the Atlas.
- Preparation of metadata: a metadata description would accompany all datasets included in the Atlas. The preferred metadata format was ISO 19115; however original metadata, wherever available, would also be used.
- Formatting the data: all data would be clipped to the geographic area of interest and, wherever required, converted to useful products such as seasonal climatological grids and contours. All products would be converted to a format suitable for use in a GIS.
- Data assembly and quality control of the final products and uploading to a central server.

The data layers in the Atlas are divided into five environmental themes: Geosphere, Hydrosphere, Atmosphere, Biosphere and Human Environment.

- Geosphere: this theme focuses on several key issues related to geohazards (including location of plates, ridges and trenches, historical earthquakes, earthquake intensity zones, tsunami events), minerals (marine minerals and core sampling sites), sediment thickness, land cover and soils.
- Hydrosphere: the oceans and fresh waters are represented in the Atlas by a number of themes including physical (temperature, salinity, sea surface temperature, average currents and sea level), chemical (nitrate, phosphate, silicate and oxygen), optics (ocean colour and turbidity) and limnology.
- Atmosphere: the Atlas is restricted to the activities within the troposphere including temperature, pressure, humidity, precipitation and cloudiness.
- Biosphere: this theme is extensive and includes biological oceanography (phytoplankton, chlorophyll, zooplankton), marine botany (algae and kelp, mangroves, sea grass), terrestrial vegetation, fisheries (fishing areas, total captures, commercial species catch and distributions for selected countries), zoology (corals, fishes, invertebrates, mammals, reptile, birds) and notable areas and species (protected areas, exotic and invasive species).
- Human Environment: this complex theme covers the area of social, economic and environmental impacts and contains data covering political boundaries (national and internal boundaries, exclusive economic zone boundaries), population densities, industrial and commercial (oil and gas, mines and quarries, oil spills), infrastructure (roads, railways, ports, airfields, power and pipelines, submarine cables) and tourism (tourist arrivals and departures, hotels).

In addition, the Atlas includes a comprehensive set of base layers that can be used as overlays to enhance the thematic datasets. These base map layers include a collection of coastline data files (of varying resolutions), relief data (bathymetry and topography), geo-referenced imagery and a gazetteer.

National, regional and continental-scale datasets were identified and edited to the agreed geographic area of interest. Much of the work to develop the Atlas, including data mining, metadata preparation, formatting and assembly, was performed by the Atlas team at their local institutions with the data uploaded to a central repository hosted at the IOC Project Office for IODE in Oostende, Belgium.

The "clearinghouse" website provides access to formatted, downloadable datasets for all environmental themes. Figure 1 shows the interface for annual and seasonal mean ocean temperature data for the surface and standard depths. The individual tiles shown in the figure are links to downloadable zip files containing geo-referenced images and gridded ASCII data for the specified time period and depth.

A dynamic web atlas (MapServer) prototype has also been developed for the Atlas to provide access to subset of the data layers, including vector and raster data. This dynamic product is built around an interactive map, shown in Figure 2, which allows the user to visualize and query relevant datasets.

Figure 1. Annual and seasonal mean ocean temperature data for the surface and standard depths (shown on the Y-axis in metres).

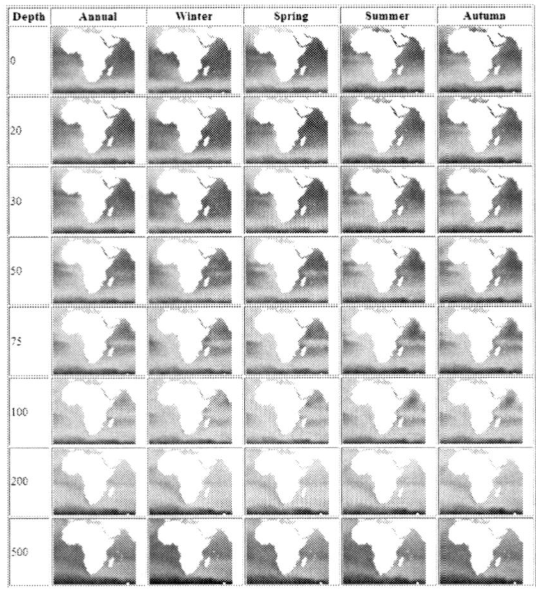

DISCUSSION AND CONCLUSION

The African Marine Atlas project has demonstrated effective capacity building for marine data management and product development in Africa through the training of a core group of GIS practitioners and having them develop atlas products. The project has also been effective at increasing access to data on the African marine environment through the establishment of a spatial data service. An additional positive outcome has been the networks and teams that have been built up through this work.

Strengths of this project are the volumes of data that were sourced and processed to a preliminary product within twelve months, with the dedicated team of marine data managers from several countries. The next step, starting in 2009, will entail completing a fully functional Web Map Service (WMS) as well as developing national atlas products that are seamlessly integrated with the African Marine Atlas. They will provide higher resolution data applicable and relevant at a scale that is useful for national and local applications.

The project for the development of the Atlas has brought benefits to participating national institutions and to Africa as a whole. The project has encouraged scientists to work together, learn new techniques and build teams that will continue to update the Atlas and develop functionality of the dissemination tools in future projects.

Challenges were sometimes posed by limited bandwidth and unreliable Internet access at the partner institution level. Some countries were facing rolling power-cuts during the project activities. This challenge was overcome by holding regular working meetings at the IOC Project Office for IODE where excellent facilities are available. The second important challenge to note is the

Figure 2. The demonstration MapServer application for the African Marine Atlas at http://www.africanmarineatlas.net

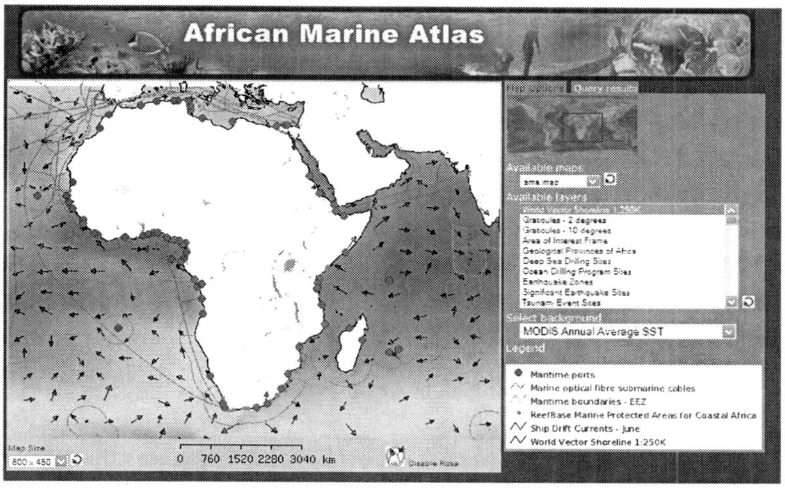

availability of data; the team was careful to use data that were in the public domain and to fully cite their use. Some data providers, however, were reluctant to share data or provide permission for their redistribution, in which case links to data sources (rather than actual data sets) have been made.

The Atlas, when moved from prototype to operational phase, will ultimately be of direct use to national institutions and many diverse user groups. The application of the data sets is dependent on the required end use, and at the moment, the relatively low resolution of most data in the atlas makes them most useful for regional and continental-scale applications.

The existing inventory of data sets in the atlas indicates gaps either in the knowledge base or in the availability of data in the public domain, and where efforts at repatriation or processing should be directed.

Future development of atlas products will aim at adding higher resolution data at national or local scales, thereby providing useful data for different applications and reaching a wider user group.

The future for the African Marine Atlas is bright; ODINAFRICA Project funding has been extended for a fourth phase of four years and partnerships with major African data providers and users, including the Large Marine Ecosystem Projects and the Nairobi Convention Clearinghouse Mechanism, have been greatly strengthened. Joint activities for the development of national data sets will add higher resolution content to the current holdings over the next two to three years and the IOC Project Office for IODE will continue to host the Pan African atlas web sites for the foreseeable future.

The African Marine Atlas, through the IODE program, is a participating member of the International Coastal Atlas Network (ICAN). Through this association the Atlas will be an important partner in broadening ICAN's reach and relevance on a truly international scale (Dwyer & Wright, 2008). The ICAN Technical Task Force, which is focussed on ways to globally integrate locally-maintained coastal atlases, has assisted with capacity development and sharing of best practices in implementing the web atlas. The African Marine Atlas will participate in the ICAN interoperability prototype as an additional node and the Atlas development team will continue to adopt the recommendations and best practices of ICAN.

ACKNOWLEDGMENT

The authors would like to thank the IOC Project Office for IODE, the African Coelacanth Ecosystem Program (ACEP), the United Nations Environment Program (UNEP) and all the editors of the African Marine Atlas: Selorm Ababio, A.K. Armah, Anis Diallo, Emmanuel Dovlo, Clousa Maueua, Desiderius Masalu, Harrison On'ganda, Zacharie Sohou and Mwangi Theuri, as well as Peter Pissierssens, Mika Odido and Murray Brown for their ongoing project support.

REFERENCES

Dwyer, N., & Wright, D. J. (2008). *Report of International Coastal Atlas Network Workshop 3 on Federated Coastal Atlases: Building on the Interoperable Approach*, European Environment Agency, Copenhagen, Denmark. Retrieved 29 July 2009, from the ICAN website: http://ican.science.oregonstate.edu/ican3_final_rpt

European Commission. (2009). *European Commission Joint Research Center African Marine Information System (AMIS)*. Retrieved 22 July 2009, from the AMIS website http://amis.jrc.ec.europa.eu/index.php

Intergovernmental Oceanographic Commission. (2003). *ODINAFRICA III: An Integrated Ocean Observation and Service Network for Africa. Project Proposal*. Intergovernmental Oceanographic Commission (of UNESCO), Paris, France. Retrieved 29 July 2009, from the ODINAFRICA web site: http://www.odinafrica.org/

Scott, L., & Brown, M. (2009). African Marine Atlas. In Odido, M & Mazzilli, S. (Eds.), *African Oceans and Coasts*. Nairobi, Kenya: United Nations Educational, Scientific and Cultural Organization (UNESCO).

Scott, L.E.P, Ababio, S., Armah, A.K., Brown, M., Diallo, A., Dovlo, E.K., et al. (2009). *The African Marine Atlas*.

UNEP. (2008). *Africa: Atlas of Our Changing Environment*. Nairobi, Kenya. Division of Early Warning and Assessment (DEWA). United Nations Environment Program (UNEP).

United Nations Environment Program. (2009). *United Nations Environment Program GEO Data Portal*. Retrieved 22 July 2009, from the UNEP web site: http://geodata.grid.unep.ch

KEY TERMS AND DEFINITIONS

Coastal Web Atlas: A collection of digital maps and datasets with supplementary tables, illustrations and information that systematically illustrate the coast, oftentimes with cartographic and decision support tools, all of which are accessible via the Internet.

ACEP: African Coelacanth Ecosystem Program.

Clearinghouse: An indexed repository of data or information available for download.

IOC/UNESCO: Intergovernmental Oceanographic Commission of the United Nations Educational, Scientific and Cultural Organization.

Metadata: Metadata in the geographical domain is structured information on a dataset, which helps the data owner to document and catalogue the data, whilst helping a data user to understand the content and fitness for use of a dataset.

NODC: National Oceanographic Data Center.

ODINAFRICA: The Ocean Data and Information Network for Africa.

UNEP: United Nations Environment Program.

Chapter 12
Caribbean

Sean Padmanabhan
Institute of Marine Affairs, Republic of Trinidad & Tobago, West Indies

ABSTRACT

The marine environment of the Caribbean Region currently faces several threats to its stability and sustainability. Mitigation of these threats requires an integrated management approach at a regional level using the best available data and information. Unfortunately, there are several shortcomings in the collection and management of marine data throughout the region. One solution is in the use and application of web atlases. In this light, the online Caribbean Marine Atlas (CMA) allows users to identify, locate, create and disseminate necessary marine and coastal information. As part of the International Coastal Atlas Network (ICAN), the CMA also promotes interoperability among atlases through practice and adherence to standards in web mapping, data and metadata generation and maintenance. This chapter deals with several of the steps in developing the CMA, as well as the challenges met and overcome throughout its growth and evolution.

INTRODUCTION

The Caribbean as a region (Figure 1) is made up of the Caribbean Sea and contains a number of coastal and water-locked states and territories each with its own distinct characteristics, concerns and cultures. Among an array of diverse geographical, environmental, social, economical and political threads, there is also a common factor that helps bind the fabric that is the Caribbean – the inalienable waters of the Caribbean Sea and the Western Atlantic Ocean upon which each coastal area in the region relies and by which they are all connected.

Within the Caribbean, coastal and marine environments are under increasing pressures from a combination of threats. First off is the challenge of dealing with (extreme) environmental events to which much of the Caribbean is exposed. These include hurricanes, tsunamis, fish kills and red tides that have hazardous and sometime disastrous

DOI: 10.4018/978-1-61520-815-9.ch012

Copyright © 2011, IGI Global. Copying or distributing in print or electronic forms without written permission of IGI Global is prohibited.

Figure 1. Index map of the Caribbean region

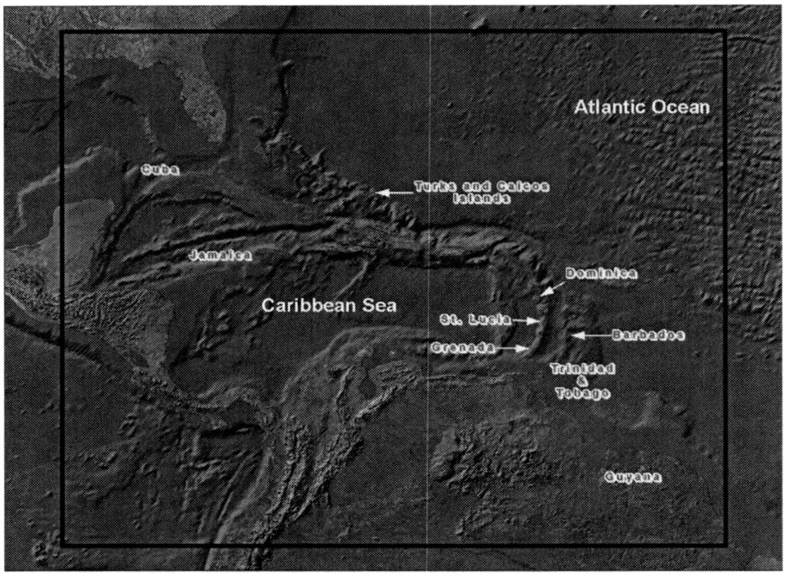

effects on human livelihood and life. Additionally, pollution, coastal erosion, biodiversity and habitat loss, over-exploitation, sea level rise and climate variability are just some of the other activities and factors affecting the coasts and waters of the region and their ecosystems. It is in managing and mitigating the effects of these events that the need to understand these environments, their resilience and how they relate to maintaining the necessary ecosystem-based goods and services provided to the region, becomes evident and unquestionable.

The management of coastal and marine areas is intrinsically difficult since it involves a complex, multi-faceted environment for which data from observation and monitoring is limited and expensive, especially in the Caribbean. Moreover, coastal and ocean management usually involves multiple users, various responsible state agencies at different levels (e.g., national, provincial or local) and, in the case of shared seas, like in the Caribbean, interactions and relations with other nations. Given the geographic distribution and proximity of landmasses in the region, a complex interplay of intersecting and sometimes undefined jurisdictions develops in an area of dynamic living resources. Furthermore, almost all of the major threats in the Caribbean have a regional or sub-regional area of effect where the events and activities occurring in local or national waters will eventually affect neighboring areas (e.g., algal bloom). The successful sustainable development of the region, its coasts and waters is therefore based both on the premise that management of these areas must be a collaborative effort among all stakeholders and on the principles of integrated management of all activities occurring in or affecting coasts and oceans.

To effectively manage the coastal and ocean environments of the Caribbean at a regional or national level, facts and understanding of the ecosystems, habitats and their components are needed to provide a foundation of information upon which good decision-making may be firmly established. In order to build such a knowledge base, appropriate data and information on living and non-living resources, an environment's health, cycles, stresses and vulnerabilities must be collected, analyzed and distributed. It is here, in the gathering, administration, application and dissemination of useful coastal and marine information that a major challenge exists for the Caribbean region and its members. There is a limited range and amount of relevant and current data and information easily available to and proficiently used by the Caribbean region and/or its constituents. Also prevalent is an overall lack of appropriate tools, techniques and expertise needed to effect timely decision-making and adaptive management of coastal and marine areas. As a result, managers at both a regional or national level frequently have outdated or deficient information on the resources that they are charged with conserving and/or preserving and find it difficult to successfully perform their mandates and functions.

The major concern about data for the Caribbean is the fact that there are several inconsistencies and limitations surrounding their collection, management, access and delivery. Over time, data on the region or parts thereof have been collected by several parties, from local and state agencies to regional and international institutional bodies, from the public sector to the private sector and non-governmental organizations (NGOs). Many of these parties, particularly at a national/local level, have taken a more sectoral approach to data gathering due mostly to limitations of their resources and/or scope of concern. At the time of collection, each of these parties may have had disparate objectives, used assorted methods and tools and sampled at different locations – all leading to incongruent data sets that are difficult to integrate with each other and that do not allow for a preferred comprehensive, regional or ecosystem-based approach to management and decision making. Further to the problems surrounding the collection of data are those that deal with their management and use. Some of the daunting circumstances surrounding existing data sets that members of the Caribbean have to face include: lack of access to data; spatial or temporal data gaps; decentralization, data redundancy and versioning; limited expertise and administrative capacities; and, lack of quality controls, policies and standards for data and metadata. Each of these issues inexorably leads to the inefficient and ineffective delivery of necessary information to decision makers and managers and retards any efforts towards better management of the coastal and marine environments of the Caribbean.

To minimize and eliminate some of these barriers facing the successful management of coastal and marine environments, a well-organized and proficient information infrastructure that capitalizes on appropriate technologies such as the Internet, geographic information systems (GISs) as well as incorporates the principles underlying spatial data infrastructure (SDI) for adequate data acquisition, integration, management, analysis and dissemination (Caribbean Marine Atlas Team, 2009) is summarily warranted.

Evolution and Overview

Fortunately, the Caribbean region was not the first to have to deal with the many issues and limitations surrounding data. Others have designed and developed various tools and methods to address many of the concerns mentioned above as well as those centering on other issues such as data interoperability and SDI at national, regional and global levels. One of the more effective solutions to these problems was the design and development of web atlases.

Building upon various existing and evolving technologies such as the Internet, GIS, remote sensing, distributable databases, Web Mapping Services (WMS) of the Open Geospatial Consortium (OGC), and content management systems (CMSs), a web atlas was an integrated and comprehensive approach to managing and solving many of the (spatial) data needs expressed above. In this case, a coastal/marine atlas for the Caribbean should "not only help identify and locate important data resources, but also create and disseminate information, principles and standards needed to ensure that future datasets be made accessible to and used by all who need them" (Caribbean Marine Atlas Team, 2009, p.

1). It was within this context of providing more effective means of data production, integration, access, sharing, visualization and interpretation that a web-based coastal/marine atlas for the Caribbean region was conceptualized.

Taking a cue from the already underway African Marine Atlas (AMA) project (Chapter 11), marine data managers and specialists from several countries within the region attended a workshop in 2007 that scoped and established the terms of reference for developing a Caribbean Marine Atlas (CMA). This workshop was facilitated through the umbrella purview and mandate of IOCARIBE – the Sub-Commission for the Caribbean and Adjacent Regions of the Intergovernmental Oceanographic Commission (IOC) of the United Nations Educational, Scientific and Cultural Organization (UNESCO).

Coming out of the workshop, there was a general consensus that many of the countries of the region shared similar concerns on several issues relative to the current and future state of marine habitats and ecosystems as well as the shortcomings surrounding the availability and management of marine information and data. In the spirit of addressing these common priorities, the purpose of the CMA was defined as one "to identify, collect and organize available spatial datasets into an atlas of environmental themes for the Caribbean region" that "will contribute to the sustainable development and integrated management of marine and coastal areas in the region" (Caribbean Marine Atlas Team, 2009, p. 1). Basic features and objectives of the CMA were also determined at the workshop and helped to define and focus the development and outcomes of the atlas (Intergovernmental Oceanographic Commission, 2007).

At the conclusion of the workshop it was decided to develop a CMA Pilot Project. This Pilot Project was to be made up of a limited/controlled number of fully committed participating member states. Fostered within the framework of the Ocean Data and Information Network for the Caribbean and South America (ODINCARSA) project and under the sponsorship of UNESCO/IOC's International Oceanographic Data and Information Exchange (IODE) and Integrated Coastal Area Management (ICAM) Programmes, the CMA Pilot Project was born as a joint initiative among eight countries in the Caribbean region: Barbados, Cuba, Dominica, Guyana, Jamaica, St Lucia, Trinidad & Tobago and Turks & Caicos (Figure 1). Suitable representatives from each of these countries were nominated to and formed a CMA Pilot Project Team, which was charged with bringing the CMA to life. With the aide and guidance of IOCARIBE and IODE, the CMA Pilot Project received fourteen months of funding from the Government of Flanders as a Small Scale Project Grant Proposal within the context of the Flanders-UNESCO Science Trust Fund (FUST). Over this period, these funds provided the resources and assisted in meeting the requirements necessary for developing the CMA Pilot Project. One of the initial resources and outputs at the conclusion of the 2007 workshop was the establishment of a CMA website (http://www.caribbeanmarineatlas.org). This website served as the launching pad for the CMA to its stakeholders and users and provided information and updates on the background and status of the atlas.

Concept, Design and Development

The lead concepts of atlas development that were adopted by the CMA came from a 2007 International Coastal Atlas Network (ICAN) Coastal Web Atlas (CWA) workshop. Addressing issues such as atlas design and usability, visual appeal, use of technology, flexible design and meeting user needs (O'Dea *et al.*, 2007), the CWA workshop provided the guidelines for creating the CMA. Consequently, the next step was to clearly evaluate and define specific targets of the pilot project within the purpose of the CMA, and how best to achieve these targets through atlas design and function.

The CMA first considered a list of critical questions posed during the 2007 CWA workshop that were meant to assist in evaluating, synergizing and focusing the goals, objectives and design of an atlas. These questions covered a wide range of topics such as the purpose of the atlas, data, functionality, skills required, required resources and standards (O'Dea *et al*. 2007). This process allowed several general parameters of the CMA to be defined as follows:

- **Aim of the CMA:** To provide spatial-relatable, environmentally relevant data and information products on the marine and coastal environment of the Caribbean through an interactive, web-based application that can be employed by users anywhere and facilitate informed decision-making and resource management.
- **Intended Audience of the CMA:** Concerned scientists, coastal zone management practitioners and natural resource managers, any of whom may be from local or national agencies to regional or international bodies.
- **Data Range/Themes of the CMA:** Coastal habitats, fisheries, environmental quality, climate variability and sea level rise, oceanography, as well as socio-economic aspects.
- **Functions of the CMA:** Identify and view relevant data and metadata; generate products (maps) composed of multiple data elements from a web-based environment; query and investigate data sets; and download data for further analysis.

Many other relevant CWA questions were also addressed and answered as the CMA developed to best meet the desired features of any coastal web atlas, such as map and data display, layer lists, atlas tools, metadata, etc. (Intergovernmental Oceanographic Commission, 2007).

By following the CWA guidelines, a framework and context to actualize the CMA was also established and led to a series of steps for atlas development. These steps covered matters such as establishing topic outlines and an area of interest, gathering, formatting and managing data and metadata, and designing, publishing and disseminating the atlas and its products via the Internet (Intergovernmental Oceanographic Commission, 2007).

The subsequent responsibility to build and develop the atlas itself based on these steps fell to its member participants — the CMA Team. Having recognized the limited and varied expertise available in the region, one of the first, most identifiable and important needs of the project was the training of its participants. One of the goals of such training centered on long-term and sustainable capacity building for member countries by first improving the technical skills and proficiencies of the participants that were necessary to develop the atlas. With an already established programme via Ocean Teacher (http://www.oceanteacher.org) and available financial and infrastructure resources, IODE provided the training to the CMA Team. Led by experts in marine data management and coastal atlas design, this IODE training effected the major aims and objectives of the CMA over a series of three technical workshops:

1. Oceanographic Data Management Course provided an introduction to data managers on several ocean data sources and the various formats that are commonly used and readily available. It also guided them on how best to explore, analyze, manipulate and integrate this type of data into useable products via GIS tools and applications.
2. Data Mining and Assembly Course aimed at familiarizing Caribbean marine data managers with the methodology required to locate, process and generate products from the themed data identified as critical to the management of the maritime space

and coastal resources of the region within the CMA framework.

3. MapServer Course brought together all of the concepts presented and data products developed at earlier workshops, as well as introduced map server applications for Windows to develop interactive web-based atlases.

Inherent to and concurrent with this training was the process of identifying and meeting the immediate needs of the project. In doing so, both the needs of the CMA and the steps necessary for its development were broken into two major categories: Data and System.

The Caribbean Marine Atlas: Data

Data Content

Using an established data classification system introduced by IODE at the 2007 CMA workshop, data content was to be separated into the following environmental themes: geosphere, hydrosphere, biosphere, atmosphere, human environment, and base maps (Intergovernmental Oceanographic Commission, 2007). While these themes helped organize data and products, the information relevant to the region that each was to contain was still outstanding. At the first CMA workshop in 2007, participants conducted an assessment of the major threats to coastal and marine areas at national and regional levels. This evaluation helped identify priority concerns and specific issues that the CMA looked to address and on which it sought to provide relevant coastal and marine information. It included topics such as coastal habitats, unsustainable resource exploitation (e.g., overfishing), environmental quality (e.g., land-based sources of pollution), and natural hazards (e.g., hurricanes, sea level rise, other vulnerabilities to climate change, etc.) (Intergovernmental Oceanographic Commission, 2007).

At the first technical training course dealing with data management for the CMA, the data needs based on these topics were re-evaluated and broken down into several more specific datasets for each of the six data themes. To maximize the resources and time available for the pilot project, the order in which these datasets were to be added to the CMA was then prioritized based on necessity as well as availability at both a regional and national levels.

Area of Interest

Establishing an area of interest (AOI) in an atlas served two major functions. An AOI centered the overall attention and focus of the atlas and its users to a specific region and separated itself from other phenomena and events outside its scope. Concurrently, an AOI assisted in spatially eliminating unnecessary or inapplicable data. Datasets that were completely outside the range of the AOI did not need to be considered, while those that were larger than but inclusive of the AOI were to be scaled to suit. Additional emphasis on identifying, prioritizing and filling data gaps for an AOI was now possible and manageable. During the first technical workshop, the bounding coordinates of an AOI were determined for the CMA by the Pilot Project Team, which was geared to covering the issues and concerns of both the individual member states and the region as a whole (Figure 2).

Data Sourcing and Gathering

The CMA employed the process of data mining in acquiring the data that made up the atlas. This entailed researching, investigating and gathering information from various sources of existing collections of data often found in different formats and scales (Intergovernmental Oceanographic Commission, 2007). The parameters of source and/or scales of information then allowed all of the data elements that were to be served up by the CMA to be segregated into three main groups.

Figure 2. Area of interest and bounding coordinates of the CMA

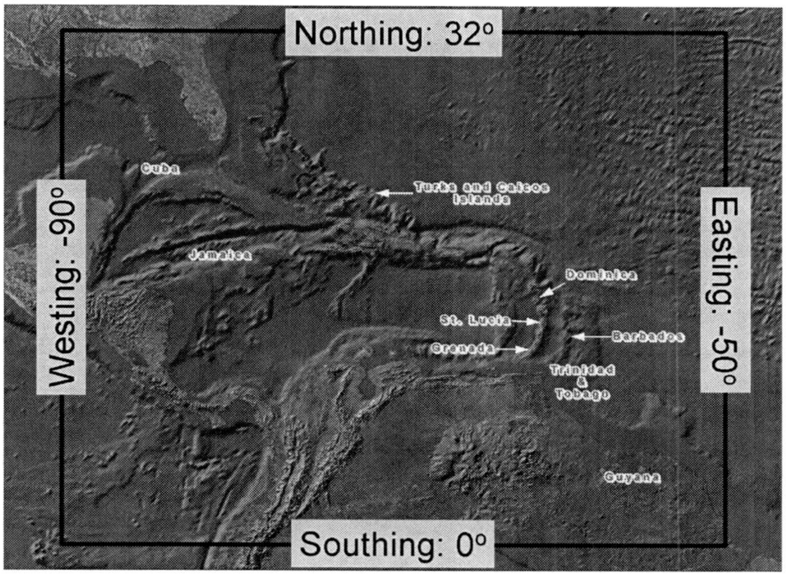

Regional-Scale Data Sources

Covering the complete atlas AOI, the majority of relevant data at a regional scale was derived from smaller scale data sets (i.e. continental or global) that were readily accessible and freely available. Much of this data provided the information for the base map layers of the atlas and was harvested from several agencies and organizations with a regional and/or international purview (Table 1).

Other regional data collections were also identified by the CMA Team, but were not as readily available. Some of these required direct permissions from their owners if they were to be used by third parties such as the CMA. While the CMA obtained owner-authorization to use a few of these datasets, it planned to continuously seek out and request approval for others over time.

National-Scale Data Sources

One of the direct tasks of the CMA team was to source and provide local and national scale datasets that were in their respective participating States' possession or national collections. Once this data was examined and reviewed, it was then possible to identify what data had to be found outside of the member States to complete a collection or theme component at a national or regional scale. As mentioned before, much of the national data collection activities within the region were conducted on a sectoral approach with singular objectives and applicable time frames in mind. As a result data within state collections contained spatial and/or temporal gaps when put in the context of a larger, national or regional scale. Combining common datasets from various local/national collections helped fill some of these gaps. One example of this was in the collection and compilation of mangrove data, where in several countries there were organizations that maintained information on local/national mangroves, with each holding at least some viable information on its own. With representatives from each country going through their own data mining exercise, these various sources of mangrove information were identified and the data itself collected and standardized in the CMA. When combined, these data sets formed a more complete picture of national mangrove distribution, which when fed into the CMA pro-

Table 1. Regional scale data sources for the CMA

Organization/Product	Source	Layers
General Bathymetric Chart of the Oceans (GEBCO)	http:///www.gebco.net	• Global Bathymetric Contours • World Coastlines
National Geospatial Intelligence Agency, World Vector Shoreline	http://rimmer.ngdc.noaa.gov/mgg/coast/getcoast.html	• World Vector Shoreline
Flanders Marine Institute (VLIZ)	http://www.vliz.be/vmdcdata/marbound/	• World Exclusive Economic Zones (EEZs) • World Maritime Boundaries
UNESCO International Hydrological Programme (IHP)	http://typo38.unesco.org/index.php?id=240	• Sea Area Names
NOAA National Geophysical Data Center (NGDC)	http://www.ngdc.noaa.gov/mgg/global/global.html	• Global Relief Model, ETOPO • Sediment Thickness
NASA JPL Blue Marble	http://onearth.jpl.nasa.gov/	• Blue Marble Imagery
United States Geological Survey (USGS)	http://www.usgs.gov/pubprod/data.html#data	• Faults & Plate Boundaries • Historic Events - Earthquakes
NOAA National Oceanographic Data Centre (NODC) World Ocean Database 2005 (WOD05)	http://www.nodc.noaa.gov/OC5/WOD05/pr_wod05.html	• Major Nutrients (Nitrate, Oxygen, Phosphate, Silicate, Chlorophyll) • Sea Surface Temperature • Salinity
The Nature Conservancy (TNC)	Acquired with permission	• Mangroves • RAMSAR Sites
World Resources Institute (WRI)	http://www.wri.org/reefsatrisk/ Acquired with permission	• Coral Reef Locations, Health & Threat Levels

vided a better overall representation of mangrove distribution for the region.

Virtual Data Sources

One recent technology to emerge and provide direct benefit to web atlases has been that of online map and feature services. Via the Internet, they offered data, map and imagery solutions to users "virtually" – that is to say, much like a web page using hypertext transfer protocol (HTTP), the data or image itself was not copied onto a users machine as a separate entity, but fully transmitted and delivered to the user and integrated into the relevant application/software, like a browser. The two most common protocols for serving up virtual data employed by the CMA were Web Map Services (WMS), which provided users with georeferenced map imagery and Web Feature Services (WFS) which served up vector features (e.g., shapefiles). The CMA sourced and integrated content from WMSs such as those provided by NASA and USGS (e.g., Blue Marble), which provided georeferenced imagery that was used to produce dynamic maps and backgrounds.

Data Formatting and Standardization

Much of the data sourced for the CMA came in a variety of formats. In collecting data on a range of topics and from several sources, it was inevitable that no one data type or format catered for all the data needs of or intended applications by its users. Consequently, with several means of packaging and distributing data to users, it was necessary to provide some level of standardization on these datasets in order to meet the specific goals and objectives of the CMA. There were four main areas of standardization that were addressed: data scale/coverage, data reference/projection, data format/conversion and metadata.

Data Scale/Coverage

As mentioned before, the intention of the CMA was to provide relevant data and information on the region. With this in mind, data relevant to the AOI was the focus and consequently, integrating datasets created at various scales raised some problems. A major issue surrounding scale for the CMA was adopting and adapting information from sources that were originally geared for either much smaller or much larger scales than the AOI. For example, oceanographic features such as bathymetry or sea surface temperature data were derived from sources that used a world scale and worked well at a regional level for the CMA, but, when looking at particular areas/countries at a sub-regional level, the detail or resolution that a user may require was sometimes reduced and lacking. Alternatively, country-specific data that were originally geared for larger scales, like coastal protection structures, became obscure and even unnoticeable at a regional scale.

Data from the multiple sources used in the CMA tended to be voluminous and varied. Often their intended use was specific to a particular user, time or place such that vertical and/or horizontal sharing among others and integration into other collective systems were limited. Certain steps in the design and execution of the CMA were thus included to deal with the range of scales (and accuracies) of data that were incorporated from several global, regional and local sources into the CMA. Primarily, the intended uses, user needs and outputs of the CMA were considered. From this, data needs and data standards were derived to set up a measure of quality control when integrating datasets such that a prescribed scale based on the AOI was set for the regional view of data to best avoid any misrepresentation or misinterpretation of information. Where possible, data was therefore re-sampled from source information to best fit this prescribed scale for the CMA, particularly in the case of grids and other raster datasets.

Data Reference/Projection

Within the CMA, no dataset was projected. Given the scale of the CMA's AOI, as well as the numerous projections used by different countries with the region, it was felt that offering data in a simple Geographic Coordinate System, which used the internationally accepted World Geodetic System (WGS) 1984 datum, best fulfilled the goals and intent of the CMA. Additionally, this approach also facilitated integration among all layers within the CMA as well as interoperability among the CMA, other sources such as WMSs and WFSs and other atlases. If and when necessary, users who downloaded data from the CMA had the option to apply specific projections on their local machine for their own further investigation or analysis.

Data Format/Conversion

Data for the CMA was mined from a variety of sources, which used a range of formats, some of which were not CMA ready to be input into the atlas. To facilitate the effective integration of data, the CMA Team was trained in and utilized a variety of tools and applications to make the necessary changes to such datasets, be it format conversions (e.g., HDF/CDF to grid and imagery), or re-scaling and re-sampling datasets. In keeping with the mission of the CMA, most of these tools used were open-source and/or freely available (directly or indirectly) through the Ocean Teacher portal of the UNESCO-IOC International Oceanographic Data and Information Exchange (IODE) programme.

Concurrently, the CMA Team chose the following options for formatting and standardizing the data, which it offered to users:

- vector data were converted to and available as shapefiles;
- raster data were converted to and available as ASCII grids; and

- imagery was georeferenced, converted to and available as geoTIFFs.

The objective here was to have all the data within the CMA in standard GIS formats or layers that were readily accessible, manageable and useable by the CMA system. These formats also met the accepted standards and practices of the wider geospatial community (e.g., OGC) and therefore facilitate interoperability between the CMA, its users and other spatial systems.

Metadata Standards

Inadequate metadata and inconsistent metadata standards existed for many of the datasets pertaining to the CMA. If the CMA was to present a more complete and coherent means of information delivery, especially in the datasets that it served up to users, standardized metadata was essential. It was therefore crucial that useful and relevant metadata be utilized and/or generated by the CMA to present sufficient descriptions on data to users. This in turn facilitated some of the intended functions and features of the CMA user interface, such as exploratory and search capabilities for data, as well as increased the interoperability of the atlas.

While the CMA provided a direct web-link to the metadata provided by the owner/publisher of a dataset that it had not altered in any way, it was responsible for generating metadata for those data layers, which it either created itself or derived from another source. While there were many international standards for metadata (e.g., the US Federal Geographic Data Committee or FGDC) available to the CMA, they consisted of numerous elements of information and often required a level of detail and attention that was very time-consuming. Weighed against the unacceptable options of providing incomplete or no metadata, a compromise arose. Using the examples and lesson learnt from other atlases, in particular recommendations from the Marine Irish Digital Atlas (MIDA) and ICAN, as well as an assessment of the metadata standards used by others for its unaltered datasets, the CMA decided upon and produced two forms of metadata based upon the International Standardization Organization (ISO) 19115 standard.

The first, called "abstract metadata", contained very basic descriptive information on a dataset and was the minimum requirement for any dataset, whether created by the CMA or not. It was a simple text file containing five parameters: title, abstract, extent, owner, and publication data. The primary purpose of abstract metadata was to quickly and succinctly describe a dataset to a user upon exploration. It was not to be used in the search capabilities of the CMA nor provide any interoperable functionality. In cases of data that were not created by the CMA and for which it was not responsible, a link to the metadata record of the original owners or creators was included.

The second form of metadata, called "discovery metadata" was based on the twenty core elements as prescribed by the ISO 19115 standard. These core elements provided the foundation upon which the CMA generated an expanded number of descriptive parameters and a greater level of detail about each dataset for which it was responsible. Additionally, in following the ISO 19115 direction, the CMA then had the ability to produce ISO 19139 eXtensible markup language (XML) tags which allowed for digital integration of its data and metadata. This in turn facilitated better search potential and functionality within and out of the atlas as well as provide for interoperability, data sharing and distribution. By adopting these parameters and applying these standards, the CMA not only ensured that proper records were maintained for its data collection, but also set itself apart from the convention of leaving metadata unattended and established itself as a regional example of proper practice and standard from which others may benefit. In this case, the metadata record contained multiple data parameters as shown in the Appendix.

Data Management

After all the datasets, layers and metadata were gathered, processed and formatted, they were then arranged in a database based on the various CMA data themes and placed on a file server. A file transfer protocol (FTP) site was established and allowed the CMA Team access to the database via the Internet for data update and management from any location. Using simple, freely available FTP software, the CMA Team was able to secure the exchange of files via a username and password, which limited access and changes to the files at the FTP location. Concurrently, FTP was the means by which files from the CMA were served up from the file server to users. Where available, CMA users could locate and download the appropriate dataset onto their own machines. Alternatively, the atlas itself ran on a Hyper Text Transfer Protocol (HTTP) that allowed the transfer of files from a Web or Map server onto a browser in order to view a Web page that is on the Internet. Although HTTP was also an option for file delivery from the CMA to users and allowed for direct links to files through the atlas web page, the separate FTP site allowed for files to be securely managed and available independent of the web/atlas site, which was particularly advantageous during atlas upgrades when the web site and/or interface was temporarily offline and inaccessible to users. Essentially, both HTTP and FTP were separately employed by the CMA, with the former governing the main functions and features of the atlas in pulling data from a map server, and the latter being the primary means of direct delivery of data and information from the CMA to its users.

The Caribbean Marine Atlas: System

To meet its goals and objectives, the CMA relied upon a system that not only presented the data and information organized in its database to its users, but also offered various utilities and tools that assisted in the discovery, exploration and analysis of data in a multi-functional and effective environment. The system employed by the CMA was broken into two categories: the back-end – the various (sub-)components that made up the system and allow it to function; and the front-end – the interface that visualized the information to users and communicated their requests. During development, the functions and utility of the front-end user interface were first considered and the appropriate back-end technologies were then adopted to meet these needs.

Front-End Interface

As mentioned before, the CMA incorporated many features of a web-based digital atlas. Some of these features allowed users to both explore and access relevant information and also provided supplementary services and functions. The main features of the CMA front-end interface were as follows (Intergovernmental Oceanographic Commission, 2007):

- Map display area:
  - Interactive/Dynamic
  - Overview map of the area of interest
  - Scale bar
  - Geographic coordinates (spatial reference)
- Data display area:
  - Both vector and raster data
  - User control (turn on/off layers)
  - Multiple layers viewed together
  - Data grouped/viewed by themes
- Legend Area:
  - Symbols colours and palettes used to represent data
- Layer List:
  - Contained all the layers available to users
- Atlas tools:
  - Zoom to another area of interest
  - Search for specific datasets
  - Advanced queries or spatial analysis

- Attribute data:
  - Identify tool
  - Table results
- Metadata:
  - Standardized basic and/or advanced
- Information:
  - Descriptions of related topics and issues
  - Links to key web sites, organizations, documents
  - Resources for specific users

These basic tools, functions and services of the CMA front-end interface now guided the CMA Team in building a back-end system that facilitated as many of them as possible.

Back-End System

According to O'Dea *et al.* (2007), several factors such as available financial and technical resources, audience needs and limitations, system architecture, web design and content management were taken into account where powerful server and software technology were warranted and used to support the hosting of CWAs. In developing the back-end system of the CMA, these factors and the various options/solutions available for each were considered and weighed against the costs and benefits of similar decisions made for other CWAs (such as the AMA). In the end, the CMA Team made the following choices that best suited both the circumstances in which the CMA was to develop/exist as well as the features and functions that it was to perform:

- The CMA was hosted by the IODE located in Oostende, Belgium, whose machines and network ran mainly on a Microsoft Server platform and whose infrastructure best facilitated the network and server needs of the CMA with greater distribution and bandwidth capability;
- The open source Apache Web Server was used;
- The open source University of Minnesota MapServer package was used as the online GIS and WMS;
- Other than the functions provided by MapServer, the CMA did not rely upon another Database Management System and where necessary it provided direct access to data files via FTP;
- The CMA was accessible through HTTP and a variety of browsers;
- The CMA best adopted and complied with international standards for both metadata creation with ISO 19115/19139 and web mapping using the Open Geospatial Consortium (OGC) web mapping standards.

Once the technologies and applications that ran the CMA were established, the next step was to design and customize them to meet user needs and provide the desired features and functions of the CMA. Using MapServer's scripts, all the relevant datasets and sources for the CMA were input into various mapfiles that controlled data use and utility. Next, simple hypertext markup language (HTML) and Java scripts and files were written to define the look, parameters and controls of the CMA available to users through their browsers. Together with the front-end system, these back-end implementations best incorporated the datasets and information, tools, presentation, interface, features and functions intended for the CMA into a useable application that could now be uploaded and distributed via the Internet.

The CMA Prototype

Given the potential number and volume of data that the CMA covered, the time consuming nature of data mining, procurement, formatting and integration, together with the limitations of the CMA Team in terms of human resources, potentials and capacities, and lessons learnt from previous atlas exercises like the AMA, the CMA decided that a smaller product created and completed in a given timeframe was of greater benefit and interest to all. With that in mind, a CMA prototype was initially developed as a more of a realistic and effective deliverable of the Pilot Project.

In essence, the prototype was the foundation upon which the CMA was to grow and expand. Accordingly, data for the prototype was prioritized by availability, use restrictions (if any), readiness (formatting/standardization required), and utility to the project. Some of these priority datasets consisted of coastlines and topography for the base maps; physical and chemical oceanographic parameters for the hydrosphere theme; coral reef and wetlands for the biosphere theme; earthquake incidents for the geosphere theme; precipitation and hurricane records for the atmosphere theme; and political and maritime boundaries for the human environment theme. Concurrently, the target functions of the prototype were a subset of the overall CMA, with basic atlas features such as map display, zoom, layer selection, standardized legends and metadata retrieval taking precedence. Other functions such as data search and query, analysis and other tools were to be developed after the prototype was completed as part of the evolution of the CMA.

As of 2009, the prototype was still incomplete and was being hosted on a closed server for testing. All of the processed data was incorporated into the mapping application and the prototype functioned as a simple viewer for the assembled datasets. Finalization of the prototype by the CMA Team was to be facilitated through the delivery of outstanding datasets, completion of metadata records and updates to Minnesota MapServer files and scripts. To best describe the state of the CMA prototype, Figure 3 shows a sample image taken from the interface.

Limitations and Challenges of the CMA

From conception to development, the overall CMA initiative and its prototype encountered a range of challenges. As in many other CWA stories, these problems covered a wide spectrum ranging from limited human resources and expertise to data inadequacies and restrictions. Some of these issues were anticipated, prepared for and pre-empted,

Figure 3. CMA prototype showing point and line vector information together with WMS image background, layout and interface including the map window, layer control and map control.

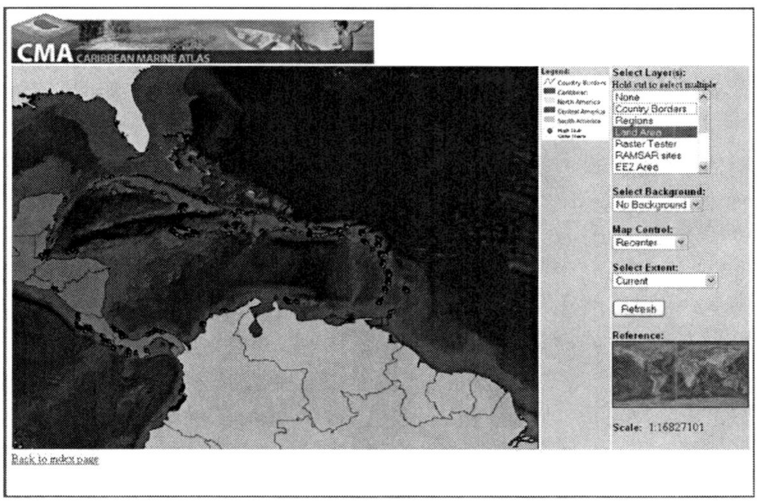

while others that were unavoidable were dealt with on the fly. This section highlights some of the main challenges of bringing the CMA to life in relation to data and systems development.

Data

Having experienced the processes of data mining, preparation, integration and management, there were several issues that the CMA Team faced and solved and some that still persist. Keeping in mind that the CMA was to grow with time, it was still the goal of the team to launch the CMA (prototype) with an acceptable number and range of datasets. At minimum, this initial offering by the CMA was to provide some data for each of the proposed themes. With this in mind, there were four main data hurdles that the CMA team confronted: data shortages, data sharing restrictions, data quality, and metadata issues.

Data Shortages: For some of the desired categories, data at a regional level was not readily found or obtained. After some discussion, the CMA Team decided on a three-tiered approach to this problem.

- Consistently search for data, both new and updated, where missing datasets were given the highest priority in this search;
- For those categories where regional data were unavailable, but local data existed for most of the participating countries, datasets were combined as best as possible to generate a temporary regional offering. It was also noted that such an offering may have contained spatial and/or temporal gaps that were to be identified in any respective metadata;
- In developing its network of partner organizations and affiliations, the CMA identified and highlighted many of these data gaps as prioritized areas of study and research in the hope that data produced from any future exercise find its way to the CMA.

This three-tiered solution generally accommodated and addressed the issue of data shortages, where one or any combination of the three recommended an effective means to successfully yield useable data.

Data-Sharing Restrictions: With the team collectively searching for data via the Internet, as well as approaching local, regional and international organizations and agencies for assistance, data collection centered on finding information that was freely available. When data was sourced, this stipulation raised two main issues dealing with data restrictions: what to do if the data had a financial cost attached and, how to accommodate for restrictions placed on the data by its creators/owners.

When faced with the choice of paying for data or having no data whatsoever, there was much deliberation among the CMA Team. Ultimately, the approach taken by the CMA to overcome this matter attempted to stick to the essence and principles upon which the CMA was conceived and developed – a spirit of cooperation, capacity building and information exchange. Where necessary, the CMA established a formal arrangement with these data providers via either a Memorandum of Understanding and/or Licensing Agreement. In this manner, the CMA initially approached the owners of data for sale and offered to act more as data promoters rather than redistributors or retailers. Since the CMA did not want to have any charges or fees to users associated with any of its content, it agreed to list or display these datasets either via a web link or WMS/WFS, where the data itself never left the owners.

Such an arrangement was also applied to situations where the restriction on the data was not necessarily associated with a cost, but with use and distribution. Several public sector entities and NGOs that owned/held datasets relevant to the CMA were willing to share their information but only under certain conditions. These conditions to deal with data restrictions are summarized as follows:

- Provider gives CMA the data to use as desired and without restrictions; users have complete access to view and download data via CMA;
- Provider gives CMA the data but CMA can only list and display the data and/or images; users cannot obtain/download files via CMA but may go to provider for actual dataset;
- Provider does not give CMA the data, but CMA can list the data as a layer and allows its display via WMS/WFS; CMA users must go to provider for actual dataset;
- Provider does not grant any access to data through CMA, but CMA lists the data and provides a link; CMA users must go to provider for actual dataset.

Therefore, an agreement between the owners and the CMA was established such that data were not to be redistributed to or downloaded by CMA users, but made available for listing or viewing only, with the appropriate credit and links back to the owners. One example of such an agreement was between the CMA and World Resources Institute (WRI), which held comprehensive datasets on coral reef distribution and health throughout the Caribbean. With the permission of WRI, the CMA included its reef information in the atlas for viewing, but provided credit and links back to WRI for anyone wanting to obtain the data themselves.

This approach allowed the CMA to not only provide data directly from its file servers, but also act as a link to other sources of information or as a shopping window for users where it promoted/advertised datasets to a wider market who may in turn go to the owners themselves to purchase or request the data. Failing this type of arrangement, the CMA's fallbacks were to wait for another freely available source of the data, generate the data itself as best as possible or budget for the purchase of data at a later stage.

Data Quality: In general, the CMA took a practical approach when addressing the issue of variable qualities of data with particular respect to formats, standards and scale. Recognizing the inevitability of integrating datasets of lower quality, which were either less accurate or incomplete in some way, the CMA Team incorporated both proactive and reactive mechanisms to deal with potential issues.

Implementing a system of quality control during data mining was primary in preserving the overall integrity and utility of the atlas when adding data. It was in scrutinizing data, data sources and metadata records that the team was able to predetermine and select what datasets met the criteria and standards of the CMA before incorporating them into the atlas and offering them to users.

Alternatively, several datasets that were harvested for the CMA were not immediately integrated and required varying degrees of additional preparation and processing. Some of the more common problems regarding these datasets were that data were not GIS ready. To resolve some of these issues, the CMA Team spent much time on data assimilation and processing. Using the data formatting and conversion tools received from the IODE Ocean Teacher technical workshops, as well as applying OGC standards for web mapping, several of the datasets mined for the CMA were converted and standardized into GIS-ready formats (e.g., shapefiles, grids, etc.) that were then uploaded into the system. It should be noted that while this step in assimilating data into the CMA was most productive and effective in yielding relevant datasets, it remained however extremely time consuming.

In other cases, where datasets to be used in the CMA had inappropriate or incomparable source data scales such that re-sampling was not a viable option and representational accuracy and potential analysis were affected and/or limited, a two-tiered approach was taken: national versus regional levels of data and update of metadata. Datasets that were only relevant to smaller areas at a sectoral (national or local) level and that were not critical to the regional aspects of the CMA were to be handed down to and incorporated into the development of a national marine atlas for an individual country (described below). As these national atlases were to fall under the umbrella structure and protocols of the CMA and eventually be linked or amalgamated into the CMA, there was no need to discard or ignore these datasets and their utility and potential was maintained. Secondly, for all datasets, which presented a continued issue with scale and/or accuracy, but, were still critical to and included in the CMA, their metadata records were adjusted accordingly and as necessary to clearly state the specifics of their incompatibility and overall quality to any potential user.

Metadata: Due to the direct relationship between data and metadata, any problems or limitations to data were usually accompanied with some metadata issues. One of the goals of the CMA was to provide comprehensive and accurate data and information to users where appropriate metadata was a critical component. Additionally, metadata provided the foundation upon which several of the potential tools and functions of the CMA, such as data searches, were directly dependent.

On several occasions while data mining, the CMA Team encountered the situation where there was little to no metadata attached to or associated with a dataset that was to be included in the atlas. Given the various categories of data incorporated into the CMA, and based on their ownership and accompanying restrictions, the CMA Team had to ensure the validity of the metadata provided to users. It was decided that for all data listed in the CMA, an abstract metadata record was to be created by the CMA Team to allow for an element of metadata standardization no matter what the data source. For those datasets that lacked sufficient metadata, the following approach was applied across the different categories of data to create discovery metadata records:

- CMA generated and owned the data – CMA was responsible for creating metadata;
- CMA obtained the data from provider with/without restrictions – CMA was responsible for creating/updating metadata;
- CMA did not obtain the data from provider but either listed it via a WMS/WFS or provided a link to where it was found – CMA was not responsible for any metadata assuming an adequate metadata record already existed;

To solve the problems of absence of good or any metadata, the CMA Team, now responsible for creating/updating the metadata records, often conducted some research into the dataset itself. In many cases, digging a little deeper and tracing the origins of datasets and methods of data generation, yielded the information necessary to create adequate and appropriate standardized (discovery) metadata.

System

Challenges to the CMA system resided mainly in the development and update of the back-end and front-end user interface. After datasets were mined, assimilated and loaded into the MapServer application, the tasks of writing scripts and generating appropriate map files for the back-end, while simultaneously creating HTML and JavaScript files for the front-end, became somewhat daunting. Much of the CMA Team had little to no prior experience of writing the type of codified language used in scripting. As such, it took quite a while, much troubleshooting and many revisions to generate a working prototype of the CMA that aptly represented the data within it as well as properly performed the functions it offered.

One of the most difficult stages in the development of the CMA continued to be in enhancing the system (back- and front-ends) to provide many of the expected (advanced) functions of a CWA. The search and querying aspects of the CMA, for instance, have proved to be somewhat of a challenge to add to the code of the system.

Greater emphasis on Database Management Systems, Content Management Systems, HTML and Java-based scripting and MapServer functions were necessary to eventually add in more advanced functionality. This in turn raised the issue of potential system upgrades and updates beyond the current scope of the CMA. To best deal with this, the CMA Team planned for further relevant training and future support in its agenda. One way forward was in further technical assistance from IODE and ICAN members. As many other atlas developers have faced and successfully solved several of these issues and problems already, it was only rational that the CMA learns as much as possible from past examples and cases.

Additionally, there was concern regarding the update capacity of the CMA, where new and revised datasets released by owners and/or generated by the CMA Team will have to be regularly updated on the system. Again, this led to the thinking of the CMA having to maintain human resources with the technical wherewithal to accommodate periodic revisions to the database, back-end and front-end systems.

Response to the CMA

As of 2009, the CMA has not been published nor opened to the public. However, its exposure and popularity have been on the rise. From its inception at the first workshop held in 2007, members, sponsors and others affiliated with the CMA have been ardently promoting it at every opportunity in various forums.

At regional and international levels, the CMA has been presented at several conferences and workshops to help advance awareness as well as garner support where possible. Similarly, each participating country in the CMA has put forth a national agenda to several relevant agencies, organizations and offices, as well as private sector entities, on how best the CMA can assist in meeting some of the needs of that country. Together with other means, like the already active CMA static website (http://www.caribbeanmarineatlas.org), which outlines and documents the goals, strategies and progress of the CMA, there has been an overall positive response to the CMA at all levels.

Many regional and international bodies have already offered assistance in the form of data, as seen in some of the data sources listed above. Local and national bodies have already forged a commitment to the CMA through their active participation in its development as well as via in-kind contributions. It has been well recognized by all that the objectives and functions of the CMA will ultimately serve many both within and outside the Caribbean Region. Consequently, support for this initiative is expected to grow over time.

Future of the CMA

The CMA will be set to launch in the third quarter of 2010, when it will be made available to interested parties with the knowledge that it is still only a prototype with limited capacities and functions. In addition, due to pending agreements with data providers, the downloading functionality of the atlas may also be restrictive at the start.

The next steps in development involve the addition of querying and sharing capabilities to the prototype mapping application. Following this phase (and with additional funding from the Government of Flanders already secured), the atlas project will be expanded to include other Caribbean countries, datasets and related information. The interface of the atlas will also be improved to make it more user-friendly and to provide additional functionality such as more sophisticated tools that will draw upon the informational resources of the atlas and that may ultimately be used in decision making and management of resources.

Equally important are the equivalent national atlases of the participating countries that will be simultaneously designed and implemented alongside the regional CMA. During the evolution of the atlas, the CMA Team recognized and addressed the fact that many of the problems of the Caribbean Region are geo-specific and localized both in their source and solutions. While a regional atlas may be able to provide data and (eventually) assist in decision-making for broader matters, a significant number of requests for data and information at a national or local level will inevitably arise. Consequently, larger-scaled local atlases will be required to meet these requests, each of which will contain specific data and capabilities relevant to that area and its respective concerns.

These national atlases will be set up and maintained by respective members of the CMA Team from each participating country. Utilizing the training and lessons learnt from the CMA, each national atlas will adhere to the umbrella rules and guidelines of the CMA with similar standards used for data formats, conversion, scaling, overall data quality and metadata. National atlases will provide the basic functions of data and information display, together with features such as zoom, layer lists and legends, but specific tools and functions incorporated into each atlas will be determined and designed based on the specific needs and limitations of each country. To facilitate local use restrictions, each country will be given autonomy and control over what data their atlas will contain and share. It is also hoped that each national atlas, and as much of its contained data as possible, will be integrated into the CMA to allow for the regional atlas to act as a single portal or entryway to users into what will be a network of national atlases. With similar standards for data and metadata used by all, interoperability among the atlases should therefore be maximized and consequently allow for relevant sectoral/local data from national atlases to be accessed by the regional CMA where possible. In this way local decision-making, data sharing, data analysis and communication, and regional collaboration will all be enhanced to promote effective marine and coastal resource management.

The CMA and ICAN

One of the main supporters of the CMA has been the International Coastal Atlas Network (ICAN). As a major champion of CWAs, ICAN through its network of members has on several occasions lent its experience and knowledge to the CMA. From its conception, the CMA adopted many of the paradigms and guidelines set forth by ICAN when it came to establishing a CWA. The data and metadata formats and standards, system design and infrastructure, software application options and concepts of interoperability that have been incorporated into the CMA have in some way been influenced or endorsed by ICAN. More directly, during training workshops, ICAN provided the CMA Team with training and expertise in metadata management and information management systems.

Becoming one of the nodes of ICAN, the CMA joined the effort towards fostering an association of web atlases with similar goals and objectives, as well as developing and promulgating web atlas guidelines. Advancing such a network meant that ICAN and its members like the CMA, AMA and MIDA endeavor towards common goals such as interoperability among atlases by adopting and promoting the use of standards in all aspects of atlas creation. It is therefore in the best interest of the CMA to continue to cultivate and strengthen ties to initiatives such as ICAN that continue to lead the way in the growth and evolution of CWAs.

CONCLUSION

With the process of developing the CMA having been outlined and some of the more pertinent challenges having been described, it is worth mentioning here that one of, if not the, most critical component that influenced the growth of the CMA was the sustained availability of adequate human resources, training and support. As mentioned earlier, a significant shortcoming of the Caribbean region was the lack of technical expertise and/or maintaining this expertise. The CMA sought to meet the goal of capacity building for the region by allowing participants to broaden and increase their skills and proficiencies in dealing with the management and dissemination of coastal and marine data through training and hands-on experience in developing this atlas. Given their roles and positions in their respective home countries, each participant will then impart their newly acquired knowledge and experiences to others. This in turn helped satisfy the need to strengthen both regional and national capacities, as well as promote and facilitate the development of similar solutions and technologies through initiatives like the national/local CWAs.

Additionally, the role of data and information to the CMA cannot be over-emphasized. Both the CMA and ultimately local CWAs rely upon data, data integrity and data standards to provide the backbone and foundation upon which many of the objectives of a CWA, and in this case the CMA, are satisfied. For the first time in the Caribbean, relevant coastal and marine data from multiple sources and several locations have been gathered and organized in a single site. While the system employed by the CMA facilitated the management and delivery of data and information to users, it is the data itself, its validity and its relevance that is essential to addressing issues and determining solutions in the course of managing resources. Although there remained a significant amount of data that CMA has still to collect in order to meet the needs of many of the themes and sub-categories identified as priorities and relevant to the region, it is equally important that it ensures that the data that it offers is of the best integrity and validity. As described above, some of the means to achieve such a goal include employing a system of quality control and standards for data and metadata. In this way, confidence in data and data integrity is maximized where users of a CWA capitalize on utility from what is available. This will also facilitate interoperability among systems and atlases that increases resources and information available to users and thus leads to better management and decision-making.

While the CMA is still in a young phase of development compared to some other CWAs, through the dedication and commitment of the CMA Team and the continued support of organizations such as IODE, ODINCARSA and ICAN, it will become an invaluable resource to national, regional and global entities. The sustainability of marine and coastal resources that provide the populations of the region a variety of goods and services is a common banner under which it is hoped that varied groups, interests and actions of the nations of the Caribbean can all get behind and support. The CMA has very ambitious and laudable goals when it comes to addressing some of the more prevalent environmental and marine issues in the Caribbean Region. By attempting to fill some of the more significant gaps in data availability, administration, accessibility and distribution, as well as provide valuable tools and resources for decision-making, the CMA hopes to become an integral part of the overall sustainable management of marine and coastal resources for the region for years to come.

REFERENCES

Caribbean Marine Atlas. (2009). Retrieved October 31, 2009 from the Caribbean Marine Atlas web site: http://www.caribbeanmarineatlas.org.

Caribbean Marine Atlas Team. (2009). *Project Proposal for Submission to the Government of Flanders (FUST 2009-2013) Version 2.1*. UNESCO/IOC.

Intergovernmental Oceanographic Commission. (2007). *IODE/ICAM Workshop on the Development of the Caribbean Marine Atlas (CMA), IOC Workshop Report No. 205*. Oostende, Belgium: UNESCO IOC.

O'Dea, L., Cummins, V., Wright, D., Dwyer, N., & Ameztoy, I. (2007). *Report on Coastal Mapping and Informatics Trans-Atlantic Workshop 1: Potentials and Limitations of Coastal Web Atlases*. University College Cork, Ireland, Coastal & Marine Resources Centre. Retrieved October 31, 2009, from ICAN web site: http://ican.science.oregonstate.edu/node/47.

Ocean Teacher. (2009). Retrieved October 31, 2009 from the Ocean Teacher web site: http://www.oceanteacher.org.

KEY TERMS AND DEFINITIONS

Caribbean Marine Atlas (CMA): An unrestricted coastal web atlas to manage and offer spatial data and related information relevant to marine and environmental issues identified for the Caribbean Region.

Caribbean Region: Area containing and immediately surrounding the Caribbean Sea that contains several islands and coastal states.

IOCARIBE: The Sub-Commission for the Caribbean and Adjacent Regions of the Intergovernmental Oceanographic Commission (IOC) of the United Nations Educational, Scientific and Cultural Organization (UNESCO).

IODE: UNESCO/IOC's International Oceanographic Data and Information Exchange.

National Marine Atlas: A sub-atlas of the CMA geared towards the data and information relevant to marine and environmental issues of a particular member country of the CMA.

ODINCARSA: Ocean Data and Information Network for the Caribbean and South America.

Training Workshops: Technical courses designed to introduce the tools and skills necessary to design and develop a coastal web atlas.

APPENDIX

Figure 4. Example of Discovery Metadata Record for the CMA.

Dataset Title: World Vector Shoreline of the Caribbean Marine Atlas Area
Owner: Caribbean Marine Atlas Project
Filename: cma_wvs.shp
Publication Date: 2008-04-25

Description
Abstract: The World Vector Shoreline dataset is a digital data file at a nominal scale of 1:250,000 and referenced to the World Geodetic System (WGS-84) datum, containing worldwide coverage of shorelines and international boundaries. This dataset is a subset for the Caribbean Marine Atlas region.
Keywords: OCEANS > COASTAL PROCESSES > SHORELINES
Data Type: vector
Scale (vector): 1:250000
Grid Spacing (raster):

Geographical Information
Description: Global coverage, clipped to CMA area of interest
Geographic Extent: Caribbean Sea, Gulf of Mexico, North Atlantic Ocean
Bounding Coordinates: North: 32.0
South: 0.0
East: -90.0
West: -50.0
Reference System: WGS_1984

Dataset Source
Creation Date: 1990
Lineage Statement: The main source material for the WVS was DMA's Digital Landmass Blanking (DLMB) database which was derived primarily from the Joint Operations Graphics and coastal nautical charts produced by DMA.
Related URL: http://rimmer.ngdc.noaa.gov/mgg/coast/getcoast.html
Description: NOAA NGDC - Vector shoreline coastline extractor program

Dataset Status
Progress: complete
Maintenance and Update Frequency: not planned

Access and Use
Format: shapefile
Format Version: ESRI v9.2
Medium: digital
Constraints:
Fees and Terms: data is free to use.

Contact Information
Organisation: Caribbean Marine Atlas Project
Person: Project Manager
Mail Address:
Telephone:
Electronic Mail: caribbeanmarineatlas@gmail.com
Website: http://www.caribbeanmarineatlas.org

Metadata Information
Creation Date: 2008-06-25
Point of Contact: Caribbean Marine Atlas project
Full Metadata:
Website:

Chapter 13
United Kingdom

David R. Green
University of Aberdeen, UK

ABSTRACT

This chapter briefly examines the origins and evolution of electronic coastal and marine atlases, and online mapping and GIS in the United Kingdom (UK). Beginning with some early examples, such as the UK Digital Marine Atlas (UKDMAP), initially distributed on floppy disk (MS-DOS) and later CD (MS-Windows), consideration is then given to some of the first online Internet-based information systems e.g., The UK Coastal Map Creator, some of the current systems now available e.g., MAGIC, MESH, and UKSeaMap, and finally the potential of Google Earth (GE) and Google Ocean (GO) to provide a framework for the development of simple local scale coastal and marine atlases. In each case, attention is paid to the origins of the atlas, its development, the user-interface, functionality, data and information content, and the target audience. Some of the advantages and disadvantages of electronic atlases are also discussed together with some of the problems, and possible solutions.

INTRODUCTION

This chapter briefly examines some of the origins and the evolution of Internet or Web-based marine and coastal atlases, and online mapping and GIS in the United Kingdom (UK). Using some of the early disk and CD-based marine atlases first available in the 1990s as the earliest examples of electronic atlases that became widely available in the UK, it also explores some of the earliest web-based equivalents when the potential of the Internet was first realized. Subsequent developments, paralleling the rapid evolution of computer technology, have provided the platforms for more recent and more widely available examples of web-based atlases. The chapter will firstly introduce the electronic atlas as a modern day equivalent of the traditional paper-based atlas. Reference to the rapid developments in computer hardware and software, coupled with increasing familiarity with

DOI: 10.4018/978-1-61520-815-9.ch013

Copyright © 2011, IGI Global. Copying or distributing in print or electronic forms without written permission of IGI Global is prohibited.

the tools and skills required to construct web-based applications, provides a contextual framework to the origin and development of web-based atlases and online map information systems. Some of the problems faced by potential atlas authors are also highlighted, including access to data and information and copyright issues that frequently inhibit the development of such atlases and their maintenance.

FROM PAPER TO ELECTRONIC: THE CONTEXT

Electronic or digital atlases are a natural evolution of the paper-based traditional paper atlas, albeit for use in a computer environment. There are many examples of electronic atlases available, some of which cover the marine and coastal environment. Electronic atlases originated at a time when computer technology (hardware and software, storage, processing, and display capabilities) had reached a stage when it was possible to design, create, and present maps and charts in an electronic or computer-based environment using computer-aided or computer-assisted cartography (CAC). While some early examples relied upon very basic black and white computer displays, the arrival of high resolution color monitors greatly enhanced the possibilities to display increasingly better quality maps on the computer screen. These took advantage of the additional dimension of color for mapping. In recent years, desktop computers (PCs) and small mobile geographic information system (GIS) hardware have evolved so rapidly that it is now possible to provide very high resolution color displays of maps and charts on the smallest of mobile platforms, including mobile phones. There are also many examples for displaying electronic chart displays on the modern yacht.

Whilst computer hardware has evolved quickly, providing the storage and memory capacity and processor speed to handle large volumes of geographical data on the desktop and mobile platform, so too has computer software. Software applications now provide the tools to design, create, and display very sophisticated map and chart displays. At the same time, software has become more user-friendly and so many more people are now empowered to create their own electronic maps, aided by software that provides the necessary cartographic, design, layout, and publishing tools. Map and graphic design software (e.g., Golden Software's MapViewer and Adobe Illustrator) can be used to create maps and present them as dynamic examples with the aid of MS-Powerpoint-based slideshows.

The potential of the electronic atlas as a medium to access, view and share coastal and marine information, has been further enhanced with the development of the Internet and a range of website development software tools, as well as GIS and online GIS-based mapping software that has ultimately led to the development of a wide range of different web-based examples of marine and coastal atlases. A sample of these can be accessed from the International Coastal Atlas Network (ICAN) website: http://ican.science.oregonstate.edu/atlases. The ICAN initiative (http://ican.science.oregonstate.edu) seeks to document best practice in the development of coastal web atlases (CWA) around the world by coordinating the sharing of international knowledge and expertise. It therefore represents a significant step in the evolution of Internet-based electronic atlases and may in turn have considerable influence on the development of data models and spatial data infrastructures (SDIs) that may underpin electronic atlases in the future.

In this chapter, the term electronic or digital atlas covers a wide range of examples ranging from electronic mapping to online GIS. The overview presented and the subsequent discussion necessarily also includes a number of other examples that are based around a searchable online catalogue of datasets, make use of maps as the interface to such a catalogue, or provide an online map/image-based service. Whilst not strictly fulfilling the definition of an electronic atlas they are nevertheless closely related and in effect form the basis of an electronic atlas.

EARLY ELECTRONIC ATLASES IN THE UK: SOME EXAMPLES

In the UK, some notable marine and coastal atlases were first developed in the 1990s by the British Oceanographic Data Centre (BODC) at Bidston. The United Kingdom Digital Marine Atlas Project (UKDMAP) and the UK Current Meter Series Database are two good examples. Other examples from the same source were the Global Bathymetric Chart of the Oceans (GEBCO) atlas and the North Sea Project Database. Together these represented a combination of software, database and visualization packages that were initially distributed on one or more 3.5" floppy disks and subsequently on CD. These were some of the first examples of electronic marine and coastal atlases in the UK, and were very similar to a number of other examples that became available at approximately the same time in North America such as the Coastal Ocean Management, Planning and Assessment System (COMPAS) and Atlas T. COMPAS (Alexander and Tolson, 1990) was developed by NOAA for the Apple Macintosh using Hypercard software and was accompanied by a Guided Tour for user familiarization Atlas T written by Rhines in 1992, was a micro-computer-based ocean atlas for tracer and hydrographic sections based on ship tracks. Running under MS-DOS, and controlled by keyboard input, Atlas T provided visualizations of ocean data including depth profiles and contour maps. It was also accompanied by a detailed Help file. All of these examples of computer-based atlases reflect the growing availability of digital data, mapping and visualization software, and the state-of the art computer technology at the time.

Some UK examples will now be briefly examined to provide some evolutionary context for the subsequent development of UK Internet-based electronic map information systems and atlases.

The UK Digital Marine Atlas

The UKDMAP (http://www.bodc.ac.uk/products/bodc_products/ukdmap/) was probably the most notable and widely available example of a UK coastal and marine atlas, and certainly the most popular and familiar (BODC, 1992; Barne et al., 1994; Green, 1994a, 1994b). This electronic atlas provided the end-user with access to a wide range of marine and coastal data (originating from BODC databases) presented in the form of a digital map or chart. Initially for desktop (PCs) computers running MS-DOS (later replaced by an MS-Windows version) and a color display monitor, this electronic atlas provided the potential for end-users to interact with and map a wide range of terrestrial, coastal, and marine datasets for the UK (Figure 1). At the time, UKDMAP was quite a novel application that had considerable potential as an information and educational resource and was targeted at people involved in research, conservation, engineering, fishing, transport and leisure, education, recreation, and planning (Green, 1994b). One advantage of the electronic format was the ease with which updated data could be made available to the end-user. Customization of the maps was also possible and there was some limited GIS functionality (Green, 1994b). One novel feature of UKDMAP was the capability to assemble a selection of maps into a slide show to examine changes over time, trends, and the exploration of specific themes (Green, 1994b). For various reasons (e.g., lack of funding, changes in computer technology etc.), after several releases of updated versions of UKDMAP, it was no longer maintained, although it is still available from BODC. This is unfortunate as the overall concept of a UK information system rapidly became very popular and the product was well-packaged; easy to use, low cost, and a very useful information and educational resource for the UK.

Figure 1. Screen snapshot of the UKDMAP (courtesy of BODC, UK)

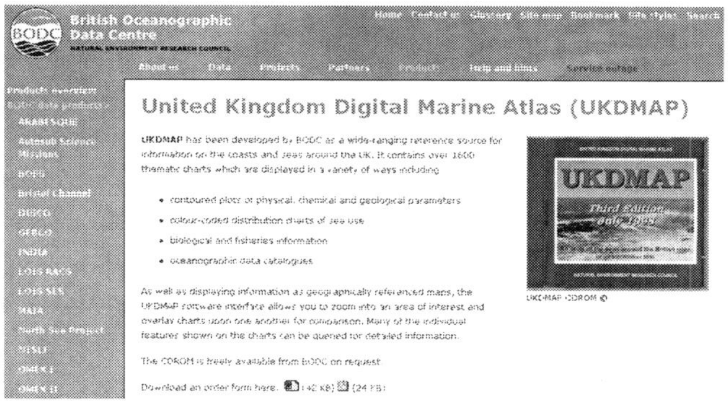

British Oceanographic Data Centre Inventory of Moored Current Meter Data

Another example, also developed by the BODC, was the Current Meter Series Database (BODC, 1991). It was developed in co-operation with international marine organizations and comprised a database and visualization software with a similar user interface of pop-up/drop-down menus found in UKDMAP. It included moored current meter data collected by sixteen countries from around the World. Although not solely focused on the UK, a large part of the BODC Current Meter Series database (part of their National Oceanographic Database) contained current meter data for the UK. The end-user was able to browse the database, select a sample of records, and to display or visualize the selected locations on a map outline. Records could be selected according to various criteria such as the start date, country, and laboratory of origin, and sorted, retained, printed to file and previous record selections reloaded for review.

Whilst initially also distributed on floppy disk, the Current Meter Series Database is now accessible via a simple map interface based around the Environmental Systems Research Institute or ESRI's ArcIMS map server software (https://www.bodc.ac.uk/data/online_request/current_meters/). It is easy to use and comprises a descriptive web page and a help file, together with some simple navigational tools e.g., zoom and pan, and previous extent. This map interface allows the end-user to explore the data holding prior to making an online request for the data. In its current guise it is probably one of the most basic examples of an electronic interface to spatial data and utilizes only a small part of the full functionality of ArcIMS software. A similarly configured product is the Wave Data Series which also includes a web-based ArcIMS interface (https://www.bodc.ac.uk/data/online_request/waves/).

North Sea Project Database

The North Sea Project Database (Lowry et al., 1992) is a comprehensive collection of physical, chemical, and biological data collected from survey cruises and process study cruises containing data on temperature, salinity, and dissolved oxygen. It also included satellite AVHRR (Advanced Very High Resolution Radiometer) data from NOAA, metadata, the BODC Moored Current Meter database, and a North Sea Bibliography. As with the other BODC products the North Sea Project Database is now supplied on a CD and was designed as an electronic publication of data complete with visualization software. A range of tools was also provided to help visualize the variety of different datasets e.g., time series, underway data and the analysis of satellite imagery. The latter tools include UNESCO's Bilko image processing software (UNESCO, 1989), Image Display

Program in Interactive Data Language (IMDISP), and PicLab (a freeware graphics package). Data could be subset, listed to screen, printed as both black and white and color hardcopy, and images saved to the GEM .IMG format facilitating use of the information in a slide show, and integration in text documents. Cruise tracks could also be plotted. Customization of the visualization was controlled by the end-user allowing for the selection of colors and specification of contour intervals. Data could also be copied to a computer's hard disk to facilitate faster access (Lowry et al., 1992).

By comparison to other BODC products this was a less user-friendly source of data, comprising a number of different interfaces to different software used to view, analyze and visualize the data, and was aimed more at the specialist audience than the layman.

GEBCO Digital Atlas

The General Bathymetric Chart of the Oceans or GEBCO Digital Atlas (http://www.gebco.net/data_and_products/gebco_digital_atlas; Jones et al., 1994) was another BODC product, first released in 1994 on 3.5" floppy disk (software) and CD (data). Complete with a comprehensive 70 page printed Supporting Volume (including Annexes) to the GEBCO Digital Atlas (GDA), it contained the following: digitized bathymetric contours and coastlines; digitized trackline control; the existing digitized bathymetric charts from the Intergovernmental Oceanographic Commission (IOC) regional ocean mapping projects; a copy of the computerized gazetteer of geographic names of undersea features as maintained by the International Hydrographic Bureau; a standard world coastline (the US Defense Mapping Agency's World Vector Shoreline); and a trackline inventory of digital sounding data (Jones et al., 1994). The end-user interface and functionality were very similar to that of UKDMAP and the Current Meter Series Database. Running under MS-DOS, with an interface comprising drop-down menus, the software provided a range of functionality to visualize the data on a map using overlays, to extract and save data, provide printout, and to generate slideshows. Current information about the latest release of GEBCO and the software is now available online at http://www.gebco.net/

In summary, all of these BODC products were clearly quite unique when first made available. Furthermore, they were well developed, executed, and subsequently supported, which at the time, placed the UK at the forefront of electronic access to coastal and marine data and information and electronic atlases. Although UKDMAP was not developed further by BODC, the closest equivalent now is the Marine Irish Digital Atlas (MIDA) (http://mida.ucc.ie/) which is discussed in Chapter 7 and utilizes web mapping technology.

HelFal

With the arrival of Windows-based operating systems, first pioneered by Apple and later pursued by Microsoft as a replacement for MS-DOS, came a new generation of micro-computers and software. As mentioned earlier in this chapter, NOAA developed their COMPAS system for the Apple Macintosh using Hypercard (a hypermedia application program) software. A similar approach was used at the University of Aberdeen by Green (1994c) to develop a simple demonstrator for coastal management of a UK estuary in Cornwall, South West England. HelFal (the Helford Passage and Fal Estuary Community Information System) was used to convey the concepts and potential of developing a multimedia GIS-based information system, using maps, text and imagery for the two estuaries. A version of this demonstrator was later also developed in Toolbook, multimedia training software program for the PC. The objective was to take advantage of the opportunities to construct a simple intuitive interface to a wide range of data and information about an estuary including maps, imagery and text documents using interactive multimedia software.

THE WORLD WIDE WEB AND SOME EARLY EXAMPLES

In the mid-to late-1990s, with the then emerging Internet, web browsers (e.g., Mosaic and Netscape), and accompanying software, attention soon turned to the potential the web offered as a means to construct online mapping and information systems and electronic atlases, and to provide access to spatial data and information by a wider end-user community via both network and communications software. A number of early web-based examples were soon developed, more often than not as demonstrators of the potential the Internet offered as an interface to a wide range of coastal and marine information from multiple disparate formats and sources.

Over time, progress towards web-based GIS has been greatly aided by the development of Internet map- and image- server software and the concept of an Internet map service (IMS). The best known software initally was perhaps ESRI's ArcIMS which has been widely used as the basis for many online GIS projects. Indeed, nearly all commercial GIS software packages now have their own IMS as part of the software or as an add-on. Besides commercial software products such as ArcIMS, there are a number of open source software examples such as GeoServer (http://GeoServer.org) and Minnesota MapServer (http://MapServer.org). Web Mapping Services (WMS), an Open Geospatial Consortium (OGC) standard, are also now supported by Google Earth (http://earth.google.co.uk/), NASA's World Wind. (http://worldwind.arc.nasa.gov/), and Microsoft's Virtual Earth (http://www.microsoft.com/maps/).

Proposal for a UK Network-Based Spatial Information System: The United Kingdom Marine and Coastal Information System (UKMCIS)

The success of UKDMAP clearly revealed the demand for a simple means to access UK marine and coastal data and information. Green (1995) therefore proposed a national coastal and marine information system based around the world-wide web. At the time this represented a unique and potentially very useful way to translate the work and content of products such as UKDMAP into a format offering many advantages over distribution on a CD. UKMCIS (the United Kingdom Marine and Coastal Information System) was a conceptual idea to integrate existing disparate coastal and marine data and information held by many UK organizations into a widely accessible national information system based around the net, which would provide and enhance opportunities for more people to access and use coastal and marine information. The rationale for this approach lay with the following advantages: establishing a resource for multi-user access; promoting awareness, contact and discussion; providing improved communication links; providing opportunities for direct sharing of data, information and expertise; providing opportunities for integration of data; developing a multidisciplinary approach to research and problem solving; minimizing duplication of data and work; providing cost-effective solution to problems; enhancing planning and decision-making. Additional benefits included: overcoming the isolation of data and information sources; promotion of data standardization and documentation; encouraging regular updating; and an emphasis on data quality and error checking.

The UK Coastal Map Creator

As with all computer technology the Internet evolved very rapidly and by the mid- to late- 1990s software to generate map-based or GIS-based map delivery across the Internet was well developed. One of the most notable early examples was ESRI's ArcIMS, a product that allowed a developer to create web-based portals providing an opportunity for map and image based data and information to be accessed, and shared across a computer network.

Figure 2. The UK Coastal Map Creator

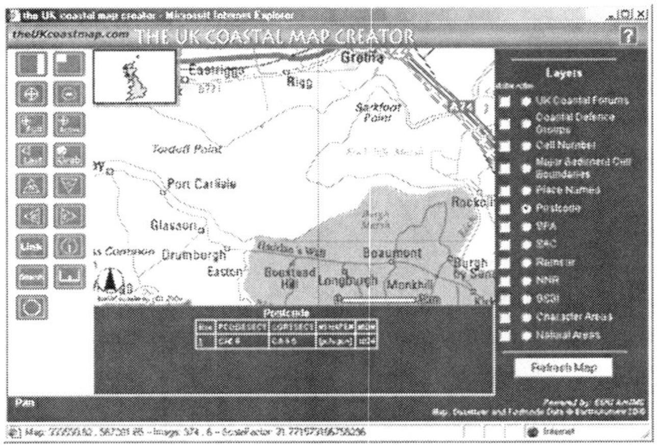

One early example that took advantage of the potential offered by the world-wide-web was the UK Coastal Map Creator (Green and King, 1998; Figure 2), the outcome of some research undertaken at the University of Aberdeen in Scotland, UK using ESRI's ArcIMS 3 software. Supported and hosted by ESRI UK, the UK Coastal Map Creator was one of the first examples of a web-based coastal zone information system in the UK. Building upon the ideas first presented by Green (1995) for UKMCIS, the UK Coastal Map Creator took advantage of the functionality of ESRI's ArcIMS to develop a framework to deliver coastal and marine information to the UK coastal community. Initially developed as a demonstrator to enhance access to UK coastal and marine information, the UK Coastal Map Creator sought to encourage potential data providers to make their data and information more widely available.

The rationale for the development of the UK Coastal Map Creator was that access to data and information is an essential component of day-to-day coastal planning and decision-making in integrated coastal zone management (ICZM). In the United Kingdom (UK), coastal zone management was undertaken by a wide variety of government and non-government agencies and institutions guided and co-ordinated by voluntary forums or partnerships for specific areas of the UK coastline. Although many of these organizations gathered data and information relating to the coast, often it was not stored in a standard or accessible format for people to use outside of a particular agency or to share easily. The web environment therefore provided an opportunity to overcome such constraints.

The Bartholomew 1:200,000 raster data provided the navigational base map for the system, developed with their postcode and gazetteer data, providing the end-user with the ability to search the database either by town name or postcode. Other datasets included were English Nature's (now Natural England) boundaries of national and international conservation site designations in England. Proposed additions to the system included aerial photography, satellite imagery, and Lidar, some aided by commercial agreements with Infoterra (http://www.infoterra.co.uk/). Considerable care was taken to develop a user-friendly and aesthetically pleasing interface for the end-user, bearing in mind the growing and ever widening end-user community for such information. Customization was easily achieved using the ArcIMS software tools.

However, whilst successful as a conceptual framework, the UK Coastal Map Creator ultimately proved practically difficult to develop into a national resource for a number of reasons, even despite an attempt to commercialize the system. One reason was the difficulty in getting map data suppliers to contribute data into the system. Commercial agreements to both list and show an example of available information on the website with a pointer to the data supplier's contact details overcame some of these difficulties but did not lead to wider population of the system with spatial datasets of interest to the coastal and marine community. Another was the ever present problem of copyright issues in the UK which usually made the base datasets prohibitively costly to purchase and use in an online system such as this.

Geo-Information Decision Support Processing and Dissemination System

An extension of the UK Coastal Map Creator idea was the GDSPDS (Geo-Information Decision Support Processing and Dissemination System) (http://www.abdn.ac.uk/~geo402/index.htm) developed by King at the University of Aberdeen in the Centre for Marine and Coastal Zone Management (CMCZM) (http://www.abdn.ac.uk/cmczm) between 1999 and 2001. Constructed using ESRI's ArcIMS 3 software, the GDSPDS sought to extend the basic and typical concept of a GIS and mapping portal into one that included links to a wide range of additional geographical data and information representations, thereby providing a multimedia information system for planning and decision-making in the coastal zone. The system focused on the Solway Firth in Scotland, UK.

In addition to ESRI's ArcIMS software used for the mapping component, the system offered the end-user access to digital satellite and airborne imagery, panoramic photography, mobile GIS data, and video, all within a single system. A key component of the work involved emphasis on the end-user interface, with a focus being placed on the aesthetics, functionality, and usability of the interface by the coastal manager and coastal practitioner, and development included some end-user tests. Some additional experiments were undertaken to explore the utilization of image compression software e.g., LizardTech's Mr. Sid (http://www.lizardtech.com/) and ER-Mapper's ECW compression formats (http://www.erdas.com), and a range of software utilities to process and visualize data and information, as well as some more experimental ideas utilizing software to develop a 3D realistic navigational interface to the coastal zone information.

As an experimental decision-making framework, based around recognition of the importance of the ease of access to coastal data and information for coastal managers, this research revealed how off-the-shelf software and hardware could easily be utilized and customized to create the framework for an online GIS mapping, data and information system for coastal zone management.

SOME RECENT UK WEB-BASED EXAMPLES

With greater availability of spatial data and developments in web-based technology, coupled with a desire to develop institutional and national online information systems using map servers, more and more organizations have been placing their marine and coastal data and information online in one format or another. Whilst some simply provide access to catalogues of data holdings with accompanying search tools, others have provided online mapping systems with customized interfaces, metadata, tutorials, manuals, help files and data download facilities. Recognition of the potential of the Internet-based marine and coastal information system is finally beginning to take shape particularly now with the UK Marine Bill (http://www.defra.gov.uk/) and the Scottish Marine Bill (http://www.scotland.gov.uk/Publications/2009/09/28115722/0). The growing use of GIS and the need for spatial datasets is also being influenced by the greater emphasis now placed on the development of Marine Protected Areas (MPAs), marine national parks, and the overall enthusiasm for marine or maritime spatial planning. A few different examples of such developments are summarized below to illustrate some of the progress that has been made to date.

Data Archive for Seabed Species and Habitats

The Data Archive for Seabed Species and Habitats (DASSH; http://www.dassh.ac.uk/) is an archive centre for digital benthic data, as well as images and video, and an online catalogue of data and metadata, providing a long-term store and a data and information resource. Funded by Department of Environment, Food and Rural Affairs (DEFRA), DASSH works with the Marine Biological Association, the Marine Life Information Network, the National Marine Biological Library, the Marine Environmental Digital Information Network (also known as MEDIN), and the Marine Data Archive Centers regarding data standards. Although primarily a search-based interface, there is a simple map interface search window in SEArchable BEnthic Data or SeaBED. This is interactive and allows one to draw a bounding box around an area on a map and returns the coordinates to the search tool and a list of datasets. These can then be downloaded in the comma separated value or CSV file format for use in a spreadsheet and subsequently imported into a GIS such as ESRI's ArcView or ArcGIS. Links to other organizations are provided where the data are not held by DASSH. Registration of a user's details allows continuing access to any datasets retrieved from the database.

Marine Environment Data and Information Network

MEDIN (http://www.oceannet.org/) is hosted at the BODC and comprises a partnership of UK organizations whose aim is to improve access to marine data. The web interface is being tested at present and provides links to marine data held by a number of other organizations e.g., DASSH, British Geological Survey (BGS), and the United Kingdom Hydrographic Office (UKHO). It provides hyperlinks to access a data discovery portal, a data search by theme, and advice on marine data standards, mainly for the UK. Searches return a series of links to detailed metadata files in Extensible Markup Language (XML).

ICES FishMap

The International Council for the Exploration of the Sea, Denmark (ICES) Fishmap (http://www.ices.dk/marineworld/ices-fishmap.asp) is an online atlas of fifteen fish species in the North Sea based on data collected from trawl surveys collected between 1983 and 2004. It was created by RIVO (Institute of Fisheries Research, The Netherlands), Center for Environment, Fisheries and Aquaculture Science, UK (CEFAS), and ICES and is partly funded by the European Commission's 6th Framework Program. Fishmap provides an online web portal together with background information, some downloadable as PDFs and includes data from GEBCO (see earlier). Fishmap has two modes: Basic and Advanced. In Basic mode, the end-user is provided with a simple, largely static, interface which provides a legend of the fifteen fish species in the North Sea. Selecting each fish species changes the map accordingly. Contextual information on each species is also provided. This is a very basic user-interface and functionality is limited to changing the fish species and the selection of Basic or Advanced mode. Advanced mode, by comparison, provides a more interactive interface with an array of zoom, measure, re-center, select area, and select polygon functionality. In addition, a pane provides the end-user with reset, export, select, query, clear and help functionality. The legend pane provides access to the data. The data layers can be selected and displayed. This is a simple web atlas with basic functionality. It operates in interactive Spatial Explorer and Administrator (iSEA), CEFAS's spatial data explorer and webGIS interface for UK data (http://www.cefas.co.uk/isea).

WaveNet

Another web-based interface provided by CEFAS is WaveNet (http://www.cefas.co.uk/wavenet-mapping) which provides access to DEFRA's wave buoy locations and wave height data. This is a very simple interface to a map, where interaction of the mouse with each symbol also reveals the data for that buoy, which includes wind and temperature information. As with ICES Fishmap, the CEFAS web interface includes a Basic and Advanced mapping capability. The Advanced mapping capability provides access to a similar map but with more functionality and many layers, including wind farms, buoys, and waveriders, that can be toggled on and off, selected, deselected, and redrawn. Clicking on a symbol takes the end-user to a separate web page with further options to display the data. Links to satellite and text data, as well as Google Earth KML files are also provided.

Mapping European Seabed Habitats (MESH)

The Mapping European Seabed Habitats (MESH) marine mapping and metadata portal is one of the outcomes of the MESH project (http://www.searchmesh.net/). Funded by Interreg IIIB NW Europe, MESH was developed over three years, starting in 2004 and finishing in 2008. It involved twelve European partners led by the Joint Nature Conservation Committee (JNCC). Accompanying the webGIS mapping interface is a very comprehensive website providing both the scientist and the non-scientist with information about benthic habitat mapping and the process of producing a map, as well as details on a data model, metadata, and a series of case studies. The webGIS was developed using the open source GIS software Minnesota MapServer, which can be freely downloaded and used to develop an online mapping system. The portal contains a wide range of marine datasets.

UKSeaMap

The UKSeaMap webGIS (http://www.jncc.gov.uk/page-3663) provides a map portal (described as a mapping data delivery service) giving access to information on seabed landscapes and seasonal water column features. Additional contextual layers (e.g., geology, bathymetry, tide stress, and salinity) are also provided. It was funded by the following organisations: the Countryside Council for Wales (CCW), DEFRA, Department of Trade and Industry (DTI), Natural England, The Royal Society for the Protection of Birds (RSPB), the Scottish Executive, the Crown Estate, the JNCC, and the World Wildlife Fund (WWF). UKSeaMap was developed as part of MESH (see above). The web GIS is launched using an initial layer chosen from a drop down menu list (e.g., a base map). The web-mapping interface was developed by the JNCC and exeGesIS SDM Limited and made use of Minnesota MapServer and the Javascript language. The interface comprises a typical web-mapping view on a map window, an overview map, a legend and a number of basic navigational tools, complete with tabs for Location, Legend, Info, Layers, and Help. These can all be selected and the information is displayed in the right hand pane of the view. Layers can be toggled on and off. The webGIS can be launched from a number of points in the website pages that accompany the mapping portal, and the water column data and the seabed landscape data can also be downloaded as zipped data files.

Multi-Agency Geographic Information for the Countryside (MAGIC)

The Multi-Agency Geographic Information for the Countryside or MAGIC (http://www.magic.gov.uk/) is a website with an interactive map service providing public access to geographic information on the environment and designations. A detailed overview of MAGIC is provided by Askew et al. (2005). As part of this online resource, a Coastal and Marine Atlas was added in 2005. This is promoted as a tool for marine planning and contingency response. It is funded by the following organizations: the Maritime and Coast Guard Agency (MCA), DEFRA, the Scottish Executive (SE), Scottish Natural Heritage (SNH), the Energy Institute (EI), JNCC, the Environment Agency (EA), English Nature (EN; now Natural England), the Department of Trade and Industry (DTI), the Hampshire County Council (HCC), Essex County Council (ECC), Kent County Council (KCC), and the British Geological Survey (BGS). Much of the data are from the Ordnance Survey (http://www.ordnancesurvey.gov.uk) and also SeaZone (http://www.seazone.com). Compared to many of the other online mapping systems discussed earlier in this chapter, MAGIC has more GIS functionality accessible via the user-interface. There are a number of navigational functions including zoom, pan and the addition of an Ordnance Survey map backdrop, grid co-ordinates, identify, measure distance and area, print, save (in GIF format), re-project, and bookmark. The view window includes a map pane, a scale and scale bar, a legend pane, and an overview map. Up to fifteen thematic layers can be displayed at any one time. Tabs provide access to map tools, and other themed resources. Aside from the online map-based information the website also includes a map tutorial, dataset information, some static maps, a training manual, and work in progress (Figure 3).

Western Channel Observatory

The Western Channel Observatory (WCC: http://www.westernchannelobservatory.org.uk/) is hosted by the Plymouth Marine Laboratory (PML: http://www.pml.ac.uk/) and provides access to marine data e.g., ocean color, sea surface temperature, chlorophyll, weather data and webcams for the Western English Channel. Data, including remotely sensed data are integrated within a GIS. At present the interface is basic and provides web links to data and graphic visualizations, as well as the satellite data. The latter utilizes the NERC Earth Observation Data Acquisition and Analysis Service (NEODASS) (http://www.npm.ac.uk/), a Java image viewer which includes some zoom and GIS data overlay capabilities. An additional, and unique, feature of this portal is the interactive

Figure 3. Screen snapshot of MAGIC (courtesy of DEFRA)

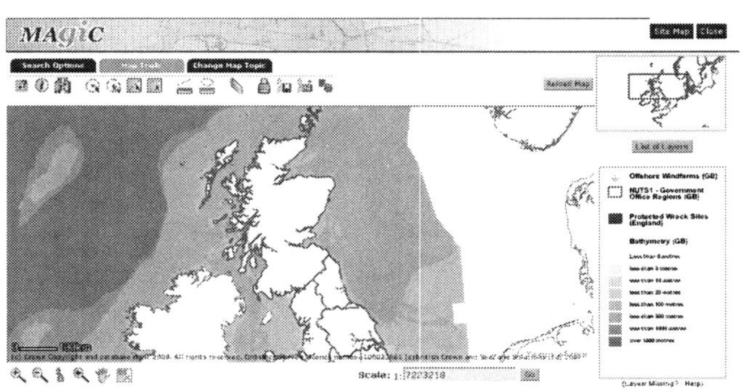

image stretch function as well as the pixel by pixel enquiry tool – both of which open in separate viewing windows – allowing for some elementary digital image processing and visualization to be undertaken.

Channel Coastal Observatory

By comparison, the Channel Coastal Observatory (CCO) (http://www.channelcoast.org/) has an online map viewer and data search tool. This provides a web-based interface to a catalog of a wide range of marine and coastal data for part of the UK, largely focused on the South Coast and the English Channel. The web view provides the end-user with a map pane (including a summary of the functionality of the map tools), an overview map, and a list of the map layers as clickable folders. These include remotely sensed data (ortho-rectified and non-rectified aerial photography, color infrared (CIR) aerial photography, and light detection and ranging (LIDaR) data, topographic and topographic model data, hydrographic and hydrographic model data, photogrammetric data, sediment distribution data, beach profile change data, real time wave buoy and tide gauge data, and other data including the Ordnance Survey (OS) grid, management units, coastal places, and GPS points. The web-map interface provides the end-user with an interactive geographical search tool (Figure 4). Following the initial area search it is possible to refine the search for the datasets available for an area by choosing from a range of formats for each data type (e.g., a recommended format such as ECW (Enhanced Compression Wavelet) for aerial imagery, and PDF (Portable Document Format). Selected datasets are then placed in the 'basket' and can be downloaded providing they are not over 300 Mb. The web interface includes a Help file and Feedback form. This is an aesthetically pleasing and navigable interface which uses the map server software to identify and display the location and coverage of the data holdings. One weakness, perhaps, is that the datasets cannot be displayed, only downloaded. The reason for this is probably to minimize storage requirements of the data. Logically, however, the end-user expectation is to be able to display and examine the data e.g., the imagery and the provision of an electronic atlas instead might be beneficial to the end-user community in the longer term.

Godiva

Godiva is a NERC project (http://www.nerc-essc.ac.uk/godiva) hosted by Reading University's e-Science Center and uses GADS (Grid Access

Figure 4. Screen snapshot of the Channel Coastal Observatory (courtesy of CCO)

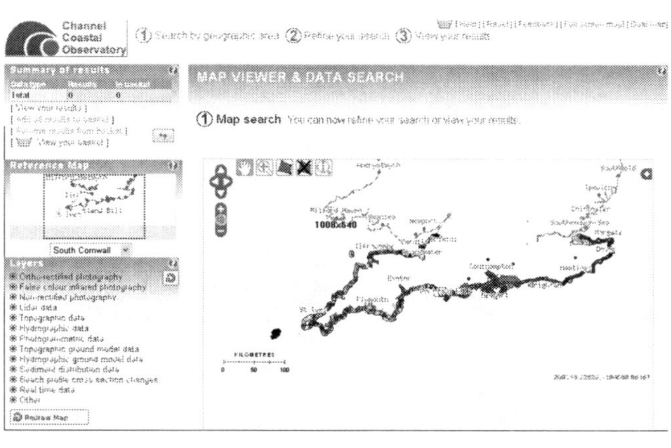

Data Service) It was designed for scientists so as to facilitate analysis and visualization of data from various sources including climate, atmospheric and oceanic models. Godiva provides a map-based interface (Godiva2) for a range of marine spatial datasets from the ESSC Web Map Service for environmental data for a number of different areas and from different origins e.g., UK Met Office. One such area is the Irish Sea. Available datasets include: sea water velocity, sea surface height, sea water temperature, and wind speed. Data can be selected for viewing with the aid of a calendar and drop down time (clock) menu. The data are visualized against a map outline. The interface also includes some basic navigational tools for the map data e.g., Pan and Zoom, together with a legend. A choice of linear or log displays can also be selected. A Transect tool allows transect graphs to be shown in a separate window. The basic interface is simple and practical, aesthetically appealing, and functional, allowing the end-user to locate and display multi-temporal marine spatial datasets, and to focus on the information, thereby providing a useful browse service. It also allows for the comparison of different datasets from different EU projects e.g., MERSEA. The accompanying User Guide gives details about using Godiva and the ways in which it can be customized by the end-user e.g., navigation, changing the color palette, creating an animation, and changing the map projection, as well as saving and emailing a view.

SeaDataNet

SeaDataNet of the Pan European Infrastructure for Ocean and Marine Data Management (www.seadatanet.org) has been developed by a number of EU partners including the British Oceanographic Data Center (BODC) in the UK. The web portal provides access to marine datasets from thirty-six countries. The viewing services are in the form of a Catalog and Maps, with point, track and area entries. Developed in compliance with OGC (Open Geospatial Consortium) standards and in co-operation with my-Ocean (see below) it includes a Common Data Index (CDI) Metadata Search interface. This includes a Geographical Box selection criteria function with a map view, simple navigational tools, and additional information resources. More details are provided at the BODC website: http://www.bodc.ac.uk/projects/european/seadatanet/.

MyOcean

At the time of writing, the MyOcean Service (2009-2012 - http://www.myocean.eu.org) is a new project (a contribution to Global Monitoring Environment and Security or GMES) that aims to provide a catalog of worldwide and European regional ocean products (in the form of real time observations, analysis and forecasts) that have been made available from previous projects e.g., MarCoast. It will include the INfrastructure for SPatial InfoRmation in Europe (INSPIRE) functionalities e.g., discovery, visualization and downloading tools and includes marine and coastal environments.

InterRisk

InterRisk (http://interrisk.nersc.no) is an EU funded project (FP6-IST) designed to provide GMES information services for environmental risk management in marine and coastal areas of Europe. It has been built on existing solutions from other GMES projects e.g., MASS GSTP and the Data Integration System for Marine Pollution Water Quality or DISMAR (http://www.nersc.no/dismar/) The pilot service is based on open system architecture of GIS and web service protocols and has been implemented for several European regional seas, one of which is UK/Irish waters (Plymouth Marine Laboratory). It is also based on a number of recognized standards including the World Wide Web Consortium (W3C), the OGC, the International Organization for Standardization (ISO), and follows INSPIRE recommendations.

The Open-source Project for a Network Data Access Protocol (OPeNDAP) and software tools have also been investigated.

MarCoast

MarCoast (Marine and Coastal Environmental Information Services: http://gmes-marcoast.com/) was a three year (2005-2008) GMES (Global Monitoring for Environment and Security) project funded by the European Space Agency (ESA). It comprises a network of marine and coastal information services e.g., the MarCoast Service Portal and provides a catalog of the services provided by the MarCoast partners at a European scale, including the UK. These are in the form of surveillance, forecast, monitoring and alert, and assessment services. This includes access to oil spill, water quality, algae bloom, water quality assessment and Met-Ocean data. The basic map interface is simple and offers navigational and search tools, and an overview map. A 'map' link provides access to Google Map as a backdrop and a Google Earth file that displays a map of links to service providers. In addition, some data are available for download and services can also be added.

Proudman Oceanographic Laboratory Observatory

Included as part of the Proudman Oceanographic Laboratory's (POL) Observatory for the Irish Sea is a MapViewer link (http://cobs.pol.ac.uk/gmapper/). The map interface is based on Google Maps, together with an interactive legend, instructions, and links to a map and satellite imagery backdrop. Additional map data layers include near real time data on Coastal Tide Gauges, WaveRider Buoys, SmartBuoys, as well as and RiverFlow monitor data, and model output. Switching to satellite provides an image backdrop instead of the map. The interactive legend can be hidden from view. This website takes advantage of the Google Maps interface as a backdrop to other layers. Clicking on the symbols gives a pop-up window which provides further links to data. For example clicking on the tide gauge symbol for Lowestoft leads to web links for near real time tide gauge data and seven day predictions. The interface is aesthetically pleasing and easy to navigate and use, with its interactive legend, and is perhaps closest to a combination of a GIS and the Google interface, something that makes it potentially quite familiar.

Google Earth and Local Information Systems

With the availability of Google Earth and a range of software utilities it has become possible to consider this as yet another way to provide access to and share data and information relevant to the marine and coastal environment. As seen earlier some of the web-based services are already making use of Google Maps and Google Earth. However, while Google Earth, in its base form, provides access to geospatial information for anywhere in the World, the additional element explored in this section of the chapter lies with the potential to create a local information system allowing for the addition of additional data and information (vector and raster) as well as data collected with mobile systems to the base layers provided in Google Earth. With the aid of software utilities e.g., Shape2Earth (http://shape2earth.com/), MapWindow (http://www.mapwindow.org/) and a number of Google blogs (e.g., OgleEarth (http://www.ogleearth.com/) it is possible to add more information to the Google Earth view that can then be shared as a KMZ (KML-Zip file) file and opened by anyone who has access to Google Earth. These provide a powerful toolbox to create a simple online GIS system.

The latest release of Google Earth, including Google Ocean, has extended the terrestrial information resources to the coastal and marine areas, allowing for the addition of bathymetric data. This combination can then be used to develop local GIS for small areas of the coast and, in an

Figure 5. Virtual Fieldtrip of Ythan Estuary in Google Earth

educational and training context, can be used as the basis for developing virtual fieldtrips or as a ground-truthing resource. Two simple examples are used here for illustration. The first is a local information system developed for the Ythan Estuary in North East Scotland, UK, which is also used to provide a virtual fieldtrip and tour of the estuary and the sand dune system. The second example is the use of Google Earth to create a local information system for Nigg Bay to the south of Aberdeen that is also used as the basis to create a multimedia coastal footpath walk, and a visual fly-through of the Bay as the basis for visualizing a proposed waterfront regeneration proposal (Figure 5). Although Google Earth has some limitations as a Geographical Information System (GIS) mainly in terms of its functionality, it has considerable potential to provide a relatively simple way to bring together and display spatial data and information, and to share the data and information without the need to have a GIS. Google Earth also allows for layering on top of an image back drop, the addition of more layers, and incorporates a DEM (Digital Elevation Model). The popularity of Google Earth is increasingly evident from the number of data providers who are now willing to share their GIS data layers in the KMZ format; for example, Scottish Natural Heritage (SNH), and the Maritime Coastguard Agency (MCA), and by the number of GIS software packages that now allow the export of data in a GIS to the KML format. Perhaps the biggest drawback to using Google Earth is the apparent limitation in file sizes that can be handled, and in the UK problems that might arise when sharing data if the source originates with, for example, the Ordnance Survey (OS).

ISSUES AND DEVELOPMENTS: PAST, PRESENT AND FUTURE

Green (1994b) raised some general issues concerning ease of access to coastal and marine data and information for the wider community. Whilst progress towards overcoming some of the constraints has clearly been made over time, there still seem to be a number of the very same issues continuing to hinder progress in this area.

Access to data and information about the coastal and marine environment has clearly improved greatly over time. This has largely been aided by the development of web portals that allow end-users to find, locate and explore the data, to visualize database holdings in the form of an interactive map (some with simple GIS functionality), and in some cases to download the digital datasets for use in a GIS. In the UK there have been a number of initiatives to collect and make more widely available data for the marine and coastal environment. Two have been the result of collaborative projects to develop links between the

Ordnance Survey (OS) terrestrial data and marine chart datasets from the UK Hydrographic Office (UKHO).This resulted in two map demonstrators. The first produced a single paper sheet map of the coastal zone for the area around Southampton and the Isle of Wight. The second, Integrated Coastal Zone Map (ICZMap), a collaborative project of the Ordnance Survey, the UKHO and the British Geological Survey (BGS) was completed in 2003 and resulted in an example demonstrator digital dataset. Latterly this led to the creation of SeaZone thematic GIS layers. These include: hydrospatial, digital survey bathymetry, charted raster, and charted vector data. These have since been successfully used within some of the web map portals described earlier. Recent developments for academic use (teaching and research) have been Marine Digimap (http://edina.ac.uk/digimap/description/marine_overview.shtml) which makes the SeaZone datasets available for in-class exercises and project work. This has been some time in coming but is definitely a very positive development for up and coming generations to be able to use the data in their education and research.

Interfaces to information systems have also clearly evolved, and whilst there is considerable variety in the end result and ultimately the functionality and usability of each one, the Internet (and some of the web mapping software) has had a huge impact upon improving access to information through more aesthetically appealing, navigable, intuitive, and usable interfaces.

In nearly all cases, the online portals also provide considerable contextual background which is focused on the likely potential end-user community and usually include resource pages with additional information, tutorials, links, manuals, and help files. This is not unlike earlier systems provided on disk and CD. What is still not included, however, for those systems where it is possible to create a new map, with the end-user having control over the choice and use of color, shading and symbols, and overall design, is guidance or best practice guidelines on how to produce the most effective map as output. This is especially important where maps are to be exported to a document or used in a slideshow presentation for communication to a wider audience. While some systems do provide information about the mapping process (e.g., MESH has a link to a Mapping Guide), there is a still a fundamental need to direct end-users of these online information systems towards the basics of map design through books and tutorials etc., particularly as Green (1994b) points out that: cartographic design is far from trivial and it is easy for those lacking knowledge and experience to end up with a poor map. This should not really be a problem as there are now many books and resources to aid the end-user (e.g., Brewer, 2005; Brown & Feringa, 2002; Dent, 2002; Erle et al., 2005; Krygier & Wood, 2005; Madej, 2000; Memorial University, 1998; Peterson, 2009).

There is clearly also a wide range of different digital examples of online mapping systems, online GIS, and electronic atlases, that provide interfaces to either databases via search tools, and/or interactive map searches, as well as providing online mapping tools, and even online GIS. Whilst many examples are simply little more than information systems, other examples allow one to download the actual map data for further analysis in a desktop GIS. Although not every end user requires the data, most simply requiring access to the information, in some cases the demand for datasets for spatial analysis in a GIS can be very advantageous for education, training, and research applications

Problems once associated with being able to read different data formats have largely disappeared and where data are available for download this is usually available in a number of well recognized and accepted formats that can be easily used directly in GIS software. These are typically ESRI shapefiles and coverages. More recently Google Earth KML Keyhole Markup Language (KML) and KMZ (KML-Zip) files are now often an option, finding favor as the popularity and functionality of Google Earth and Google Ocean continues to grow.

More emphasis is now also being placed on the development of data models e.g., the Arc Marine data model (http://dusk.geo.orst.edu/djl/arcgis/), and metadata standards for documenting spatial datasets. In Europe the INSPIRE initiative is becoming increasingly relevant to the future of UK (as part of the EU) datasets. However, although much has also been written about national SDI and marine data infrastructures over the years (see Longhorn, 2009) there still seems to be some problems and perhaps even reluctance with translating words into practice, and there is much work still to be done in this area. As noted earlier, this is perhaps one area where the work and collective expertise of ICAN can begin to help to contribute to and drive progress in the future.

The main issue remaining, however, in the UK still lies with copyright and charging associated with digital datasets. On the whole most UK datasets are still relatively costly to buy and use and this tends to restrict their use by all the interested parties ultimately hindering attempts to produce Internet-based mapping and GIS systems at the local, regional, and national scale. While it is perfectly acceptable that data should be charged for (after all data costs money to collect, capture, create, and maintain) identifying what seems to be a 'reasonable' and 'acceptable' cost for use still appears to be quite difficult to achieve. Certainly substantial progress has been made over the years, as can be seen from the many online examples now available. But it would seem that despite the best efforts of people and organizations, many of the online websites are either very specific or are at best are still only partial solutions, despite the general availability of the software tools to create the resources. This is somewhat disappointing given the timeframe over which these ideas have been discussed, and the speed with which the technology has been developed to provide technical solutions. However, at the time of writing, recent Government-led developments at the Ordnance Survey (OS) have launched a new initiative called OS OpenData designed to provide users with free and unrestricted access to most of its map data (http://news.bbc.co.uk/1/hi/8597779.stm). This is an interesting and welcome step forward and one that will hopefully address many of the issues that have prevented more widespread development of the online examples of marine and coastal atlases in the UK.

SUMMARY AND CONCLUSION

Rapid developments in computer hardware and software have provided considerable opportunities for accessing and sharing geospatial data and information about the marine and coastal environment in the last twenty years. While early examples of marine and coastal atlases provided opportunities for more people to explore and utilize such data and information, the Internet and associated software in the form of map and image servers have provided the biggest step forward in providing access to both data and information for the marine and coastal environment. Although still not plug and play solutions, Internet-based map server software is more widely available and can now be more easily configured by virtually anyone with the help of various manuals, tutorials, books and blogs, which means that more people can set up their own web portals and create electronic atlases. With the development of products such as GeoServer, MapServer, TimeMap and more recently Google Earth, and Google Ocean, there is now even more scope to share geospatial datasets across the net, and to develop digital coastal atlases. The University of Aberdeen, for example, is currently investigating the development of a coastal atlas for the Aberdeen and Aberdeenshire coast (http://www.sacrp.info), and another for the Living North Sea (LNS) (http://www.livingnorthsea.eu) Project using Google Earth, TimeMap and MapServer.

An historical overview reveals the development of the hardware and software, networking and communications technology, the Internet, and the growing freedom for the individual to upload and share their geographical data. In the UK though it has been relatively slow to develop and take off. Over the years there have been many examples of proposals, demonstrators, and project-based examples of Internet-based electronic atlases. In the UK, some of the potential and the will seems to have been lost in all the continuing discussions about data availability, data models, spatial data infrastructures, and a general lack of recognition of the value of the Internet as an information resource, as well as continuing to involve people who lack the vision of the potential, underfunding, lack of commitment, and to some extent the degree to which it is all become 'bogged down' in words rather than actions.. This perhaps goes some way towards explaining the origins and success of Google Earth! In other parts of the World there would appear to be far fewer constraints and barriers to accessing digital map and chart data which has, not surprisingly, led to the development of some very comprehensive and very appealing online examples of electronic atlases which provide the sort of resource that is desirable by the end-user community, including that in the UK.

Where to now in the UK? ICAN has revealed how useful, important and powerful web-based mapping, atlas, and GIS tools are and, from the list of examples provided, just how successful they have become. The technology is also proven and more and more digital spatial data are becoming available. It is a pity then that there are so few UK examples which truly mirror the earlier success of UKDMAP. The resources now available range from the basic to the comprehensive, simple to aesthetic and intuitive interfaces, and basic to sophisticated functionality. Certainly there are some good UK examples and much has been achieved. Yet it would seem, from looking at other examples originating in North America and Europe that we could do so much more and better in the UK. Perhaps the work of ICAN will provide the required stimulus?

Finally, a thought. Despite the advantages of all this computer-based spatial technology to provide access to and share digital coastal and marine data and information, there are still a number of printed, paper-based examples of atlases being produced around the World (albeit with the use of electronic data and the help of computer software and usually provided on an accompanying CD/DVD for use in GIS software). While there are clearly advantages and disadvantages of both media, and a certain appeal and practicality associated with the electronic format, it is hard to beat the look and feel of information presented in the form of a bound paper volume. Maybe there is a challenge here to consider new ways of presenting and visualizing coastal and marine information in the electronic atlases of the future that tries to capture some of the visual attraction of the printed atlas? Assuming of course the data are available!

ACKNOWLEDGMENT

The Living North Sea (LNS) Interreg IVB North Sea Region Program: Investing in the future by working together for a sustainable and competitive region.

REFERENCES

Alexander,C., &Tolson, J.P.(1990). COMPAS - NOAA's Coastal Ocean Management, Planning, and Assessment System National Ocean Service, National Oceanic and Atmospheric Administration. Unpublished. Galveston Bay Information Center, (pp. 213-214).

Arc Marine Data Model. (2009). Retrieved October 31, 2009 from the OSU Davey Jones' Locker web site: http://dusk.geo.orst.edu/djl/arcgis

Askew, D., Evans, S., Matthews, R., & Swanton, P. (2005). MAGIC: a Geoportal for the English Countryside. *Computers, Environment and Urban Systems*, *29*, 71–85.

Barne, J., Davidson, N. C., Hill, T. O., & Jones, M. (1994). Coastal and Marine UKDMAP Datasets: A User Manual. *JNCC Report No. 209*. Joint Nature Conservation Committee, Coastal Conservation Branch.

Brewer, C. A. (2005). *Designing Better Maps: A Guide for GIS Users*. Redlands, CA: ESRI Press.

British Oceanographic Data Centre. (1991). *Manual of Inventory of Moored Current Meter Data*. Plymouth, UK: British Oceanographic Data Centre.

British Oceanographic Data Centre. (1992). *United Kingdom Digital Marine Atlas. Version 2.0. User Guide*. Plymouth, UK: British Oceanographic Data Centre.

British Ordnance Survey. (2009). Retrieved October 31, 2009 from the British Ordnance Survey web site: http://www.ordnancesurvey.gov.uk.

Brown, A., & Feringa, W. (2002). *Colour Basics for GIS Users*. New York: Prentice Hall.

Centre for Marine and Coastal Zone Management. (2009). Retrieved October 31, 2009 from the University of Aberdeen web site: http://www.abdn.ac.uk/cmczm.

Channel Coastal Observatory. (2009). Retrieved October 31, 2009 from the Channel Coastal Observatory web site: http://www.channelcoast.org

Current Meter Series Database. (2009). Retrieved October 31, 2009 from the British Oceanographic Data Centre web site: https://www.bodc.ac.uk/data/online_request/current_meters

Data Archive for Seabed Species and Habitats. *(2009). Retrieved October 31, 2009 from the Data Archive for Seabed Species and Habitats web site:*http://www.dassh.ac.uk.

Data Integration System for Marine Pollution Water Quality. *(2009). Retrieved October 31, 2009 from the* Nansen Environmental and Remote Sensing Center *web site:*http://www.nersc.no/dismar

Dent, B. D. (2002). *Cartography with ArcView GIS Software: Thematic Map Design: With ArcView GIS Software*. New York: McGraw-Hill Higher Education.

Directory of Coastal Web Atlases Worldwide. *(2009). Retrieved October 31, 2009 from the International Coastal Atlas Network website:*http://ican.science.oregonstate.edu/atlases.

Erdas Imagine. *(2009). Retrieved October 31, 2009 from the Erdas, Inc. web site:*http://www.erdas.com.

Erle, S., Gibson, R., & Walsh, J. (2005). *Mapping Hacks Tips & Tools for Electronic Cartography*. O'Reilly.

General Bathymetric Chart of the Oceans GEBCO Digital Atlas. *(2009). Retrieved October 31, 2009 from the Global Bathymetric Chart of the Oceans website*http://www.gebco.net/data_and_products/gebco_digital_atlas

Geo-Information Decision Support Processing and Dissemination System. *(2009). Retrieved October 31, 2009 from the University of Aberdeen web site:*http://www.abdn.ac.uk/~geo402/index.htm

GeoServer. *(2009). Retrieve dOctober 31, 2009 from the GeoServer website* http://GeoServer.org

Global Bathymetric Chart of the Oceans. *(2009). Retrieved October 31, 2009 from the Global Bathymetric Chart of the Oceans web site:*http://www.gebco.net

Godiva. *(2009). Retrieved October 31, 2009 from the Natural Environment Research Council of the UK website*http://www.nerc-essc.ac.uk/godiva

Google Earth. (2009). Retrieved October 31, 2009 from the Google Earth web site: http://earth.google.com.

Green, D. R. (1994a). The United Kingdom Digital Marine Atlas: A review. GeoCal, (June), 280-30.
Green, D.R. (1994b). Geographic and marine information systems: Some recent developments and progress in the United Kingdom. In Proceedings Hydro '94, The Ninth Biennial International Symposium of the Hydrographic Society, UK Branch (pp. 13-15). Aberdeen Exhibition and Conference Centre. Special Publication No. 33.

Green, D. R. *(1994c). Using GIS to construct and update an estuary information system. In* Proceedings of Management Techniques for the Coastal Zone. *Centre for Coastal Zone Management, University of Portsmouth. October 24-25, (pp. 129-162).*

Green, D. R. (1995). Internet, the WWW and Browsers: The Basis for a Network-Based Geographic Information System (GIS) for Coastal Zone Management. In *Proceedings of AGI'95 Conference*. Expanding Your World. 21-23 November 1995. (pp. 5.1.1- 5.1.12).

Green, D. R., & King, S. D. (1998). The UK Coastal Map Creator: ArcIMS provides a solution to coastal map and data delivery for the United Kingdom's coastal community. In Proceedings of the ESRI User Conference, San Diego, CA.

Infoterra. (2009). Retrieved October 31, 2009 from the Infoterra website http://www.infoterra.co.uk/.

interactive Spatial Explorer and Administrator. *(2009). Retrieved October 31, 2009 from the Center for Environment, Fisheries and Aquaculture Science, UK website*http://www.cefas.co.uk/isea

International Council for the Exploration of the Sea. Denmark ICES Fishmap. *(n.d.). Retrieved from* http://www.ices.dk/marineworld/ices-fishmap.asp.

InterRisk. *(2009). Retrieved October 31, 2009 from the* Nansen Environmental and Remote Sensing Center *website*http://interrisk.nersc.no

Jones, M. T., Tabor, A. R., & Weatherall, P. (1994). *Supporting Volume to the GEBCO Digital Atlas. British Oceanographic Data Centre (BODC).* Swindon, UK: Natural Environment Research Council.

Krygier, J., & Wood, D. (2005). *Making Maps: A Visual Guide to Map Design for GIS*. New York: The Guildford Press.

Longhorn, R. (2009). Coastal and marine data infrastructure . In Green, D. R. (Ed.), *Coastal Zone Management* (pp. 206–225). London: Thomas Telford Limited.

Lowry, R. K., Cramer, R. N., & Rickards, L. J. *(1992).* North Sea Project CD-ROM User's Guide. *British Oceanographic Data Centre. Swindon, UK: Natural Environment Research Council.*

Madej, E. (2000). *Cartographic Design Using Arcview GIS*. Albany, NY: OnWord Press.

MapWindow. *(2009). Retrieved October 31, 2009 from the MapWindow website*http://www.mapwindow.org

Marine and Coastal Environmental Information Services. *(2009). Retrieved October 31, 2009 from the Marine and Coastal Environmental Services website http://gmes-marcoast.com*

Marine Digimap. *(2009). Retrieved October 31, 2009 from the Marine Digimap website*http://edina.ac.uk/digimap/description/marine_overview.shtml

Marine Environment Data and Information Network. *(2009). Retrieved October 31, 2009 from the Marine Environmental Data and Information Network website*http://www.oceannet.org

Marine Irish Digital Atlas. *(2009). Retrieved October 31, 2009, from the Marine Irish Digital Atlas website:*http://mida.ucc.ie

Memorial University. (1998). *Guide to Map Design and Production Using CorelDRAW 7 and CorelDRAW 8*. St. Johns, Newfoundland, Canada: Memorial University of Newfoundland.

Microsoft Virtual Earth. *(2009). Retrieved October 31, 2009 from the Microsoft Virtual Earth web site:*http://www.microsoft.com/maps/.

Minnesota MapServer. (2009). Retrieved October 31, 2009, from the Minnesota MapServer web site: http://www.MapServer.org.

Mr. Sid. *(2009). Retrieved October 31, 2009 from the LizardTech website*http://www.lizardtech.com

Multi-Agency Geographic Information for the Countryside. (2009). Retrieved October 31, 2009 from the Multi-Agency Geographic Information for the Countryside web site: http://www.magic.gov.uk.

MyOcean Service. 2009-2012. *(2009). Retrieved October 31, 2009 from the My Ocean web site:*http://www.myocean.eu.org.

NERC Earth Observation Data Acquisition and Analysis Service. *(2009). Retrieved October 31, 2009 from the Natural Environment Research Council web site:*http://www.npm.ac.uk.

OgleEarth. *(2009). Retrieved October 31, 2009 from the OgleEarth web site:*http://www.ogleearth.com.

Peterson, G. N. (2009). *GIS Cartography: A Guide to Effective Map Design*. Boca Raton, FL: CRC. doi:10.1201/9781420082142

Plymouth Marine Laboratory. *(2009). Retrieved October 31, 2009 from the Plymouth Marine Laboratory web site:*http://www.pml.ac.uk.

Proudman Oceanographic Laboratory Observatory. *(2009). Retrieved October 31, 2009 from the Proudman Oceanographic Laboratory Obervatory web site:*http://cobs.pol.ac.uk/gmapper/

Scottish Marine Bill. *(2009). Retrieved October 31, 2009 from the Scottish government web site:*http://www.scotland.gov.uk/Publications/2009/09/28115722/0.

SeaData Net. *(2009). Retrieved October 31, 2009 from the British Oceanographic Data Centre web site:*http://www.bodc.ac.uk/projects/european/seadatanet/.

SeaZone. *(2009). Retrieved October 31, 2009 from the SeaZone Solutions web site:*http://www.seazone.com.

Shape2Earth. *(2009). Retrieved October 31, 2009 from the Shape2Earth web site: http://shape2earth.com.*

UK Digital Marine Atlas. *(2009). Retrieved October 31, 2009 from the UK Digital Marine Atlas web site:*http://www.bodc.ac.uk/products/bodc_products/ukdmap.

UK Marine Bill. *(2009). Retrieved October 31, 2009 from the Department of Environment, Food and Rural Affairs web site:*http://www.defra.gov.uk.

UKSeaMap. *(2009). Retrieved October 31, 2009 from the Joint Nature Conservation Committee web site:*http://www.jncc.gov.uk/page-3663.

UNESCO. *(1989). Bilko Software. Retrieved-October 13, 2009, from the UNESCO-Bilko web site:*http://www.noc.soton.ac.uk/bilko.

Wave Data Series. *(2009). Retrieved October 31, 2009 from the British Oceanographic Data Centre web site:*https://www.bodc.ac.uk/data/online_request/waves/.

WaveNet. (2009). Retrieved October 31, 2009 from the Center for Environment, Fisheries and Aquaculture Science, UK web site: http://www.cefas.co.uk/wavenetmapping.

Western Channel Observatory. *(2009). Retrieved October 31, 2009 from the Western Channel Observatory web site:*http://www.westernchannelobservatory.org.uk.

World Wind, N. A. S. A. *(2009). Retrieved October 31, 2009 from the NASA World Wind web site:*http://worldwind.arc.nasa.gov.

KEY TERMS AND DEFINITIONS

Coastal Web Atlas: A collection of digital maps and datasets with supplementary tables, illustrations and information that systematically illustrate the coast, oftentimes with cartographic and decision support tools, all of which are accessible via the Internet. Also known as web atlas, digital atlas, digital coastal atlas.

Coastal and Marine Information: Both spatial and non-spatial information regarding the coastal and marine areas of a region or country.

Metadata: Metadata in the geographical domain is structured information on a dataset, which helps the data owner to document and catalogue the data, whilst helping a data user to understand the content and fitness for use of a dataset.

Spatial Data Infrastructure or SDI: A framework via an organization of people or government agencies, via the Internet, or via a series of guiding policies or standards to assist people with acquiring, processing, using, and preserving spatial data. The spatial data are often in geographic information system (GIS) format, are not, but not limited to this.

Spatial Data Visualization: The ability to view digital data with a spatial dimension in a computer environment containing a coordinate reference system.

Web GIS: A geographical information system which can be accessed over the Internet and allow visualization and interaction with spatial data via a map as well as providing analysis functionality such as spatial analysis, querying and buffering.

Virtual Globe: A three-dimensional visual rendering of the Earth or of other planets, usually available as online software application, or as an application for the desktop. A virtual globe provides the user with the ability to navigate throughout the virtual world freely in all directions, by changing viewing angle, azimuth, and heading. A virtual globe is primarily a visualization tool (oftentimes open source) and is not completely synonymous with a geographic information system in terms of spatial analysis capabilities. Examples include Google Earth, NASA World Wind, and Microsoft Virtual Earth.

Chapter 14
Spain

Alejandro Iglesias-Campos
Government of Andalusia, Spain

Gonzalo Malvarez-García
University of Pablo de Olavide, Spain

José Ojeda-Zújar
University of Seville, Spain

José Manuel Moreira-Madueño
Government of Andalusia, Spain

ABSTRACT

The Regional Ministry for the Environment of the Regional Government of Andalusia launched the Environmental Geographic Information System of Andalusia (SinambA) in 1984. The main aim then was the development of tools for environmental planning and the management of natural resources in the region. Subsequently, the Environmental Information Network (REDIAM), established in May 2000, produced standardized protocols and still disseminates information on the environment in Andalusia, providing a useful tool for management, research, decision-making and environmental knowledge in general. In the last five years the REDIAM has undergone a major redesign to accommodate the European Union's INSPIRE Directive and to provide a comprehensive service to the public, decision makers and researchers. The strategy follows the development of a series of OGC-compliant subsystems (biodiversity, climate, coastal, etc.) that aim at facilitating accessibility to and robustness of complex geodatabases that were mostly designed for Internet use through 2D and 3D viewers by ELIMCO Systemas. The SIGLA - Sistema de Información Geografica del Litoral Andaluz or Coastal Information System of Andalusia is now under development as a REDIAM subsystem for refining access protocols and procedures to utilize the vast data sets available as well as formalizing agreements to maintain external services from other administrative bodies that gather and maintain spatial databases with coastal information.

INTRODUCTION

Since the creation of the Andalusian Government, the environmental authorities of the Autonomous Region have facilitated the creation and development of an information system that meets the needs of development planning and management of natural resources in the region. The Environmental Geographic Information System of Andalusia

DOI: 10.4018/978-1-61520-815-9.ch014

Copyright © 2011, IGI Global. Copying or distributing in print or electronic forms without written permission of IGI Global is prohibited.

(SinambA), launched in 1984, when the previous Environment Agency of Andalusia was created, was an instrument that integrated georeferenced databases (land use, climate, socioeconomic data, data on the emission of pollution, etc.) graphical or visual information (maps, photographs, etc.), and information obtained through remote sensing, with a wide range of related developments.

In SinambA, which was organized in a systematic manner, the most up to date environmental data available from multiple sources were fed into a local spatial database at the Regional Ministry and were subjected to continuous processes of quality assurance and updates. The relationship among the global media, information elements and the multidisciplinary team which designed and operated the System (computer scientists, geographers, engineers, physicists, economists, etc.) aimed at providing a robust and effective tool to support decision makers in the environmental administration. Its operating structure, as a system capable of simulating complex real life situations, aimed at allowing the evaluation of policies or actions likely to impact on the environment, as well as to enable the implementation of standards and legal instruments of high relevance, such as environmental impact assessments, programs for the promotion of disadvantaged areas, and monitoring of environmental issues throughout the region (Moreira, 1991).

The original SinambA was organized for the preparation of an alphanumeric and georeferenced database (topography, soil, climate, water use and existing vegetation) in relation to existing socio-economic variables, environmental quality indices and all the actors (humans, flora, fauna, etc.) present. One of the main goals of the SinambA focused on the development of techniques for digital processing of satellite imagery (Cornaet & Moreira, 1992). This last aspect was vital since there was a comprehensive database for all of Andalusia that could be used at reconnaissance (<1:100,000), semi-detailed (1:50,000) and detailed scale (>1:10,000).

However, the SinambA was a generic system for all environmental data and no system was developed for specific issues or complex environments, such as coastal systems. It was as part of a parallel project, mainly fed by data already gathered by SinambA that a coastal system was developed, aimed specifically at the spatial relevance of the information. The system was the SIGLA - *Sistema de Información Geografica del Litoral Andaluz* or Coastal Information System of Andalusia, a joint initiative of the Regional Ministries of Environment and Public Works in the 1990s.

Following the experiences gained through development of SinambA, the Regional Ministry for the Environment of Andalusia developed a generic environmental data information system called REDIAM (*Red de Información Ambiental)* or Environmental Information Network, in May 2000. The Regional Ministry leads the coordination of all the regional agencies that collect environmental data and focuses on the creation of harmonized procedures for generating and manipulating the geoinformation, so it can be used to generate further geodata. The focus is to promote inter-sectoral cooperation, by creating synergies with other interested parties such as universities and private research institutions. In July 2007, the Act on Integrated Environmental Quality Management officially established the REDIAM to standardize and disseminate information on the environment in Andalusia, providing a useful tool for management, research, decision-making and knowledge in general.

THE ENVIRONMENTAL INFORMATION NETWORK (REDIAM) AND COASTAL INFORMATION SUBSYSTEM (CIS)

As a result of the preliminary implementation of the Andalusian Strategy for Integrated Coastal Zone Management and the special need to establish information and management tools for the governance of these areas, the Regional Ministry for the Environment of Andalusia is developing the Coastal and Marine Information Subsystem, based on the already existing SIGLA Project and integrated into the Environmental Information Network of Andalusia (REDIAM).

The main aspiration of this initiative is to standardize a core set of information on the coastal and marine environments of Andalusia, accessible to all users, by collecting, structuring and putting into service the information according to a consistent data model with an environmental information network.

The dissemination of information on coastal and marine resources is undertaken in parallel with a variety of activities on this topic; however, the Andalusian Government has responsibilities for coastal management, consistent with the principles of ICZM promoted by and for all the administrative levels (European, National, Regional and Local). As part of this co-ordination role for ICZM, the REDIAM network is developing cooperation agreements between the Regional Ministry for the Environment of Andalusia and other administrative bodies and institutions that produce information, share databases, methodologies and GIS tools.

The REDIAM aims to ensure the provision of public environmental information of interest to citizens, under rules that support the right of access to environmental information. The information distributed by the Regional Ministry for the Environment of Andalusia on its web page (www.juntadeandalucia.es/medioambiente) and in the REDIAM Channel (http://www.juntadeandalucia.es/medioambiente/site/web/rediam), that was launched in February 2009, is understood to be official and reliable data.

The REDIAM aims to become a permanent source of environmental information for society, and in particular for scientists, experts and managers to ensure continuous improvement and sustainability of the environment. As with the previous SinambA, this network consists of alphanumeric databases, digital maps and images captured by remote sensors which are interconnected in space and time and are a key source of information for environmental management and analysis.

In terms of accessibility and dissemination, the REDIAM Channel is the permanent portal that facilitates access to environmental information, its main objectives being to:

- be the showcase for information for environmental management and a portal for connection with its information management systems;
- monitor compliance with obligations under European rules on access to environmental information and the INfrastructure for SPatial InfoRmation in Europe (INSPIRE) directive;
- allow sharing of environmental information resources (scientific, technological and quality assurance) generated by different institutions involved in research and / or environmental management aspects in Andalusia.
- encourage development of environmental research by using new information technologies and scientific knowledge and expertise;
- disseminate data and information to all citizens;
- create an instrument for improving access to information which allows not only consultation but download services as well; and
- ensure the security of the system by restricting data access for the general public whilst allowing full access through Extranet for members of the REDIAM and Intranet for users, principally from the environmental administration.

Dissemination Strategy in REDIAM

Research institutions and public administrations that are associated through collaborative agreements use the REDIAM Channel as the best "window" for accessing all available environmental information, as well as the technical methodologies, dictionaries, environmental variables, viewers and other kind of software for the management of such a large amount of information.

Environmental information is channeled to both technical users and the public. The Regional Ministry makes every effort to process the information in such a way as to make it accessible to the public at large. The outcome of these efforts is included in multiple environmental information products of mass dissemination produced by the Regional Ministry, of which three deserve a special mention - the Annual Report on the State of the Environment, the Collection of maps and statistics, and the web site of the Regional Ministry.

The web site of the Regional Ministry for the Environment of Andalusia has been available since 1995, with the purpose of providing the public with updated and reliable information on the state of the environment in Andalusia and on the policies implemented by the Regional Ministry. Over the years, the quality and quantity of the contents have improved as larger segments of society used this web site as their channel of communication of choice, and as the information producing centers became increasingly aware of the ability of this new, user-friendly medium to reduce distances and costs.

The web site has led the way in providing information in real time, *inter alia*, of air quality and coastal water quality measurements, frequently updated information on environmental protection legislation and, in particular, of legislation enacted by the Regional Ministry for the Environment of Andalusia. It is the vehicle for presenting documents to citizens during the public information stage of consultations on environmental procedures, contains numerous digital publications, from newsletters to the latest publications of the Regional Ministry, such as the Annual Report on the Environment, or draft documents open for comment. Moreover, it also offers a comprehensive compendium of those maps and statistical documents produced by the Regional Ministry, and summarizes all the environmental documentation available in more than 8,000 hypertext pages.

In addition to these means and media for the free dissemination of environmental information, the Regional Ministry publishes a publicly available catalogue of products that are for sale. Such sale is pursuant to the principle set out in the Law 38/1995 on free access to environmental information which provides that a charge may be levied but such a charge should be reasonable and subject to a schedule of standardized and updated charges and fees. Since 2000, EGMASA, a public body, is responsible for product distribution.

Subsystems of the REDIAM

As the REDIAM network grew, it was decided to provide a series of Geographic Information Subsystems (e.g., biodiversity, climate, coastal) that aimed at facilitating access to complex geodatabases that were mostly designed for internet use through 2D and 3D viewers. This approach allowed the provision of a wider range of scales and details for specific resources and covered local, regional and national extents. A common concern in the design of the subsystems was to ensure easy access and user focused procedures for the subsystems' web portals.

To integrate standardized information, the REDIAM is based on a structure that ensures the interrelation of environmental datasets and variables referenced on a space-time matrix (Figure 1). Within this large information system thematic areas are defined with specific management units consistent with the study of environmental variables and phenomena. Currently, there are three fully functional information subsystems, dedicated to Water, Climate and Biodiversity respectively,

Figure 1. The REDIAM structure

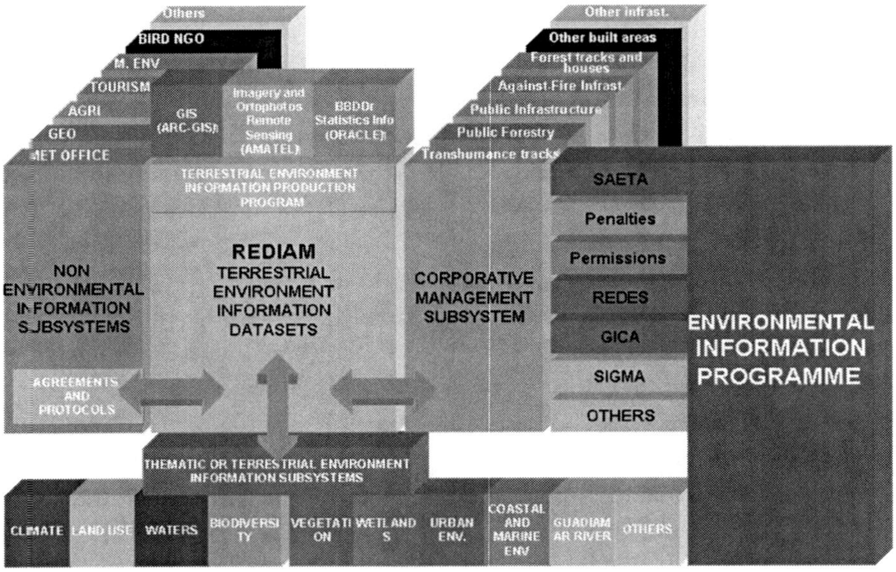

and there is a version for the Network of Protected Natural Areas in Andalusia (RENPA) and Wetlands. Each subsystem responds to a different stakeholder need and the regional administration and thus the design varies greatly. Each subsystem is described below.

Water Subsystem

The Regional Ministry for the Environment of Andalusia, through the Andalusian Water Agency, has commissioned the preparation of studies and plans that allow the definition of water policy and management in the region, the promotion of partnerships between Government and the promotion of sustainable water use. To do so it is considered essential to have an information system to store, manage and analyze spatial databases and topics related to water, which will be used as a working tool and source of information for such complex matters as supply, sanitation, regulations, customs, disaster prevention and environmental protection.

The overall goal of the Water Subsystem is the development of a geographic information system in its broadest sense, in order to manage all the relevant information for territorial water management. This organized set of information must be accessible and easily understandable to all competent technicians and users involved in water management. The REDIAM Water Subsystem (Figure 2) includes geographic information related to water management: physical and biotic data, human environment and infrastructure. All the thematic applications are accessible through the Water Subsystem for the storage, handling and modeling of the Geographic Information as well as procedures and methods for the incorporation, management and updating of the information. A list of experts related to the working groups on water is also accessible.

Products of the subsystem (interoperable with other subsystems) are:

- General Overview for Groundwater: alphanumeric and graphical information on hydrogeology, inventory of water points, aquifer systems, hydrogeological units, permeability, piezometry, water quality

Figure 2. The REDIAM Water Subsystem interface which can be accessed at http://bit.ly/3T6iEQ.

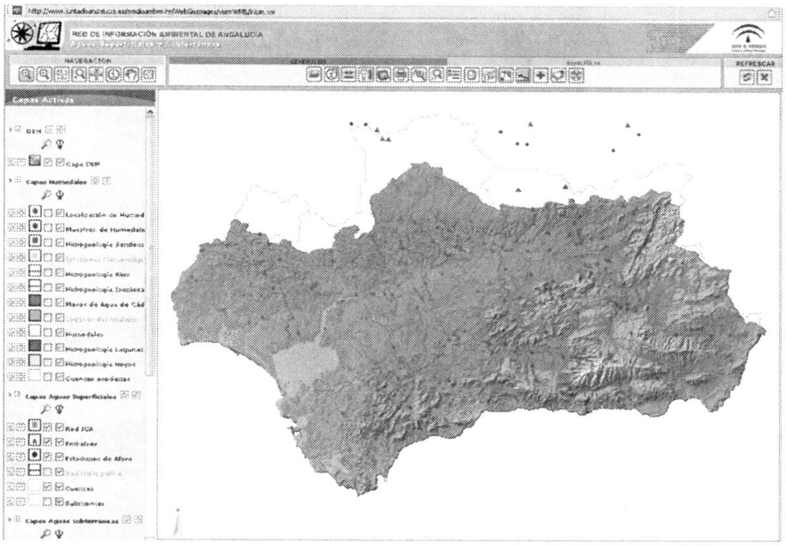

networks, hydrogeological interest areas, vulnerability, etc.

- General Overview for Surface Water: database with alphanumeric and graphical information on watersheds, river-gauging stations, monitoring networks and associated data.
- Wetlands: alphanumeric and graphical information for all the Andalusian wetlands. General information, including inventory and monitoring networks, data sampling and analysis, etc.
- Reference Cartography for Hydrological baselines in Andalusia: Maps of the drainage network extracted from an elevation model (20m), related basins and subbasins for each section and their physical characterization. Associated statistics on land use, permeability, hydrological units, soil and lithology.

Biodiversity Subsystem

The main objectives of the Biodiversity Subsystem are to upload, to manage and to maintain all the information that is generated on biodiversity in Andalusia, as well as to provide consultation to all potential users: practitioners, researchers, educators and the general public. This is an open system which covers all information related to biodiversity in Andalusia, as well as all groups that normally work with this information.

The Biodiversity Subsystem contains three different modules, each of which has considerable autonomy but which are also linked. *The Core* stores basic information for each biological entity, *the Management Module of Plans and Studies* consists of a repository of documentation, data and images that are obtained from various sources and *the Monitoring Module* manages all data and trends for time series of species' population included in the plans.

All entities that are integrated in *The Core* are classified and ordered as a general dictionary for entities, an organized inventory that compiles information on living organisms, habitats and communities located in Andalusia. As of January 2009, there are about 20,700 records and there is a constant expansion and updating process that ensures consistency and uniformity of all the

Subsystem's contents. The subsystem can be accessed at http://bit.ly/4exovu.

Environmental Climatology Subsystem (CLIMA)

The Environmental Climatology Subsystem (CLIMA; Figure 3) includes an extensive network of meteorological stations belonging to different agencies. It employs a software application that enables data integration, quality control and shared management.

The main objective of the CLIMA is the homogenization of the quality-checked data from weather stations. The CLIMA is mainly used to obtain environmental indicators to assess the state of the environment and especially the potential impacts and consequences of climate change in the region.

The CLIMA Subsystem was established following an agreement between the Andalusian Regional Ministries of Environment, Agriculture and Fisheries and the Spanish State Agency of Meteorology, integrating a total of five monitoring networks consisting of nearly 2,300 meteorological stations. These include those of the Spanish State Agency for Meteorology (AEMET), the Network of automatic weather stations (EMA), the Regional Ministry for Agriculture and Fisheries of Andalusia (CAP) stations, the Phyto Sanitary Information Alert Network (RAIF), the Institute for Agricultural Training, Fisheries, Food Production and Environment (IFAPA) and the Agro-climatic Information Network (RIA).

THE SIGLA: COASTAL INFORMATION SYSTEM OF ANDALUSIA AND THE REDIAM'S COASTAL SUBSYSTEM

The SIGLA, or Coastal Information System of Andalusia was the result of previous collaboration between the Regional Ministries for the Environment and Public Works of Andalusia. These projects, which also involved the University of Se-

Figure 3. Screen snapshot from the CLIMA Subsystem home page at http://www.juntadeandalucia.es/medioambiente/clima

Figure 4. Map showing changes of the Andalusian coastline between 1956 and 2006

ville and different regional administrative bodies, included the production of the *Mapa Fisiografico del Litoral Andaluz* series (Physiographic Map of the Littoral of Andalusia) at a scale of 1:50,000 and a more generic *Physiographic Map of the Coastal Zone of Andalusia* (1:200,000).

The main objectives of the SIGLA were: (1) to integrate and geo-reference all the data needed for the management of the coast of Andalusia; and (2) to provide planners with a versatile, "live" tool to inform the decision-making process and to monitor data.

A vast database was generated and organized into a file-based GIS, enabling various applications such as direct and easy access to information, cartographic representation and data management and dissemination. From a thematic point of view, the SIGLA has a structure as follows.

Basic and Historical Cartography

The vector (scales 1:100,000 and 1:10,000 for topographic maps of Andalusia) and raster (scales 1:50,000 and 1:10,000 for topographic maps) cartography available for the Andalusian coastal area has been compiled enabling simultaneous viewing of different data layers (Figure 4). Also included is satellite imagery such as SPOT-panchromatic (1995-96) and IRS-panchromatic, Landsat and NOAA images. For the nearshore region, charts of the Andalusian coastal area are incorporated. (Junta de Andalucía – Consejería de Medio Ambiente, 2009).

On the other hand, a series of cartographic products has been created including shadow maps or map frames to serve as a background to cartographic production, and sets of ArcView vector files to define map areas for quick access (Figure 5). The process of building the spatial database required geometrically consistent integration of data from charts (Mercator projection) and topographic maps (UTM, zones 29 and 30). Finally, the spatial database of the GIS uses UTM (zone 30) projection as its reference co-ordinate system.

There are also different sorts of historical documents included in the SIGLA: (a) first editions of the National Topographic Map, (b) a set of 19th century charts, (c) a collection of "mapas singulares," historical landmarks in the cartographical representation of Andalusia, and d) a set of maps

Figure 5. Intervisibility analysis tool operating on DTMs. The circled area indicates where the user has chosen to launch the intervisibility tool.

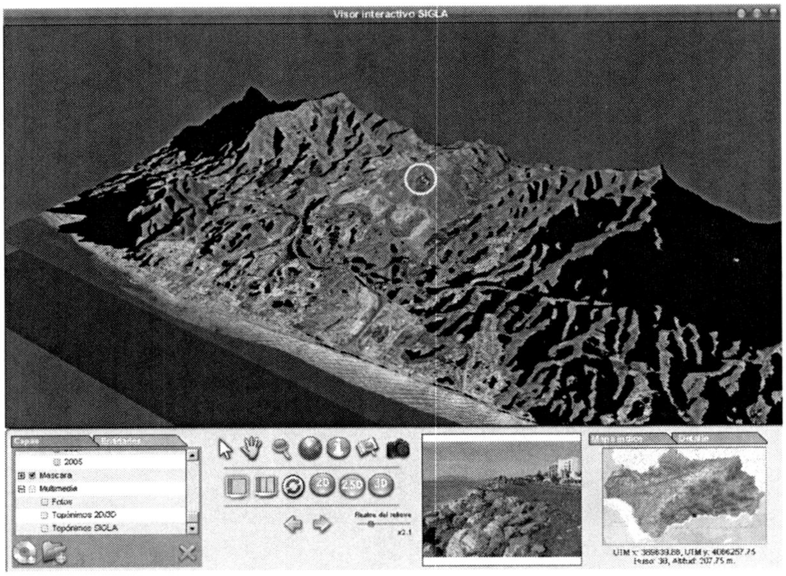

and charts from the *Cartoteca Histórica (Library of historical maps of Andalusia).*

Digital Terrain Models (DTMs)

The main concern in the process of building the DTMs was to guarantee the continuous representation of emerged topography and submerged bathymetry (Figure 6). Existing DTMs have been compiled in the SIGLA and some have been transformed to suit different projects. The DTMs in SIGLA are of different scales and resolutions: National (350 m), Regional (200 m), emerged areas of the Andalusia region (80, 40, 20 and 10 m) and Physiographic Map of the Andalusian coastal sheets (20 m).

Figure 6. Blended emerged and submerged topographies

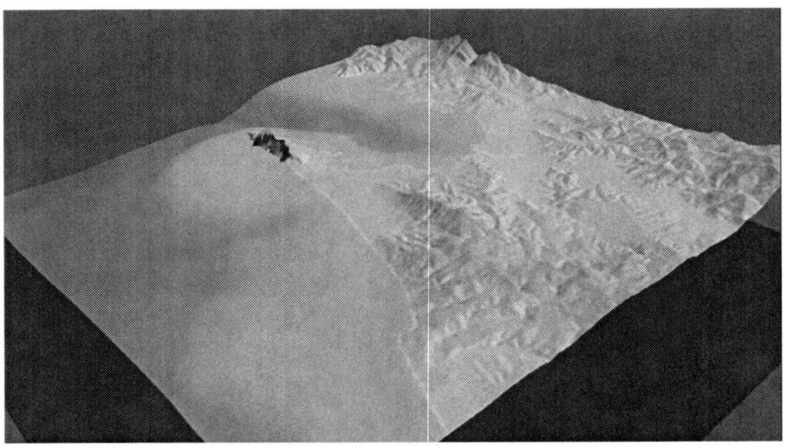

Aerial Photography, Multimedia, Thematic Maps

There is a wide diversity of aerial photography included in the SIGLA, including the early 1956 *Vuelo Americano*, black and white 1999 series, and orthophotos from 1995 to 2006. The most recent effort includes the acquisition of oblique aerial photographs, a vast collection of photographs (7000) and videos (nearly 700) of the Andalusian coast taken during 2001 and positioned by GPS. 3D images (1:200,000 scale) have also been generated.

Thematic maps provide the most specific information in the SIGLA. It includes physiographic cartography units at four levels, shoreline maps with a large amount of linked information, land use changes from 1956 to 2007 (Regional Land Cover Projects, Lacoast and Corine-Land Cover) and hydrological information (including basin and river catchment maps) (Moreira *et al.*, 1997).

Maritime Climate and Related Data

Special attention has been given to the maritime climate data due to its great importance in the natural processes and economic aspects of the region. The following variables (daily records) have been compiled: wind from five Andalusian meteorological coastal stations, wave data, sea level and tidal records, temperature and rainfall.

The SIGLA also includes a wide variety of vector maps related to complex information gathered by the different regional authorities in coastal areas. This includes variables affecting the coastal environment, such as jurisdictional maritime limits, sand movements and coastal vulnerability indices (Figure 7).

The ongoing process of data gathering, management and classification to conform with the SIGLA format continues to expand this already important resource from the various regional governmental agencies and administration.

In the new format, SIGLA is becoming The Coastal Subsystem of REDIAM, different in nature to the other subsystems since the spatial,

Figure 7. Example of viewer showing a three-dimensional (3D) view with vulnerability maps of the Andalusian Coast.

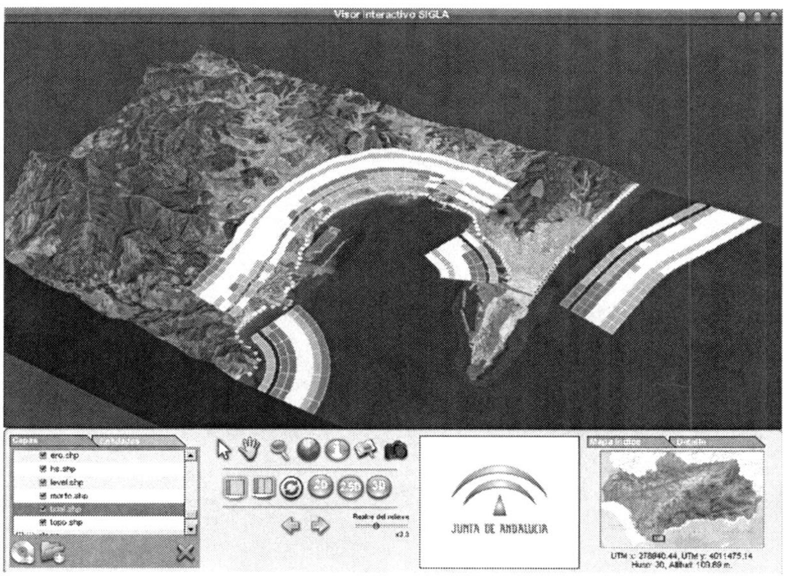

rather than thematic dimension is paramount in its design. The differences reside in the fact that multi-sectoral policy has to be taken into account and thus the focused approach of other subsystems is not valid for the coastal areas. Hence, the main thrust is to ensure inter-sectoral coordination, data harmonization and establishment of robust agreements between the data providers to ensure interoperability and long-term scalability.

The coastal subsystem benefits from the existing databases. However, the main hurdle is to achieve a new design that moves SIGLA from the file-based GIS to a fully OCG compliant geodatabase where the concept of service networks is at the base of the structure. At this point, the coastal subsystem has begun the process of development based on this conceptual and technical migration of SIGLA. This implies that all the thematic and baseline information (listed above) becomes available for the subsystem.

THE 2D/3D GEOGRAPHICAL VIEWER

Although the migration process is ongoing, the development of a two-dimensional/three-dimensional (2D/3D) geographical viewer (the ELIMCO Sistemas Viewer) is facilitating access to the information for the two main users: technical users from the regional administration by intranet and general public through a simplified version on Internet using a plug-in similar to the one used by Google Earth.

The viewer allows users to graph and visualize the information in 2D, 2.5 and 3D interactive flights (Figure 8). It is possible to synchronize spatially the different visual windows, which facilitate access to a wide collection of operations including consultation and comparison of the information from different periods and the ability to move synchronously several windows via a user-friendly interface. The user has the option to add vector formats with their thematic attributes (cad, shape) and raster formats (tiff, geotiff, jpeg, bmp, ecw, bil, bip, etc.) (Ojeda-Zújar *et al.*, 2007).

Figure 8. Thematic cartography for land use changes between 1956 and 2005.

For the 2.5D views and the 3D interactive flights a Digital Terrain Model of Andalusia (10 m) is needed to offer different options to the users like the production of hypsometric maps of elevations, the interactive production of contours, hill-shade images, slope maps with interactive control of visualization parameters, interactive inter-visibility analysis and the production of topographic profiles. One of the main attributes of this viewer is the possibility of drawing (points, lines, polygons) or digitizing synchronously (into a shape format) in any of the supported views (2D, 2.5D, 3D), facilitating the technical work and the comparison of spatial information in time periods and themes.

FUTURE TRENDS AND THE COASTAL SUBSYSTEM AS A GATEWAY OF REDIAM

REDIAM's Coastal Subsystem offers a unique opportunity to provide online access to different types of coastal data from many sources, and provides tools and services that facilitate interoperability. The technology used for the development of this allows the different systems to operate in a compatible way, despite each of the subsystems having a different architecture.

The trend is to develop a service architecture that interacts by connecting the repositories of information, which store and maintain the digital information, usually databases and geographic information layers from various institutions. One of the main interests for the subsystem in the future is the development and integration of the following web services:

- Management Service User Network: provides a single point of access to different REDIAM services and a centralized management of users and groups.
- Service Catalog and Information Services: search engines allow users to filter and query in a structured way other environmental information that is available at the Regional Ministry.
- Download Information Service: system, which allows the download of all the available environmental information in the REDIAM channel by users. It includes a download manager, which provides guidelines on how to access the various data files.
- Service registration and management of information requests: allows citizens to make requests for environmental information and provides a tool for monitoring their application. It also allows the Administration to manage, track and reply to requests for information.
- Service to the Citizen: Citizens' mailbox, the Citizen Service Center (CAU).
- Frequently Asked Questions (FAQs) and public participation tools
- Spatial Information Viewers and Map servers: development of a web GIS viewer; Development of a map server that acts as a client and a server to share spatial information and operate with remote data. Following the OGC standard it is interoperable as it includes Web Map Service (WMS) and Web Feature Service (WFS), and the possibility of integrating the data into Google Earth and Google Maps.
- Service registration and management agreements for Partners: enables the processing of high level global management and publishing agreements, subscription of members, requests for specific information, common spaces for the exchange of information, publication of environmental information, etc.
- Thematic sub-channels: development of thematic areas where the major work lines of the Regional Ministry are grouped.

CONCLUSION

Using the data infrastructure already created by the REDIAM, the Coastal and Marine Subsystem integrates the data of interest related to the seas and coasts of Andalusia, especially those with a strong spatial component. In addition to the socio-economic data related to fisheries, tourism, regional planning and urban pressure, the purely environmental issues (biodiversity, climate, physiographic, environmental planning and quality), as well as other topics, provide a clear environmental component. Information on marine hydrodynamics (currents, tides and waves) is also provided which is indispensable to understand the importance of the seas in the coastal dynamics and the sediment transportation as part of coastal ecosystem maintenance.

Facilitating accessibility of data by the public and all the partners of the REDIAM is as important as the collection and structuring of the information. This requires development of appropriate platforms and advanced viewers to enable users to query and use the data and the information.

It is imperative to ensure data quality, avoid duplication and redundancy and ensure that the data is up to date. The data model should also be consistent with those models already established. For this reason coordination amongst the institutions that generate information in Andalusia (the National Ministry of Environment of Spain - Directorate General of Coasts, the Spanish Institute of Oceanography, Spanish Naval Institute of Hydrography, Universities, etc.) is required.

Achieving all of these objectives, the REDIAM and the Coastal Subsystem represents a valuable resource for managers and researchers, as well as a meeting point for citizens interested in improving knowledge of their coastal environment.

REFERENCES

Cornaert, M. H., & Moreira, J. M. (1992). La coordination de l'information sur l'environnement à la C.E.E. au travers de projets de systèmes d'informations géorepérées. L'experience CORINE-SinambA. Rev. Regions d'Europe. *Special Environnement et Monde Rural,* 5, Besançon, France.

Environmental Climatology Subsystem (CLIMA). (2009). Retrieved October 31, 2009 from the CLIMA web site: http://www.juntadeandalucia.es/medioambiente/clima (in Spanish).

Environmental Information Network of Andalusia (REDIAM) Biodiversity Subsystem. (2009). Retrieved October 31, 2009 from the REDIAM web site: http://bit.ly/4exovu (in Spanish).

Environmental Information Network of Andalusia (REDIAM) Channel. (2009). Retrieved October 31, 2009 from the REDIAM web site: http://www.juntadeandalucia.es/medioambiente/site/web/rediam (in Spanish).

Environmental Information Network of Andalusia (REDIAM) Water Subsystem. (2009). Retrieved October 31, 2009 from the REDIAM web site: http://bit.ly/3T6iEQ (in Spanish).

Junta de Andalucía – Consejería de Medio Ambiente. (2009). *Compendio de Cartografía y Estadísticas Ambientales de Andalucía. 2 Edición 2008* (de Información Ambiental de Andalucía, R., Ed.). Sevilla: REDIAM.

Moreira, J. M. (1991). SinambA: an information system for the environment of Andalusia. Evaluation of the results of the CORINE Programme 1985-1990- Lessons for the European Environment Agency. *Workshop organized by the Commission of the European Communities*, Luxembourg, 28-30 May. 1991.

Moreira, J. M., Fernández-Palacios, A., & Rodriguez, M. (1997). Land use and land capability. Pressure and state indicators of the evolution of ecosystems. *European Topic Centre on Land Cover. Workshop on Land Cover Applications. Needs and Use*. Copenhagen, 12-13 May, 1997.

Ojeda-Zújar, J., Cabrera-Tordera, A., Fraile-Jurado, P., & Loder, J. (2007) Design and functionality of a 2D/3D viewer for a Coastal management oriented GIS. In *Proceedings of CoastGIS 2007*, Cantabria, Spain.

Regional Ministry for the Environment of Andalusia. (2009). Retrieved October 31, 2009 from the Regional Ministry for the Environment of Andalusia web site: http://www.juntadeandalucia.es/medioambiente (in Spanish).

KEY TERMS AND DEFINITIONS

Andalusia: An autonomous community (region) within the Kingdom of Spain located in the south of the Iberian Peninsula.

Coastal: Area between the seaward limit of terrestrial influence and the landward limit of marine influence.

Integrated Coastal Zone Management or ICZM: A dynamic, multi-sectoral approach to managing the coast which takes into account social, economic and environmental concerns of all parties with an interest in the coastal space.

Information System: The interaction between people, processes, and technology.

Marine: The oceans, seas, bays, estuaries, and other major water bodies, including their surface interface and interaction, with the atmosphere and with the land seaward of the mean high water mark.

Management: The act of getting people together to accomplish desired goals and objectives.

Public Participation: The principle of public participation holds that those who are affected by a decision have a right to be involved in the decision-making process. Public participation implies that the public's contribution will influence the decision.

REDIAM: The Environmental Information Network of Andalusia (Spain) led by the Regional Ministry of Environment – Junta de Andalucía.

Sectoral Cooperation: The process of working or acting together, which can be accomplished by both public and private agents from different sectors.

Spain: Officially the Kingdom of Spain, member state of the European Union, located in southwestern Europe on the Iberian Peninsula.

Section 3
Coastal Web Atlas Management and Governance Issues

Chapter 15
The International Coastal Atlas Network

Dawn J. Wright
Oregon State University, USA

Valerie Cummins
University College Cork, Ireland

Edward Dwyer
University College Cork, Ireland

ABSTRACT

This chapter introduces coastal web atlas (CWA) management and governance issues by way of a summary of the International Coastal Atlas Network (ICAN). ICAN is a newly-founded, informal group of over 30 organizations from over a dozen nations who have been meeting over the past two years to scope and implement data interoperability approaches to CWAs. The strategic aim (or mission) of ICAN is to leverage the expertise of its members to find common solutions to CWA design and implementation (e.g., user and developer guides, handbooks and articles on best practices, information on standards and web services, expertise and technical support directories, education, outreach, and funding opportunities, etc.). It also seeks to encourage and facilitate global operational interoperability between CWAs in order to enhance data and information sharing among users, and assist in the translation of coastal science to coastal decision-making.

INTRODUCTION

It is important to note further that while individual atlas efforts are important, as evidenced by the chapters in Section 2 of the book, collaboration between atlases on a regional and international scale can be critical as well. As a springboard to understanding the many dimensions of such collaboration, including both the inherent potentials and limitations, a trans-Atlantic workshop was held in Cork, Ireland, in July 2006. This gathering enabled participants from Europe and North America to assess the potential and the limitations of selected CWAs from the United States and Europe (O'Dea *et al.*, 2007). The catalyst for the workshop came initially from bilateral discussions between atlas developers in Oregon State University (OSU) and in University College Cork.

DOI: 10.4018/978-1-61520-815-9.ch015

Copyright © 2011, IGI Global. Copying or distributing in print or electronic forms without written permission of IGI Global is prohibited.

A community of interested parties grew as the two university groups facilitated the introduction of larger networks from both continents. One of the central outcomes of the workshop was recognition of the value of networking among the CWA workshop participants, and a desire for further opportunities to share valuable data and information. Thus, the seed was sown for the development of the International Coastal Atlas Network (ICAN), the subject of this chapter.

The strategic aim of ICAN, as initially established in an initial workshop (described in the next section), is ***to share knowledge and experience among atlas developers in order to find common solutions for coastal web atlas development whilst ensuring maximum relevance and added value for the users***. These atlases can be local, regional, national and international in scale. This is a mutually beneficial international activity with complementary strengths in evidence from North America, Europe, Africa, the Caribbean and Australia. ICAN participants can play a leadership role in forging further international collaborations of value to participating nations. This has tremendous potential for global spatial data infrastructures and Internet mapping projects. Benefits of participation in ICAN include: (1) receiving guidance and best practices for your own local CWA development from a community of experts in the information technology, GIS, data management and coastal/marine governance domains; (2) making your CWA interoperable with a larger universe of resources and communication channels that are needed for effective marine spatial planning, resource management, and emergency planning; and (3) participation in teaching and learning activities organized by ICAN or other organizations in the CWA domain; and (4) collaborative research proposal development (Wright *et al.*, 2009a & 2009b).

Given our long-term strategic goal of encouraging and facilitating the development of digital atlases of the global coast based on the principle of distributed, high-quality data and information, ICAN continues to convene workshops to:

- Ensure that ICAN has representation from coastal web atlas development and user groups from across the world.
- Develop technical and policy guidelines to assist coastal web atlas developers in acquiring data and engaging with data providers. Accordingly, collate and publish a set of best-practice guidelines for the development of coastal web atlases.
- Highlight the benefits of interoperability and standards-based systems to the coastal atlas developer communities.
- Develop collaborative projects for the sharing of know-how, implementation of technical solutions and demonstration of atlas benefits to users.
- Align the atlas efforts of the Network partners in order that interoperability can be facilitated.
- Engage with other relevant international projects and developments.
- Involve representatives of the relevant user communities to help in tailoring coastal web atlases to their needs.

FORMATIVE WORKSHOPS: 2006-2007

The first workshop, entitled "Potentials and Limitations of Coastal Web Atlases," was hosted by the Coastal and Marine Resources Centre (CMRC) at University College Cork in Ireland from July 24th to 28th, 2006 (O'Dea *et al.*, 2007). This workshop brought together over 40 participants from academia, government agencies and conservation organizations from Europe and North America to share technological knowledge and lessons learned from the development of national, state and regional CWAs. Among the aims of this workshop were the identification of state-of-the-art approaches to marine and coastal mapping and informatics and the development of guidelines as a resource for developers and decision makers on CWA projects. (O'Dea *et al.,* 2007).

Four key issues were examined: institutional capacity, technology, atlas design and data issues. Table 1 provides a summary of the top five "strengths, weaknesses, opportunities and threats" (SWOT) for each issue. This process helped to identify themes to be collectively addressed via future networking. In particular it helped to refine the selection of a topic for a second workshop on "Coastal Atlas Interoperability" on the campus of OSU in 2007 (Wright *et al.*, 2008).

Oregon workshop participants recognized that the terminology used to describe similar data can vary between specialties or regions, which can complicate data searches and integration. Current inventories within coastal atlases are insufficient for the purposes of networking *between* them. Each atlas has different classifications of data and information (e.g., critical information on coastal erosion that may be needed across a broad geographic region as supplied by several different atlases). Use of the word *seabed* in Europe versus use of the word *seafloor* to describe the same feature in North America is a good example of this scenario, as is the interchangeable use of *coastline* versus *shoreline* in both regions. Agreements on content can help to eliminate such problems, leading to a condition of "semantic interoperability" that will make searches between disparate, but mutually beneficial, projects feasible. Semantic interoperability is defined in Graybeal (2009) as "the ability of multiple systems to exchange information in useful ways; in particular, the ability for each system to 'understand' the terms of the other sufficiently to use those terms correctly." The topic is also discussed in much more detail in Chapter 4 (Lassoued *et al.*, this volume).

Ontologies provide the mechanism to enable semantic interoperability. An ontology, as defined in Lassoued *et al.* (this volume), is an organized list of terms that defines a set of concepts along with their meanings, the relationships among the concepts, and the logical rules that these meanings and/or relationships need to follow within a broader catalog of maps and data. Workshop participants learned how to use controlled vocabularies and ontologies in order to build a common approach to managing and disseminating coastal data, maps and information (Wright *et al.,* 2007), and concluded with the aim of designing and developing a demonstration interoperability prototype using the metadata catalogues of two mature atlases. The Oregon workshop also helped to crystallize the aforementioned long-term aims and objectives for ICAN through the formation of a strategy working group.

ICAN PROCESS

ICAN has grown from an idea at the first workshop to the cusp of a formal "virtual organization"/ community of practice, which has captured the interest of local and state governments, non-governmental organizations, universities, NOAA, the European Environment Agency (EEA), and UNESCO's Intergovernmental Oceanographic Commission. OSU in the U.S. and the CMRC of University College, Cork, Ireland have emerged as the lead universities and hubs of this nascent international network. ICAN has thus far proven to be an organic, highly motivated, productive informal group that despite its informality and lack of incorporation, is capable of generating tangible, useful products that attract and help users. Thus far these products are in the form of short workshops that ICAN members have given at various conferences (such as the international CoastGIS and the US Coastal Zone meetings) to introduce the notion of coastal web atlases, why they are important, and initial best practices for implementing them. This book is also a tangible product of ICAN. Other products in progress include user and developer guides, handbooks and articles on best practices, information on

Table 1. Results of a SWOT analysis: Top five issues relating to CWA development for atlas design, data and metadata, technology and institutional capacity. Source is O'Dea et al. (2007) and (in prep.).

Atlas Design
Strengths
- **Intuitive structure** of web sites and map pages; - Inclusion of **contextual information** in order to better understand the data; - **Hierarchical** data organization; - **Multiple user pathways** to retrieve maps and layers of interest; - **Tools** for data analysis and creating reports.
Weaknesses
- The **cartography / design** challenge of displaying many layers; - **Inadequate database management system** (DBMS) for efficient management of information, metadata and data; - Inadequate **search functions** for data and content; - Failure to meet user needs where atlas developments are **technology-driven**; - **Lack of distributed systems** to enable data owners to share and manage their own data.
Opportunities
- **Open source** technology; - **Enhanced DBMSs** to accompany open source web mapping technology to efficiently manage data, metadata and CWA content; - Improved **cartographic display** of large quantities of layers in coastal atlases; - Potential for sharing data through **distributed networks** (e.g., utilising Web Map Services and Web Feature Services); - Potential to develop **regional nodes** that tie in with larger atlases (e.g., national or statewide).
Threats
- **Funding** limitations (e.g., focus on technology rather than maintenance; staff turnover); - Keeping up with **design expectations** of users (e.g., Google Earth); - **User interpretation**: misunderstanding of how to use atlases or their components; - **Data policies, cost and Intellectual Property Rights** issues impact atlas design in data quality and accessibility, and thus atlas functionality (e.g., spatial analysis using large scale data).
Atlas Technology
Strengths
- **Improving technology** for publishing maps on the web: choice between open source (OS) and commercial-off-the-shelf (COTS) products; - Maturing **standards and specifications** (e.g., OGC specifications, ISO metadata and W3C standards); - Progress in **network capacity & hardware** (e.g., processor speeds, storage capacity and monitor resolutions); - Contribution of **other technologies and tools** (e.g., XML, UML and content management systems) to web mapping development; - Advantages of **OS tools** (e.g., broad community support, access to source code, low cost; Lack of COTS levels of technical support is possible disadvantage).
Weaknesses
- **Software support issues**: COTS software may offer more readily available commercial support, although OS software does not preclude this; - **Large datasets** can require significant disk space and are not always supported by web GIS software (e.g., raster data); - **Hardware becoming obsolete** (e.g., media obsolescence; backup software cannot deal with physical media; compatible drives no longer available); - **Inadequate metadata** may limit functionality (e.g., be incomplete, out of date and not match the data object; digital object identifiers (DOI) could be used to link data to metadata); - Web GIS is presently poor at dealing with **time series** and **3D/4D data**.
Opportunities
- **3D and 4D web GIS** riding on increased hardware and network capacity; - Simulation and **online spatial analysis**. - **Data mining**; - Widespread use of **geo-tagging** (e.g., geoRSS) to facilitate incorporation of many more items in web mapping systems; **Recommender systems** to supplement search queries; - Increased interest in CWA by policy makers and regulators as SDI initiatives become established leads to **funding potential** (e.g., EU Marine Green Paper).

continued on following page

Table 1. continued

Threats
- Difficulty in coping with **high server loading** during peak use; - **Technology evolution** can be disruptive: need to balance the exploration of new technologies against maintaining a stable and functioning system; - The challenge of **keeping data current**; - **Lack of funding** and consequent personnel turnover; - **Partners who are weak** or unwilling to co-operate.
Atlas-Related Data and Metadata
Strengths
- **Growing awareness** and acceptance of standards; - **Regulation** is driving the need for data; - Provides **publicity** for data products; - **Reduced labour costs** for routine searches; - **Widely accessible** to a broad range of users.
Weaknesses
- **Limited quantitative and analytic utility**: tools can sometimes produce suspect/alarming results; - **Data patchiness**; - Assessment of **data quality** is difficult on map presentations, original purpose and fitness for use can be hidden: 'pretty map syndrome'; - **Inadequate metadata**; - Existence of **multiple portals** to same data.
Opportunities
- Focus on the delivery of **source data** and **value-added products**, not only interactive maps; - **Identification of data gaps** and need for data collection requirements; - **Community-building** and **harmonization** among atlas providers (e.g., ontologies); - Become the **definitive reference** for certain data, if it contains current, good quality data; - Use **new, emerging technologies** for data and metadata presentation/delivery.
Threats
- **New competitive technologies** for improved data access (e.g., Google Earth); - **Intellectual property restrictions** limit data re-distribution; - Data viewed as **source of income**; - **Erratic funding** affects ability to develop and maintain atlas data as well as causes loss of skilled staff; - **Lack of incentives** for data providers.
Atlas-Related Institutional Capacity
Strengths
- **Academic CWA host institutions** have the ability to leverage additional research and education funds; - **Government CWA host agencies** may have **mandate** for CWA development; - Opportunities for **collaboration** with other institutes; - The permanent nature of government agencies ensures long-term **institutional support**; - Data and information requirements for the **Coastal and Marine sector** stimulate demands for CWA development.
Weaknesses
- Volatile and **short term nature of funding** and all associated impacts (e.g., staff turnover; difficult to maintain atlases); - Vulnerability to **political trends** and changes in priorities; - **Data access limitations**, licensing, and desire to recoup costs; - Limited experience in **marketing** and building awareness; - Tendency towards **project control** limits the formation of partnerships for data sharing.
Opportunities
- **Collaboration**: availability expertise and experience in CWA community; - Movement to E-GOV and **knowledge-based economy** (e.g., geospatial data can underpin many government activities); - Delivering on **government policy** (e.g., implement ICZM mandate); - **Economic development**: open data licenses could lead to new products; - **Leveraging data acquisition** (e.g., opportunities to pool resources to obtain more or better datasets).

continued on following page

Table 1. continued

Threats
- Changing **policy drivers**; - Perception of **'too many' databases** and mapping applications; - Credibility is affected by **poor quality data and metadata**, poor models and decision support software; - Over or **poor marketing** means user expectations not met; - **Challenges of collaboration**: partner doesn't deliver up to specifications.

standards and web services, expertise and technical support directories, education, outreach, and funding opportunities.

In 2008 the EEA invited and funded core ICAN members to present the idea of ICAN and to foster further collaborations at their headquarters in Copenhagen, Denmark (Dwyer & Wright, 2008). The value of the annual ICAN workshop event is evidenced by plans for a fourth annual ICAN workshop in Trieste, Italy, November 2009 (Wright & Dwyer, 2009), with a focus on end users; and a fifth annual workshop planned for Oostende, Belgium in 2010.

To implement these objectives ICAN has identified a range of activities in the areas of strategic planning, technical implementation, atlas assessment, outreach, training, and participation in scholarly communities, and strategic planning and funding. ICAN has established a web site to coordinate and report upon these activities (Figure 1) The web site is powered by Drupal, a powerful, flexible, open source, web-based content management system, suitable as an intranet and extranet server, as a document and image publishing system, a portal server and as a groupware tool for collaboration between entities. It follows standards for usability and accessibility, is licensed under the GNU (GNU's not Unix) General Public License (similar to Linux), and even has tools for managing multilingual content. It is therefore an

Figure 1. Example of ICAN's new web presence via Drupal hosted by the OSU College of Science, which will provide web resources in support of the workshop planning and proceedings (http://icoastalatlas.net).

ideal platform for supporting a virtual organization such as ICAN.

CURRENT MANAGEMENT AND GOVERNANCE ISSUES

There is a growing need to move beyond the demonstration of a conceptual framework towards a more formal organizational structure for ICAN. This will help to deliver the aspiration of operational interoperability in terms of a functioning web-based catalog of linked CWAs, so that in time, users will be able to conduct sophisticated and meaningful queries across a range of atlases. As a result, the long-term view is for global level operational interoperability, which will evolve as the ICAN community strives to increase awareness of the opportunities that exist for increased data sharing among policy makers and resource managers as strategic users of a CWA.

Current management and governance approaches are being reviewed to address these issues. Structures on governance (including formal procedures for receiving new members), strategic planning, and technical activities are under consideration so that ICAN can formally incorporate as an organization. The decision formally incorporate was made at the Oregon workshop, as this seems the best way to keep the momentum of current initiatives going, to attract more coastal atlas projects to use the products and services of ICAN, as well as to provide more nodes in the interoperability prototype, and to more readily secure funding. Plans are underway to assign chairs and members to a central coordination team, an advisory steering group, and within the existing technical task force, an ontology content governance group to manage storage, security, change management and serving of the ICAN interoperability ontologies (e.g., the fourth ICAN workshop in Trieste, Italy; Wright & Dwyer, 2009).

ICAN partners have had a high success rate in securing research and travel grants from national programmes in the past. However, future governance arrangements will provide a roadmap for greater diversity and innovation in securing funding for international collaboration. Examples of funding programmes to be targeted include the National Science Foundation (NSF) INTEROP (Community-based Data Interoperability Networks), NSF PIRE (Partnerships for International Research and Education), and, NSF STCI (Strategic Technologies for Cyberinfrastructure), the European Science Foundation (COST programme) and European Union Framework Programme proposals.

Future sustainability will also require collaborative partnerships with ongoing initiatives such as the European Environment Information and Observation Network (EIONET) of the EEA. For example, EIONET brings together nationally funded experts or groups of experts in organizations, which are regular collectors, or suppliers of environmental data at the national level and/or possess relevant knowledge of specific environmental issues (http://www.eionet.europa.eu/partners). Affiliation with international initiatives such as LOICZ (Land Ocean Interactions in the Coastal Zone), part of the International Geosphere-Biosphere Programme (IGBP) also provides a forum for additional exposure.

FUTURE DEVELOPMENTS

At the time of writing, ICAN still has much more work to do to: (1) formalize the governance and strategic planning structure for this new organization; (2) further its mission of establishing scientific grounding and knowledge base among the important knowledge domains of CWA development (geography, computer science, and coastal resource management); (3) develop advances in semantic interoperability of CWAs; (4) exchange lessons learned in spatial data infrastructure between the participating regions and (5) prepare collaborative submissions to funding programmes. Topics for future technical exploration include:

- Continued growth of ICAN as a formal organization.
- Implementation of a Shared Environmental Information System using ICAN's interoperability prototype (for a complete overview of the ICAN prototype and lessons learned thus far, please refer to Chapter 4).
- Applications emphasis on coastal zone vulnerability, climate change impacts.
- ICAN involvement in Ocean Teacher Academy project, "Training Academy for IODE Ocean Data and Information Networks."
- A continued focus on users: better knowledge of our users, their needs, and on continued inventory, assessment, and evaluation of atlases.
- Evaluating atlas impact, and developing analysis and decision-support tools in atlases.
- Updated guidelines to the coastal/marine research community and decision makers on the development of coastal atlas projects.
- Recommendations on the development of a joint program of work to the NSF, European Science Foundation, and the UNESCO IODE which will also provide international research experiences for the enhancement of doctoral dissertation research.

SUMMARY

In summary, ICAN seeks to further its mission of establishing scientific grounding and knowledge base among the important knowledge domains of CWA development (geography, computer science, and coastal resource management). In its formative stage ICAN has served to mobilise an international community of interest by demonstrating the added value of participation in a series of networking events. This process has been coupled with the production of a range of didactic material for technicians and coastal resource managers, which is also disseminated through a centralised web portal (by which continuing developments may be tracked at http://icoastalatlas.net or http://ican.science.oregonstate.edu). The commitment to the future enhancement of ICAN through evolving governance structures and technical plans, in addition to buy-in from multiple organisations, is evidence of the important role that ICAN has to play in fostering connectivity among geographic subfields and across disciplines in ways that enhance basic knowledge and methods that benefit all participating scholarly communities.

ACKNOWLEDGMENT

The authors wish to thank Oregon State University, University College Cork, the U.S. National Science Foundation, the Marine Institute of Ireland's Marine RTDI Networking and Technology Transfer Initiative under the National Development Plan, the European Environment Agency, Marine Metadata Interoperability, SeaDataNet and the Oregon Coastal Management Program.

REFERENCES

Dwyer, N., & Wright, D. J. (2008). *Report of International Coastal Atlas Network Workshop 3 on Federated Coastal Atlases: Building on the Interoperable Approach*. European Environment Agency: Copenhagen, Denmark. Retrieved July 29, 2009 from the ICAN web site: http://ican.science.oregonstate. edu/ican3_final_rpt.

European Environment Information and Observation Network. (2009). Retrieved October 31, 2009 from the European Environment Agency web site: http://www.eionet.europa.eu/partners.

Graybeal, J. (2009). *Achieving Semantic Interoperability*. Retrieved July 29, 2009 from The Marine Metadata Interoperability web site, Guides: Navigating the World of Marine Metadata section: http://marinemetadata.org/guides/vocabs/cvchooseimplement/cvsemint.

International Coastal Atlas Network. (2009). *About ICAN*. Retrieved October 31, 2009 from the International Coastal Atlas Network web site: http://icoastalatlas.net or http://ican.science.oregonstate.edu.

Lassoued, Y., Pham, T., Bermudez, L., Stocks, K., O'Grady, E., Isenor, A., & Alexander, P. (2009this volume). Coastal atlas interoperability . In Wright, D., Dwyer, E., & Cummins, V. (Eds.), *Handbook of Coastal Informatics: Web Atlas Design and Implementation*. Hershey, PA: IGI-Global.

O'Dea, L., Cummins, V., Wright, D., Dwyer, N., & Ameztoy, I. (2007). *Report on Coastal Mapping and Informatics Trans-Atlantic Workshop 1: Potentials and Limitations of Coastal Web Atlases*. University College Cork, Coastal & Marine Resources Centre: Cork, Ireland. Retrieved May 7, 2009, from the ICAN web site: http://ican.science.oregonstate.edu/node/47.

Wright, D., Dwyer, N., Haddad, T., O'Dea, L., Hart, D., LaVoi, T., et al. (2009b). Introducing the International Coastal Atlas Network. In *Proceedings of Coastal Zone '09*, Boston. Retrieved July 29, 2009 from the ICAN web site: http://ican.science.oregonstate.edu/node/40.

Wright, D. J., Cummins, V., & Dwyer, E. (2009a). *About ICAN*. Retrieved October 31, 2009 from the International Coastal Atlas Network web site: http://ican.science.oregonstate.edu/about_ican.

Wright, D. J., & Dwyer, E. (2009). *ICAN Workshop 4 and EEA National Reference Centers Meeting in Trieste, Italy, November 2009*. Retrieved October 31, 2009 from the International Coastal Atlas Network web site: http://ican.science.oregonstate.edu/ican4.

Wright, D. J., Watson, S., Bermudez, L., Cummins, V., Dwyer, N., O'Dea, L., et al. (2007). *Report on Coastal Mapping and Informatics Trans-Atlantic Workshop 2: Coastal Atlas Interoperability*. Corvallis, OR: Oregon State University. Retrieved July 29, 2009 from the ICAN web site: http://ican.science.oregonstate.edu/node/46

KEY TERMS AND DEFINITIONS

Coastal Web Atlas: A collection of digital maps and datasets with supplementary tables, illustrations and information that systematically illustrate the coast, oftentimes with cartographic and decision support tools, all of which are accessible via the Internet. Also known as web atlas, digital atlas, digital coastal atlas.

Capacity Building: With regard to coastal atlases, the actions involved in building local GIS infrastructure, including maintenance and update of GIS data layers, software, computer equipment and labs/offices, salaried personnel, and action items for project initiatives.

Coastal Informatics: A broad academic field encompassing the management and analysis of data collected from and representing the coast. Informatics can include spatial data infrastructure, hardware and software infrastructure, computational networking, modeling, and experimentation, the design and deployment of data portals and Internet mapping sites, as well as the creation, analysis, and understanding of data/metadata vocabularies and ontologies, metadata creation/extraction/cross-walking tools, geographic and information management systems, and grid computing.

Information Management: The means by which an organization, agency, or individual collects, documents, shares, and uses information. It often involves creating or identifying the appropriate resources to find quality information in order to fill gaps in knowledge. Coastal web atlases are therefore a key component of information management as applied to coast region of the world.

Marine Spatial Planning: A critical part of regional governance that involves the integrated, forward-looking planning and consistent decision-making regarding various uses of the coast and nearshore. Marine spatial planning must be guided by specific policies and regulations governing usage, the conditions that apply, and an eye toward what possible conflicts in use may arise. Marine spatial planning is often aided by a coastal web atlases which can provide data on habitats, species migrations, land and territorial sea use by humans, navigation, managed areas (parks, reserves, disposal sites, etc.), commercial and recreational fishing, and the like.

Ontology: A type of knowledge organization system that defines a set of concepts with meanings, relationships among them, and logical rules on these meanings and/or relationships.

Regional Governance: Policies and initiatives that allow state and local governments to pursue concrete, practical steps toward more coordinated and holistic management of ocean and coastal resources. This often includes establishing lines of communication among regions to facilitate implementation of best practices, forming partnerships among regions, and deploying proven management principles and approaches (e.g., ecosystem-based management). Coastal web atlases often provide the datasets, mapping tools, and contextual information needed for effective regional governance.

Research Collaboration Network: A collaboration of research scientists, resource managers, technical staffers, and other interested parties to develop a coordinated research network focused on a particular topic. The group fosters communication among those with common goals and interests, along with collaboration on common projects across disciplinary, organizational, institutional and geographical boundaries. The International Coastal Atlas Network is an example of a research collaboration network.

Semantic Interoperability: The ability of multiple systems to exchange information in useful ways, in particular, the ability for each system to "understand" the terms of the other sufficiently to use those terms correctly. In the computing and GIS world, this is a condition where two or more computer systems are able to exchange information and have the meaning of that information accurately and automatically interpreted by the receiving system.

Spatial Data Infrastructure (SDI): A framework via an organization of people or government agencies, via the Internet, or via a series of guiding policies or standards to assist people with acquiring, processing, using, and preserving spatial data. The spatial data are often in geographic information system (GIS) format, are not, but not limited to this.

Virtual Organization: A group of individuals, agencies, whose members and resources may be geographically dispersed, yet function as a coherent unit through the use of information technology to share knowledge, skill sets, and access to each others' resources and expertise in nontraditional ways.

Web GIS: Geographic information systems functions as deployed via a web site or series of web sites rather than just on the desktop.

Chapter 16
Coastal Atlases in the Context of Spatial Data Infrastructures

Tony LaVoi
NOAA Coastal Services Center, USA

Joshua Murphy
NOAA Coastal Services Center, USA

Gabe Sataloff
NOAA Coastal Services Center, USA

Roger Longhorn
Info-Dynamics Research Associates Ltd., Belgium

Andrus Meiner
European Environment Agency, Denmark

Ronan Uhel
European Environment Agency, Denmark

Dawn J. Wright
Oregon State University, USA

Edward Dwyer
University College Cork, Ireland

ABSTRACT

This chapter summarizes key projects and initiatives that are being implemented on very large scales (national/international) by national governments and commissions to build coastal spatial data infrastructures (SDIs). These include SDI efforts in the US and Europe that are closely related to ICAN, and as such are of great value to its mission of developing interoperable atlases, providing along the way solutions for the integration of not only technologies, but people, institutions, and institutional objectives.

DOI: 10.4018/978-1-61520-815-9.ch016

Copyright © 2011, IGI Global. Copying or distributing in print or electronic forms without written permission of IGI Global is prohibited.

INTRODUCTION

Coastal and marine spatial data infrastructure (SDI) is largely comprised of national-level datasets that are created with consistent technical standards, quality, and under adequate licensing terms, accessible for use in GIS and Web GIS, with accompanying metadata following appropriate standards, all complying with best practice (for a complete treatment of SDI in all parts of the world see Onsrud, 2007; Masser, 2007; and van Loenen *et al.*, 2009). It follows that if an SDI is the overarching framework to assist people with acquiring, processing, using, and preserving spatial data, then a coastal web atlas (CWA) is an important subset of an SDI where all things "coastal" are concerned. CWAs often provide the datasets, mapping tools, and contextual information needed for an effective SDI, and in doing so reduce valuable time, money, and effort in the process. It follows also that CWAs would depend upon an SDI where standards are concerned (such as the use of Federal Geographic Data Committee or FGDC, contents standard for metadata in the US and the adoption of International Standardization Organization or ISO metadata standards in Europe). The issue of how CWAs will contribute to existing and future SDIs remains an exciting and still-open question (e.g., Wright & Dwyer, 2009; Wright *et al.*, 2009). The case studies presented Chapters 6-14 of this book have provided many examples, all with the common theme of compiling various data sources and presenting them in a common, more understandable framework (e.g., a common clearinghouse), which in turn has fostered discussions, spatial analyses, and collaborative relationships across (and sometimes within) various jurisdictions.

This chapter reports briefly on the developing concept of a federated atlas of the entire US coastal zone as in important component of an SDI. The US federal government is instituting the federated atlas in cooperation with partners at the state and local government levels, as well as academic institutions, non-governmental organizations and private industry. Many if not most of these partners would be purveyors of a CWA. Next is a description of an emerging European Atlas of the Seas as instituted by the European Commission, based on several years of planning and due for an initial launch by the beginning of 2010. The chapter ends with descriptions of related projects focused primarily on offshore oceanographic science but still greatly relevant to the coastal zone management emphasis of CWAs: the European Coastal-shelf Sea Operational Observing and Forecasting System (ECOOP), SeaDataNet, and the projects of SeaZone Solutions Limited, a wholly owned subsidiary of the UK Hydrographic Office. All of these efforts represent important opportunities for CWAs to leverage data, methods, and approaches and are of great value to ICAN in its mission to develop interoperable atlases that would contribute to an SDI and to effective coastal/ocean governance.

A FEDERATED COASTAL ATLAS OF THE USA: EFFORTS TO BUILD THE COASTAL NATIONAL SPATIAL DATA INFRASTRUCTURE

The responsibility for mapping the US coastal zone lies with a multitude of federal, state, and local agencies, each charged with supporting their unique mission or geographic area. Historically, these organizations have worked in isolation, obtaining coastal and marine data that meet their specific needs. As a result, this valuable and often expensive information exists in a variety of formats and projections, with public access provided through an assortment of disparate on-line catalogs, clearinghouses, and Internet mapping applications. The need for a standardized approach to coastal geographic data collection and dissemination resonates with the coastal management community, which has embraced the use of geographic information systems (GISs) to support decision-making. Consequently, there are several key efforts underway within the federal government to bring this desired cohesion to coastal mapping. Integrated ocean and coastal mapping (IOCM; http://www.oar.noaa.gov/oceans/t_oceanmapping.html), the Federal Geographic Data Committee's (FGDC) Marine and Coastal Spatial Data Subcommittee (MCSD; http://www.csc.noaa.gov/mbwg/htm/multipurpose.html), Digital Coast (http://www.csc.noaa.gov/digitalcoast), and Multipurpose Marine Cadastre (MMC; http://www.csc.noaa.gov/mbwg/htm/multipurpose.html) represent management initiative that place an emphasis on partnerships, coordination, and standards. These activities support a larger federal effort to build a coastal portion of the National Spatial Data Infrastructure (NSDI; http://www.fgdc.gov/nsdi/nsdi.html), ensuring that coastal geographic data are developed and disseminated in an efficient manner. This approach, in turn, supports the concept of a "Federated Coastal Atlas of the USA," through which users have access to coastal geographic data that are standardized and easy to use.

In March 2009, Congress passed the Omnibus Public Land Management Act of 2009 (US House of Representatives Bill 146), which includes the Ocean and Coastal Mapping Integration Act (Sec. 12201 of the Omnibus Act), calling for "a program to develop a coordinated and comprehensive Federal ocean and coastal mapping plan for the Great Lakes and coastal state waters, the territorial sea, the exclusive economic zone, and the continental shelf of the United States that enhances ecosystem approaches in decision-making for conservation and management of marine resources and habitats, establishes research and mapping priorities, supports the siting of research and other platforms, and advances ocean and coastal science" (US Congress, 2009). This legitimization of the need for coastal information should contribute to the long-term sustainability of numerous coastal web atlas projects in American coastal states.

Data Coordination and Standards

Recent advances in coastal and ocean mapping technologies have made the collection of spatial data faster and more accurate. However, this upward trend in technology has resulted in increased costs. The effective coordination of data collection efforts is critical to ensuring that ever-dwindling mapping resources are used in an efficient manner. Likewise, the development and promotion of standards for data creation and documentation will ensure that, once data have been collected, it can be used by common software tools and understood by many users. The following efforts recognize these fundamental aspects of coastal mapping and demonstrate a coordinated response at the federal level.

Integrated Ocean and Coastal Mapping

The goal of IOCM is to improve coordination among federal, state and local government, nongovernmental organizations (NGOs), and private sector mapping activities. Led by representatives from the National Oceanic and Atmospheric Administration (NOAA), US Geological Survey (USGS), Minerals Management Service (MMS), and US Army Corps of Engineers (USACE), IOCM is a framework for data collection and management that will eliminate duplication of efforts, and leverage mapping capabilities and resources from various organizations. Supporting IOCM activities is the recently-passed Ocean and Coastal Mapping Integration Act, which directs the President to "establish a program to develop a coordinated and comprehensive Federal ocean and coastal mapping plan for the Great Lakes and coastal state waters, the territorial sea, the exclusive economic zone, and the continental shelf of the United States that enhances ecosystem approaches in decision-making for conservation and management of marine resources and habitats, establishes research and mapping priorities, supports the siting of research and other platforms, and advances ocean and coastal science" (US Congress, 2009). It is anticipated that IOCM, through its response to these directives, will serve as a solid foundation from which current and future mapping activities will proceed in a coordinated and efficient manner, upholding the maxim of "map once, use many times."

FGDC Marine and Coastal Spatial Data Subcommittee

The primary mission of the FGDC Marine and Coastal Spatial Data Subcommittee is to develop and promote the Marine and Coastal NSDI. Formed in 2000, the subcommittee consists of representatives from federal agencies that collect, or finance the collection of marine and coastal spatial data as part of their mission or have direct application of these data to their programs. As a member of the FGDC, the focus of the group's activities centers on the promotion of standards and interoperability. To date, this has included: the development of a National Shoreline Data Content Standard, establishment of bathymetry as a primary NSDI data theme, and the development and maintenance of the ocean and coastal community within the Geospatial One Stop (GOS; http://geodata.gov). In July 2009, following a review of the FGDC subcommittee structure by the FGDC Coordination Group, it was recommended that the Marine Boundary Working Group (MBWG), formed in 2001, be subsumed into the Marine and Coastal Spatial Data Subcommittee, to better reflect the cross-cutting issues that impacted on both these areas of interest (US Congress, 2009). As the IOCM effort moves forward, the subcommittee envisions itself as continuing to take the lead in the development of data standards, which will allow mapping organizations at various levels to develop and distribute data that are easy-to-use and understand.

Data Access and Dissemination

The impacts of the aforementioned federal efforts to coordinate and standardize coastal geographic data collection would be diminished if not for the development of mechanisms to provide effective and efficient data access. Several efforts to provide geographic data from across the federal government have been launched over the last decade, including The National Map (http://nationalmap.gov) and GOS and Data.gov (http://data.gov). As their names imply, these data portals serve a broad spectrum of data, including many national-level coastal datasets. However, their comprehensive scope means that they generally lack a focus on coastal issues and contexts. The Digital Coast and Multipurpose Marine Cadastre were developed to fill this void in coastal data access. Similar to the efforts of IOCM and the FGDC MCSD, these data access and dissemination platforms emphasize partnership building and coordination within federal, state, and local scales to ensure their relevance and utility to the coastal management community.

Digital Coast

The Digital Coast is a partnership and community resource initiated by the NOAA Coastal Services Center for use by those organizations managing the nation's coastal resources. It was developed to address the needs of its users by providing a simplified way to access coastal data. However, what separates the Digital Coast from other data access mechanisms is its focus on the tools and methods to turn that data into useful information. A comprehensive web site (http://www.csc.noaa.gov/ digitalcoast) is the primary mode of communication for the Digital Coast effort. Here, users will find access to an ever-growing inventory of imagery, land cover, elevation, benthic habitat, hydrology, and marine boundary datasets. Decision support tools are provided in the following categories: analysis, informational, simulation, data visualization, and data handlers. Training programs linked to the Digital Coast are designed to build the skills and knowledge needed to fully apply these data and tools. The "Digital Coast in Action" section of the web site combines all facets of the Digital Coast into real world examples of how end users are applying geographic information. Through short narratives, this section shows how organizations interact with the Digital Coast partners and use the data, training, and tools to address specific coastal issues.

While the web site is the public face of the Digital Coast, partnerships are critical to the success of the overall effort. A Partnership Group, composed of organizations representing users at various levels (currently consisting of the NOAA Coastal Services Center, the Association of State Floodplain Managers, the Coastal States Organization, the National Association of Counties, the National States Geographic Information Council, and The Nature Conservancy), helps determine the focus of content, form and function of Digital Coast. The group also identifies Digital Coast priorities and then works together to address coastal issues, with the goal to unify groups that might not otherwise have worked well together. Collaborations with IOCM, The National Map, Geospatial One Stop and other associated federal mapping efforts further ensure that federal groups work in unison to provide interchangeable data and information to the user. In addition, close working relationships with the private sector and academic mapping communities allows the Digital Coast to provide users with assistance in filling identified gaps in coastal data. The focus on partnerships and collaborations, coupled with an issue-specific and application-driven approach to data access and dissemination makes the Digital Coast more than a coastal web atlas; it is an enabling initiative for coastal managers.

Multipurpose Marine Cadastre

The Multipurpose Marine Cadastre (MMC) accessed through the Digital Coast http://www.csc.noaa.gov/mbwg/htm/multipurpose.html), is an example of a federal-level effort to provide geographic data access and dissemination for a specific issue area. National-level data and information contained in this Internet mapping application is used to address a range of marine spatial planning (MSP)-based issues, including the demand for alternative energy. The MMC project also provides region-specific data to address issues that involve aquaculture, submerged lands leasing, and marine conservation. The provision of marine cadastral data by the authoritative source is a fundamental concept to the MMC. As such, standardized data publishing guidelines, developed in conjunction with the FGDC Marine Boundary Working Group, are being developed to encourage additional data providers to join the effort. By providing issue-specific geographic data in context with supporting information related to decision-making, the MMC serves a model for other coastal data access efforts where focused access to issue-specific data is desired.

Assessment Thus Far of US Federated Atlas Concept

The aforementioned efforts in the US are still nascent and have not quite come together in total, but the prognosis for such is quite positive. As discussed in Dwyer & Wright (2008), factors that should weigh heavily in the ultimate success of a federated atlas include:

- Education and advocacy work to show the added value that CWAs systems bring, especially as operational decision-support systems (a key role for ICAN, Chapter 15).
- The presence of a vocal champion for a *federated* atlas. Now that the Digital Coast initiative has been launched, the time is

ripe for this in the US. NOAA CSC and the FGDC Marine and Coastal Spatial Data subcommittee will begin to make inroads here, and this is yet another key role for ICAN as well.

- Regional ocean governance is not constrained by state boundaries, and yet it is still often hard to implement partnerships between states. Although oftentimes easier, agencies must avoid taking a "silo" approach to one's atlas, worrying only about datasets, concepts, and language within your own state (and not having to worry about vocabularies or ontologies that must translate to project and initiatives in other regions).

In recent years, there has been a rising awareness that there are several critical, systemic societal issues that CWAs can address. Among these is the emerging awareness of the *impacts* of sea level rise and of global climate change on coastal communities, which are enabling agencies and directorates to obtain additional funding for high-resolution light detection and ranging (LIDAR) mapping and other surveys. Developers and administrators of atlases will be focusing on these with increasing regularity and effectiveness, especially while in pursuit of funding to maintain atlases (see Chapter 19).

A EUROPEAN ATLAS OF THE SEAS

Following the 2006-2007 public consultation on a future Maritime Policy for the Union, the European Commission focused on raising the visibility of *Maritime Europe*. One of the proposals to pursue this goal was the creation a European Atlas of the Seas, the development of which has already started by the European Commission (Commission of the European Communities, 2007), and for which the closest American counterpart is likely the Digital Coast initiative.

Much information is available about Europe's oceans and about maritime activities, but there is a lack of published material bringing it all together. As part of the Maritime Policy Action Plan proposed in October 2007 (Commission of the European Communities, 2007a), the European Commission will launch a European Atlas of the Seas, using available spatial information and building on the work of a EMODNET (Commission of the European Communities, 2009; Table 1). The development of the Atlas will also demonstrate the relevance of setting up an integrated data network (GIS-based web facility), and the importance of the cross-sectoral accessibility of such data.

An alternative view of atlas contents is represented by the publication of Vivero & Meteos (2007):

Table 1. Preliminary contents of the European Atlas of the seas (from Commission of the European Communities, 2007b).

Category	*Indicative content elements*
Sciences	Oceanography, meteorology, biology.
Governance	Spatial dimension of EU policies, marine jurisdictions, administrative units, spatial planning, cooperation agreements.
Economy	Transport, energy, extraction, fishery, telecom, tourism, research.
Society	Demography, employment, education institutions, activities
Environment	Protected areas, marine pollution, maritime accidents, climate change, waste, artificial coastline.
Cultural heritage	Coastal & underwater archaeology, classified heritage sites at coast, maritime museum, naval battle sites & submerged wrecks.

- **Generalities:** Main jurisdictional concepts; Europe in the global ocean; history and tradition.
- **Territory and maritime jurisdictions**: European waters; maritime borders; insularity; marine regions; territorial conflicts; coastal zones.
- **Uses and types of exploitation**: Marine productivity; fishing and aquaculture; sea transport; energy; waste.
- **Management and policies**: Regulatory framework and international agreements; fisheries management and environmental management; spatial planning.

Based on suggestions from public consultation, the Atlas is being developed as an on-line, free-of-charge, regularly updated, descriptive and cartographic resource, covering seas, coast and outermost regions, and based on existing data and initiatives. It should encourage the public to appreciate marine resources and be produced in several languages to meet the needs of different European maritime regions. Relevant data sources for six broad categories with more that 100 themes will initially be collected. Following progress in building the EMODNET, the Atlas will also show collected data in a way that is easily understood by scientists, but also the wider public. It will highlight European Union (EU) legislation (fishing zones, etc.) and EU research projects.

The planned geographic extent of the Atlas is beyond the continental shelf and will include as a minimum three zones: coastal zone (from the shore to the 12 nautical mile limit), shelf (from the coastal zone to the continental shelf edge), and deep sea (beyond the shelf edge to the limit of Exclusive Economic Zone or EEZ, 200 nautical miles from shore). Given the influence of fluvial input to many European seas, the Atlas might also extend upstream from the river mouths.

Work has started to examine in detail which environmental features, useful data and information are already available, such as Water Information System for Europe (WISE), which has data relevant to coastal and marine environmental issues). Existing policy-relevant information should also be included, e.g. simplified scientific advice on the basis of which decisions were prepared under the Common Fisheries Policy (i.e., advice in a generally understandable and accessible form). Some of the information layers of the Atlas may well have to be based on information systems of existing community institutions and relevant EU agencies. A working group has been established to facilitate coordination.

There are certain challenges for the Atlas as a public web-based information system. It should be noted that the semiotics (i.e. the meaning of different signs or symbologies) of on-line atlases is often different from that of paper versions. Ideally, three dimensional viewing can be implemented in an online atlas, which is important from a maritime perspective with its 3 "parallel worlds" – sea surface (space), water column and seabed. A less ambitious and more realistic approach in the first version of the Atlas may result in fewer of such three-dimensional maps being included, but will be an objective for later versions.

Conceptual development of the European Atlas of the Seas started in 2008. Various entitites with the European Commission started to develop relevant database and web interface in 2009, and the first online version is targeted for release by the beginning of 2010.

Other European Regional Marine SDI Developments

In Europe, coastal zone management is seen as a key function of the Integrated Maritime Policy for Europe and the European Union (Commission of the European Communities, 2007b) and the marine spatial planning instrument within that policy. Access to harmonized marine and coastal information also figures largely in development of the EU's Shared Environmental Information System (SEIS) (Commission of the European

Communities, 2008). Both these initiatives claim dependency on continued development of the INSPIRE European SDI (Commission of the European Communities, 2007c) to create the level of harmonization needed to be able to discover, access and integrate disparate datasets that cross borders, cross disciplines and cover multiple data themes.

The INSPIRE Directive (Infrastructure for Spatial Information in the European Community) sets legal requirements that apply to spatial datasets (any data with a locational attribute) at all levels of government. It establishes technical specifications for metadata, data and information services to current international standards set by the International Standards Organisation (ISO) and the Open Geospatial Consortium, Inc. (OGC).

The INSPIRE Directive defines 34 data themes, many of which apply to coastal, marine (oceanic and sea regions) and related data, such as marine protected sites, coastal zone management units, etc. Most of the data themes relating to the coast are will be covered by INSPIRE's technical implementing rules, as pertaining to metadata and data. These include guidelines pertaining to the harmonization of services and overall interoperability. The goal is to complete implementation between 2013 and 2019 (Commission of the European Communities, 2008). INSPIRE's 34 data themes span three "Annexes" which determine the timetable by which the Directive's various implementing rules must be brought into effect for each theme. Some themes are quite broadly defined, such as "Area management / restriction / regulation zones and reporting units," which includes coastal zone management along with waste dumps, noise restriction zones, etc.

The Directive's implementing rules are legally enforceable in all EU member states, and are developed by experts with agreement of an official INSPIRE Regulatory Committee comprising representatives from all EU states. The rules cover metadata standards and content, data harmonization and interoperability specifications for all themes, various types of services (including performance criteria), data access and sharing principles, and monitoring performance of the SDI and its impact. It is important that members of the different coastal and marine stakeholder communities be involved in creating the implementing rules, otherwise special characteristics of this otherwise generic SDI that are important for coastal SDI needs may be overlooked. This is equally true for national SDI initiatives, yet only a very few of the NSDIs now under development appear to have any special input from the coastal/marine community.

Of the 34 data themes covered by the INSPIRE Directive, Tables 2 and 3 indicate the 22 themes that are applicable to the coastal and marine communities directly or indirectly. It is important to reiterate that, unlike other regional and even many national SDI initiatives, the components of the pan-European SDI are legally mandated by the European Union institutions. Thus, European Union member states must enact legislation and regulations implementing the specifications and standards developed within INSPIRE, according to a set timetable – one that currently stretches to 2019.

In many cases, the type of input and advice needed from the coastal community, directed to the various data specification technical working groups (TWGs), relates to ensuring that coastal information needs are not forgotten, for example in setting boundaries – in three dimensions and even in time – regarding aquaculture facilities, risk zones, location of off-shore energy platforms or mineral extraction sources, etc. Hydrography, elevation (bathymetry and shoreline) and area management zones (including coastal zone management) are the most obvious data themes where direct involvement of a coastal information expert on the relevant TWG would be useful.

Table 2. INSPIRE data themes of direct relevance to coastal information needs (from Commission of the). European Communities, 2007c

Annex I
Hydrography - Hydrographic elements, including marine areas and all other water bodies and items related to them, including river basins and sub-basins.
Protected sites - Area designated or managed within a framework of international, community and member states' legislation to achieve specific conservation objectives.
Annex II
Elevation - Digital elevation models for land, ice and ocean surface. Includes terrestrial elevation, bathymetry and shoreline.
Annex III
Area management/restriction/regulation zones and reporting units - Areas managed, regulated or used for reporting at international, European, national, regional and local levels. Includes dumping sites, restricted areas around drinking water sources, nitrate-vulnerable zones, regulated fairways at sea or large inland waters, areas for the dumping of waste, noise restriction zones, prospecting and mining permit areas, river basin districts, relevant reporting units and coastal zone management areas.
Agricultural and aquaculture facilities - Farming equipment and production facilities (including irrigation systems, greenhouses and stables).
Environmental monitoring facilities - Location and operation of environmental monitoring facilities includes observation and measurement of emissions, of the state of environmental media and of other ecosystem parameters (biodiversity, ecological conditions of vegetation, etc.) by or on behalf of public authorities.
Natural risk zones - Vulnerable areas characterised according to natural hazards (all atmospheric, hydrologic, seismic, volcanic and wildfire phenomena that, because of their location, severity, and frequency, have the potential to seriously affect society), e.g. floods, landslides and subsidence, avalanches, forest fires, earthquakes, volcanic eruptions.
Oceanographic geographical features - Physical conditions of oceans (currents, salinity, wave heights, etc.).
Sea regions - Physical conditions of seas and saline water bodies divided into regions and sub-regions with common characteristics.
Energy resources - Energy resources including hydrocarbons, hydropower, bio-energy, solar, wind, etc., where relevant including depth/height information on the extent of the resource.
Mineral resources - Mineral resources including metal ores, industrial minerals, etc., where relevant including depth/height information on the extent of the resource.

Table 3. INSPIRE data themes of indirect relevance to coastal information needs (from Commission of the European Communities, 2007c).

Annex I
Coordinate reference systems - Systems for uniquely referencing spatial information in space as a set of coordinates (x, y, z) and/or latitude and longitude and height, based on a geodetic horizontal and vertical datum.
Geographical grid systems - Harmonised multi-resolution grid with a common point of origin and standardised location and size of grid cells.
Annex II
Land cover - Physical and biological cover of the earth's surface including artificial surfaces, agricultural areas, forests, (semi-) natural areas, wetlands, water bodies.
Geology - Geology characterised according to composition and structure, including bedrock, aquifers and geomorphology.
Annex III
Land use - Territory characterised according to its current and future planned functional dimension or socio-economic purpose (e.g. residential, industrial, commercial, agricultural, forestry, recreational).
Human health and safety - Geographical distribution of dominance of pathologies (allergies, cancers, respiratory diseases, etc.), information indicating the effect on health (biomarkers, decline of fertility, epidemics) or well-being of humans (fatigue, stress, etc.) linked directly (air pollution, chemicals, depletion of the ozone layer, noise, etc.) or indirectly (food, genetically modified organisms, etc.) to the quality of the environment.
Utility and governmental services - Includes utility facilities such as sewage, waste management, energy supply and water supply, administrative and social governmental services such as public administrations, civil protection sites, schools and hospitals.
Production and industrial facilities - Industrial production sites, including installations covered by Council Directive 96/61/EC of 24 September 1996 concerning integrated pollution prevention and control and water abstraction facilities, mining, storage sites.
Bio-geographical regions - Areas of relatively homogeneous ecological conditions with common characteristics.
Habitats and biotopes - Geographical areas characterised by specific ecological conditions, processes, structure, and (life support) functions that physically support the organisms that live there. Includes terrestrial and aquatic areas distinguished by geographical, abiotic and biotic features, whether entirely natural or semi-natural.
Species distribution - Geographical distribution of occurrence of animal and plant species aggregated by grid, region, administrative unit or other analytical unit.

OTHER RELATED PROJECTS

So far we have seen that in Europe and the US there is significant activity in the area of coastal and marine spatial data infrastructure building and growing momentum for federated coastal atlases. The European Commission Research Framework Program funds a number of projects to advance data handling and information along these lines, two of which ECOOP and SeaDataNet. This section, drawn from Dwyer & Wright (2008), also reports briefly on the government-funded commercial projects of SeaZone Solutions Limited, a wholly owned subsidiary of the UK Hydrographic Office.

European Coastal-Shelf Sea Operational Observing and Forecasting System

The European Commission, as part of its Framework 6 Research Program, funds the ECOOP project. It involves 71 partners in 31 European countries and the project duration is 2007 to 2010. The overall goal of ECOOP is to: consolidate, integrate and further develop existing European coastal and regional seas operational observing and forecasting systems into an integrated pan-European system targeted at detecting environmental and climate changes, predicting their evolution, producing timely and quality assured forecasts, providing marine information services (including data, information products, knowledge and scientific advices) and facilitate decision support needs. This is being attained through the following activities:

1. Integration of existing coastal and regional sea observing (remote sensing, *in-situ*) networks into a pan-European observing system.
2. Integration of existing coastal and regional sea forecasting systems into a pan-European forecasting system and assimilating the pan-European observation database into the system.
3. Assessing the quality of the pan-European observing and forecasting system.
4. Advancing key technologies for the current and next generation of the pan-European observing and forecasting system.
5. Developing and generating value-added products for detecting environmental and climate change signals.
6. Integrating and implementing a pan-European Marine Information System of Systems (EuroMISS) for general end user needs.
7. Developing a methodology and demonstrating a European Decision Support System for coastal and regional seas (EuroDeSS) that responds to the needs of targeted end users, as emphasized in the Global Earth Observation System of Systems (GEOSS) and Global Monitoring for Environment and Security (GMES) initiatives (see Chapter 5).
8. Carrying out technology transfer both in Europe and at an intercontinental level; establishing education and training capacities to meet the needs of ocean forecasters.

Underpinning the above is the development and implementation of a pan-European Data Management System (EDMS) for marine data. The EDMS is being developed in line with the INSPIRE Directive and therefore discovery, view and download services are key components.

The EDMS focuses on marine observational data including sea temperature, salinity, wave, currents, sea level and river run-off data. Currently, there are many different systems and processing chains existing in-house within the Regional Operational Oceanographic System (ROOS) centers that manage these types of data. The EDMS does not aim to re-engineer these internal systems, rather the principle of the architecture is that regional data centers make their real-time and near real-time data and metadata available in harmonized data formats. It is then possible to build INSPIRE-type services on top of these common data formats. In fact, the EDMS acts as a component that simply sits over this harmonized data and metadata. Once the regional centres conform to the harmonised formats, the EDMS component can be simply installed within each centre. Also, common quality control procedures can be more easily implemented and reused on top of this harmonized data. This generic approach reduces duplication of effort and cost, compared to each organisation individually implementing services, quality control procedures, etc., possibly in slightly different ways, which can affect interoperability.

To facilitate the implementation of this architecture, the various standards adopted, developed and maintained within SeaDataNet (discussed below) play an important role within ECOOP. Finally, open source technology, such as GeoServer (http://geoserver.org) and GeoNetwork (http://geonetwork-opensource.org), free, open-source, web-based geographic metadata catalog systems developed by the United Nations Food and Agricultural Organization (FAO) and the United Nations Environment Program (UNEP) are used in implementing this architecture.

SeaDataNet

SeaDataNet is an EU funded project that runs to 2011 and it aims to create and operate a pan-European, marine data management infrastructure (i.e., primarily offshore), accessible online through a unique portal (Maillard *et al.*, 2007; http://www.seadatanet.org). Its primary goal is the development of a system, which provides access to marine datasets and data products from 35 countries in and around Europe. It will construct a standardized, distributed system for managing the large and diverse datasets collected by oceanographic fleets and the new automatic observation systems.

Key elements in the realization of such a distributed system include common standards for data processing, communication and quality assurance. This includes the use of XML and international standards, such as ISO 19115, and shared vocabularies. These shared or governed vocabularies facilitate interoperability between the different databases within SeaDataNet. They are managed in a semantic technical infrastructure, known as the NERC Data Grid (NDG) Vocabulary Server (Lowry & Williams, in press). Permanent, machine-readable labels or uniform resource names (URNs) represent the semantics in the data and metadata. The URNs are converted into uniform resource locators (URLs), which are NDG Vocabulary Server term identifiers further used to return Simple Knowledge Organization System (SKOS) documents on the searched for concept and its mappings. SKOS provides a standard way to represent knowledge organization systems using the Resource Description Framework (RDF). Encoding this information in RDF allows it to be passed between computer applications in an interoperable way.

The SeaDataNet experience can be of great value to ICAN in its development of interoperable atlases, particularly with the SeaDataNet Vocabulary Server already in operation with existing lists and rich semantics that can be reused in an ICAN environment (e.g., Chapter 4). The content governance structure is also of great relevance to ICAN as it addresses metadata validation by ensuring that only terms from the master vocabulary are used in metadata as well as doing correctness checks on the words being entered.

Global Marine SDI Initiatives

A prototype global oceanographic SDI exists in the form of agreed data collection, formatting, exchange and policy standards and guidelines from the Intergovernmental Oceanographic Commission (IOC) of UNESCO, via its Committee on International Oceanographic Data and Information Exchange (IODE) (Longhorn, 2002). While the data interchange collaboration work of IOC's IODE programme has been ongoing for several decades, specification of equivalent coastal SDIs at national level began much more recently, as described in the following sections on individual CSDI initiatives. Marine SDI developments within the hydrographic community also received new impetus with formation of the International Hydrographic Organization's (IHO) Marine Spatial Data Infrastructure Working Group (www.iho-ohi.net/english/committees-wg/hssc/msdiwg.html) in May 2007, which held its inaugural meeting in February 2008 (Osborne & Pepper, 2008; International Hydrographic Organization, 2008). The IHO is also developing a new Geospatial Standard for Hydrographic Data (S-100), which will "support a greater variety of hydrographic-related digital data sources, products, and customers" (International Hydrographic Organization, 2008). This includes imagery and gridded data, and new applications that go beyond the scope of traditional hydrography (for example, high-density bathymetry, seafloor classification, marine GIS, etc.). It will also support the use of Web-based services for acquiring, processing, analyzing, accessing, and presenting data" (International Hydrographic Bureau, 2008)

Role of the Private Sector

The experience of the private sector can aid greatly in developing and implementing an effective marine SDI. One example (and it is an example only, not a marketing endorsement) is that of SeaZone Solutions Limited in the UK, which has for many years strived towards a coherent and coordinated approach to marine data acquisition, management and service provision supporting a wide range of activities and application. SeaZone Solutions also undertakes the collection of comprehensive and authoritative digital marine reference data, comparable to land mapping. Their expertise lies in creating combined coastal hydrographic and topographic datasets, to common vertical and horizontal datums, merging datasets maintained by hydrographic surveys (seaward) and topographic mapping agencies (landward) where they meet at the shoreline. Such harmonized datasets are very valuable to those agencies working in the coastal zone, which formerly had to acquire and integrate marine (hydrographic) and land (topographic) datasets independently.

Recently, as part of the IHO's developing interest in marine SDI, the company has been involved in a survey of International Hydrographic Organization member states to establish levels of knowledge of SDI, levels of involvements in national SDI, status of data to support SDI creation (including for purposes other than navigation), and overall willingness to support an NSDI. They will analyze the results and establish a benchmark for future IHO support, while also providing an IHO SDI Guide for Member States.

CONCLUSION

With over half of the world's population living on the coast and an ever-increasing focus on issues such as climate change, sea level rise, coastal development and overpopulation, and coastal sustainability for humans and all manner of other species alike, the demand for coastal geographic data will continue to grow. The initiatives summarized in this chapter, although limited to the US and Europe based on the experience of the authors, represent progress toward comprehensive federal- or continental-level responses to coastal geographic data development and dissemination. To reiterate, if an SDI is the overarching framework to assist people with acquiring, processing, using, and preserving spatial data, then a CWA is an important subset of an SDI, via its provision of needed datasets, mapping tools, and contextual information. In the US, this is the first step towards the development of a coastal NSDI framework and eventual "Federated Coastal Atlas of the US." For the nations of the European Union, it is the first step toward an integrated European Atlas of the Seas. In the US activities are guided by its NSDI (implemented by a single agency within the US federal government), where state and local governments beneath the federal level have developed strong data holdings and mature atlases to look after their own interests. In Europe the major initiative is INSPIRE, which is considerably more coordinated and better funded than elsewhere, due to the support of many strong governments linked together in a European Union. Hence local governments are relatively unburdened with regard to data holdings and local SDI. In Europe, the metadata standard of the ISO (i.e., ISO 19115) is in full practice whereas in the US the focus is still on the FGDC Content Standard for Geospatial Metadata (CGDSM), and as a member of the ISO the US will be required in the near future to revise the CGDSM in accord with the ISO 19115.

This chapter has essentially painted separate pictures of SDIs in the US and Europe, but in the end, SDIs are continually converging, as represented by organizations such as the Global Spatial Data Infrastructure (GSDI) Association. At the time of this writing, the GSDI, in concert with the European Commission's Joint Research Centre, the European UmbRella Organization for Geographic Information (EUROGI), the Dutch innovation program Space for Geo-Information (RGI), and Geonovum (the Netherlands SDI) had just concluded the GSDI 11 World Conference with the theme "Spatial Data Infrastructure *Convergence*: Building SDI Bridges to Address Global Challenges" (van Loenen *et al.*, 2009). We refer the reader to van Loenen *et al.* (2009) for the latest in "big picture," future visions regarding SDI, which is beyond the scope of this chapter. This event was also the Third INSPIRE Conference and the national conference on Dutch SDI Results and Challenges. The research and workshops presented at this event continue to show clearly that "one size does not fit all" with regard to building SDIs, and nations are beginning to look beyond their own borders and seeking to interoperate among their existing SDIs. This trend may continue not only from national SDI to national SDI, but from terrestrial SDI to coastal/ocean SDI (Vaez *et al.*, 2009).

With so many vested interests at national and international scales, it's no surprise that converging partnerships, collaboration and standards continue to be common themes. However, there remain large impediments to effective partnerships, hence the justification for continued work to bring these partnerships to fruition. As these efforts move forward, it will be important to continue work in these areas, and wherever possible, develop best practices and lessons learned so that other supporting efforts can be easily integrated into the larger framework (e.g., the work of ICAN). Here CWAs have the potential to play a key role as they can serve as the engines powering a successful SDI. However, this potential remains to be fully tapped and the initiatives described in this chapter are very much an ongoing work in progress. So there is still much good work to be done, to assist in the successful implementation of SDIs throughout the world.

REFERENCES

Commission of the European Communities. (2007a). *SEC(2007) 1278–Accompanying Document to the Communication from the Commission to the European Parliament, the Council, the European Economic and Social Committee and the Committee of the Regions, An Integrated Maritime Policy for the European Union.* European Commission, Official Journal of the European Community, Luxembourg.

Commission of the European Communities. (2007b). *An Integrated Maritime Policy for the European Union.* Luxembourg: Office for Official Publications, Retrieved April 24, 2009 from the European Commission web site: http://ec.europa.eu/maritimeaffairs/policy_documents_en.html.

Commission of the European Communities. (2007c). Directive 2007/2/EC of the European Parliament and of the Council of 14 March 2007 Establishing an Infrastructure for Spatial Information in the European Community (INSPIRE). *Official Journal of the European Community, 50*(108). Retrieved August 6, 2009 from the European Commission web site: http://bit.ly/21BuET

Commission of the European Communities. (2008). *Communication from the Commission to the Council, the European Parliament, The European Economic and Social Committee and the Committee for the Regions - Towards a Shared Environmental Information System (SEIS),* COM (2008) 46 final, Brussels, 1.2.2008. Office of Official Publications, Luxembourg. Retrieved April 4, 2009 from http://bit.ly/Q8hw0

Commission of the European Communities. (2009). *SEC (2009) 499 final - Building a European marine knowledge infrastructure: Roadmap for a European Marine Observation and Data Network*, European Commission, Official Journal of the European Community, Luxembourg. Retrieved August 6, 2009 from the European Commission web site http://ec.europa.eu/maritimeaffairs/pdf/roadmap_emodnet_en.pdf

Data.gov. (2009). Retrieved October 31, 2009, from the Data.gov web site http://data.gov

Digital Coast. (2009) Retrieved October 31, 2009, from NOAA Coastal Services Center web site: http://www.csc.noaa.gov/digitalcoast.

Dwyer, N., & Wright, D. J. (2008). *Report of International Coastal Atlas Network Workshop 3 on Federated Coastal Atlases: Building on the Interoperable Approach*. European Environment Agency: Copenhagen, Denmark. Retrieved July 29, 2009 from the ICAN web site: http://ican.science.oregonstate. edu/ican3_final_rpt.

FGDC Marine and Coastal Spatial Data Subcommittee. (2009). Retrieved October 31, 2009, from NOAA Coastal Services Center web site: http://www.csc.noaa.gov/mcsd

GeoNetwork Opensource. (2009). Retrieved October 31, 2009, from the GeoNetwork Opensource web site http://geonetwork-opensource.org

GeoServer. (2009). Retrieved October 31, 2009, from the GeoServer web site: http://GeoServer.org

Geospatial One Stop. (2009). Retrieved October 31, 2009, from the Geospatial One Stop web site http://geodata.gov

International Hydrographic Bureau. (2008). *IHO Geospatial Standard for Hydrographic Data, Version 0.0.0 – January 2008.* International Hydrographic Bureau, Monaco, 2008, Special Publication No. 100, Retrieved April 24, 2009 from the International Hydrographic Organization web site http://www.iho.int/COMMITTEES/CHRIS/TSMAD/S-100-Feb08.zip

International Hydrographic Organization. (2008). *Marine Spatial Data Infrastructure Working Group (MSDIWG) Terms of Reference*. International Hydrographic Organization, IHB, Monaco. Retrieved April 24, 2009 from the International Hydrographic Organization web site: http://www.iho.shom.fr/COMMITTEES/CHRIS/MSDIWG/MSDIWG-TOR.pdf

Longhorn, R. (2002). Global Spatial Data Sharing Frameworks: the case of the Intergovernmental Oceanographic Commission (IOC). In *Proceedings of GSDI-6, Budapest, Hungary, 16–19 September 2002*. Retrieved April 21, 2009 from the Global Spatial Data Infrastructure web site: http://bit.ly/yqhZM.

Lowry, R., & Williams, G. (in press). Putting meaning in SeaDataNet. *Information Sciences* (special issue on the 2008 International Conference on Marine Data and Information Systems).

Maillard, C., Lowry, R., Maudire, G., Schaap, D. & the SeaDataNet Consortium (2007). *SeaDataNet: Development of a Pan-European Infrastructure for Ocean and Marine Data Management*. Contribution to EurOcean07, theme 5.0 subtheme 5.1 - Paper 070131-026.

Masser, I. (2007). *Building European Spatial Data Infrastructures*. Redlands, CA: ESRI Press.

Multipurpose Marine Cadastre. (2009) Retrieved October 31, 2009, from NOAA Coastal Services Center web site: http://www.csc.noaa.gov/mbwg/htm/multipurpose.html.

National Spatial Data Infrastructure. (2009). Retrieved October 31, 2009, from the Federal Geographic Data Committee web site: http://www.fgdc.gov/nsdi/nsdi.html

NOAA Office of Oceanic and Atmospheric Research. (2007). *Integrated Ocean Mapping*. Retrieved October 31, 2009 from the NOAA Office of Oceanic and Atmospheric Research web site: http://www.oar.noaa.gov/oceans/t_oceanmapping.html.

Onsrud, H. (Ed.). (2007). *Research and Theory in Advanced Spatial Data Infrastructure Concepts*. Redlands, CA: ESRI Press.

Osborne, M., & Pepper, J. (2008). Marine SDI and the International Hydrographic Community. *IHO/CHRIS Marine Spatial Data Infrastructure Working Group (MSDIWG) Meeting, IHB, Monaco, 4-5 February 2008*. Retrieved April 24, 2009 from the International Hydrographic Organization web site: http://bit.ly/4mx6Yk.

SeaDataNet. (2009). Retrieved October 31, 2009 from the British Oceanographic Data Centre web site: http://www.bodc.ac.uk/projects/european/seadatanet

US Congress. (2009). *Public Law 111-11: Omnibus Public Land Management Act of 2009, Section 12201: Ocean and Coastal Mapping Integration Act*. Retrieved August 6, 2009 from the US Government Publications Office web site http://bit.ly/1Smb23.

Vaez, S. S., Rajabifard, A., & Williamson, I. (2009). Seamless SDI model -- Bridging the gap between land and marine environments . In van Loenen, B., Besemer, J. W. J., & Zevenbergen, J. A. (Eds.), *SDI Convergence: Research, Emerging Trends, and Critical Assessment, 48* (pp. 239–252). Delft, the Netherlands: Netherlands Geodetic Commission and Global Spatial Data Infrastructure Association.

van Loenen, B., Besemer, J. W. J., & Zevenbergen, J. A. (2009). *SDI Convergence: Research, Emerging Trends, and Critical Assessment* (*Vol. 48*). Delft, the Netherlands: Netherlands Geodetic Commission and Global Spatial Data Infrastructure Association.

Vivero, J. S., & Mateos, J. R. C. (2007). *The Atlas of European Seas and Oceans - Marine Jurisdictions, Sea Uses and Governance. Ediciones del Serbal*. Retrieved July 29, 2009 from the EuroOcean web site: http://www.eurocean.org/np4/file/18/Atlas_of_the_European_Seas.pdf

Wright, D., Longhorn, R., Dwyer, N., Lassoued, Y., & Boucelma, O. (2009). The International Coastal Atlas Network: An emerging spatial data infrastructure initiative. In *Proceedings of the Third INSPIRE Conference, in conjunction with the Eleventh International Conference of the GSDI (GSDI 11)*, Rotterdam, The Netherlands. Retrieved October 31, 2009 from the International Coastal Atlas Network web site http://ican.science.oregonstate.edu/node/75

Wright, D. J., & Dwyer, E. (2009). *ICAN Workshop 4 and EEA National Reference Centers Meeting in Trieste, Italy, November 2009*. Retrieved October 31, 2009 from the International Coastal Atlas Network web site: http://ican.science.oregonstate.edu/ican4.

KEY TERMS AND DEFINITIONS

Coastal Web Atlas: A collection of digital maps and datasets with supplementary tables, illustrations and information that systematically illustrate the coast, oftentimes with cartographic and decision support tools, all of which are accessible via the Internet. Also known as web atlas, digital atlas, digital coastal atlas.

Capacity Building: With regard to coastal atlases, the actions involved in building local GIS infrastructure, including maintenance and update of GIS data layers, software, computer equipment and labs/offices, salaried personnel, and action items for project initiatives.

Coastal Informatics: A broad academic field encompassing the management and analysis of data collected from and representing the coast. Informatics can include spatial data infrastructure, hardware and software infrastructure, computational networking, modeling, and experimentation, the design and deployment of data portals and Internet mapping sites, as well as the creation, analysis, and understanding of data/metadata vocabularies and ontologies, metadata creation/extraction/cross-walking tools, geographic and information management systems, and grid computing.

Digital Coast: An initiative launched by the NOAA Coastal Services Center to form a network of diverse partners working together to build an information delivery system that efficiently serves not only data, but also the training, tools, and examples needed to turn data into useful information to address a host coastal issues. Digital Coast therefore seeks to ensure the wise use and management of coastal resources. The NOAA Coastal Services center launched the Digital Coast web site (http://www.csc.noaa.gov/digitalcoast) in 2008 and provided all of the content in phase one. However subsequent phases are adding content from other sources, including other parts of NOAA and other federal agencies, as well as, state and local governments, academia, and the private and nonprofit sectors.

Federated: A condition in which it is possible to simultaneously search multiple sources (such as coastal web atlases or other online databases or resources) and to place them on a single map. A federated search often provides a single search interface, as well as single map canvas on which to place data layers.

FGDC Marine and Coastal Spatial Data Subcommittee: The Federal Geographic Data Committee (of the US Geological Survey) Marine and Coastal Spatial Data Subcommittee is made up all of the agencies in the US federal government that have responsibility for coastal and ocean spatial data. The subcommittee works to develop strategic partnerships, relevant standards, and to provide outreach that will enhance access to and utility of coastal and ocean framework data. The subcommittee is chaired by the NOAA Coastal Services Center (http://www.csc.noaa.gov/mcsd).

Information Management: The means by which an organization, agency, or individual collects, documents, shares, and uses information. It often involves creating or identifying the appropriate resources to find quality information in order to fill gaps in knowledge. Coastal web atlases are therefore a key component of information management as applied to coast region of the world.

Integrated Ocean and Coastal Mapping: The act of mapping the ocean bottom (seafloor or seabed), from the shoreline out to the deeper ocean, oftentimes at least to the extent of the Exclusive Economic Zone (EEZ), 200 miles offshore. This mapping also applies to lakes such as the Great Lakes region in the US. Integrated ocean and coastal mapping therefore involves the seamless integration of terrestrial digital elevation model data with nearshore bathymetry, including the appropriate vertical datum at the land-sea interface. The resulting maps differ from nautical charts for navigation, in that they represent multiple facets of the ocean and coast, including its physical, biological, geological, chemical, and archaeological aspects. Integrated ocean and coastal mapping is also an initiative to improve coordination among federal, state and local government, non-governmental organizations (NGO), and private sector mapping activities, as mandated by the Ocean and Coastal Mapping Integration Act of the US federal government.

Marine Spatial Planning: A critical part of regional governance that involves the integrated, forward-looking planning and consistent decision-making regarding various uses of the coast and nearshore. Marine spatial planning must be guided by specific policies and regulations governing usage, the conditions that apply, and an eye toward what possible conflicts in use may arise. Marine spatial planning is often aided by a coastal web atlases which can provide data on habitats, species migrations, land and territorial sea use by humans, navigation, managed areas (parks, reserves, disposal sites, etc.), commercial and recreational fishing, and the like.

Multipurpose Marine Cadastre: An integrated submerged lands information system of the NOAA Coastal Services Center consisting of legal (i.e., property ownership or cadastre, physical, and cultural information) in a common reference framework, especially to assist in the process of marine spatial planning and to demonstrate how marine boundary and related data are being used to support decisions on ocean uses (http://www.csc.noaa.gov/mbwg/htm/multipurpose.html). The data focused on the geographic extent of rights and interests that are benefits or enjoyment in real property that can be conveyed, transferred, or otherwise allocated to another for economic remuneration. The gathering of the data for the marine cadastre has required joint planning, interaction, and commitment by federal, state, local, territorial, and tribal agencies working through public and private partnerships.

Regional Governance: Policies and initiatives that allow state and local governments to pursue concrete, practical steps toward more coordinated and holistic management of ocean and coastal resources. This often includes establishing lines of communication among regions to facilitate implementation of best practices, forming partnerships among regions, and deploying proven management principles and approaches (e.g., ecosystem-based management). Coastal web atlases often provide the datasets, mapping tools, and contextual information needed for effective regional governance.

Research Collaboration Network: A collaboration of research scientists, resource managers, technical staffers, and other interested parties to develop a coordinated research network focused on a particular topic. The group fosters communication among those with common goals and interests, along with collaboration on common projects across disciplinary, organizational, institutional and geographical boundaries. The International Coastal Atlas Network is an example of a research collaboration network.

Spatial Data Infrastructure or SDI: A framework via an organization of people or government agencies, via the Internet, or via a series of guiding policies or standards to assist people with acquiring, processing, using, and preserving spatial data. The spatial data are often in geographic information system (GIS) format, are not, but not limited to this. Coastal web atlases often provide the datasets, mapping tools, and contextual information needed for an effective SDI.

Web GIS: Geographic information systems functions as deployed via a web site or series of web sites rather than just on the desktop.

Chapter 17
Creating a Usable Atlas

Timothy Nyerges
University of Washington, USA

Kathy Belpaeme
Coordination Centre on Integrated Coastal Zone Management, Belgium

Tanya Haddad
Oregon Coastal Management Program, USA

David Hart
University of Wisconsin Sea Grant Institute, USA

ABSTRACT

Knowing user audiences for coastal web atlases is important for designing atlas capabilities that address different user skill levels. This chapter presents guidelines about how to better understand coastal web atlas users, how to undertake user-centered design and development, and how to avoid major pitfalls with web interfaces. User groups are formed based upon understanding user characteristics. User-centered design for different user groups can take advantage of a logic model; that is, a series of steps for scoping, designing, implementing and testing the capabilities. The end result of design and implementation should be a usable system, thus software usability is an important goal. Regardless of how well designers know users, web interface pitfalls inevitably arise during the development process, some of which are discussed based on personal experience of the chapter authors.

INTRODUCTION

Coastal Web Atlas (CWA) specialists at a workshop about coastal mapping and informatics (O'Dea *et al.*, 2007) concluded that existing atlases are sometimes too complicated for general audiences. Recommendations were made to suggest that: (1) development must be responsive to user needs; (2) developers should consider designing multiple interfaces and capabilities to offer a range of services; and (3) regular user feedback is crucial for atlas success. This chapter is written in the general spirit of providing guidelines for creating a usable CWA; it is directed at designers and developers. Previous chapters have already provided insight about the capabilities of a CWA, and therefore the focus here is on what makes those capabilities more or less usable.

DOI: 10.4018/978-1-61520-815-9.ch017

Copyright © 2011, IGI Global. Copying or distributing in print or electronic forms without written permission of IGI Global is prohibited.

Design information including basic constructs for developing CWAs has been presented in previous chapters. However, it is important to reiterate that basic CWA constructs such as map data layers with various themes and tool capabilities such as pan, zoom and search form the basis for generating information for a user. Users develop an understanding of a CWA in their heads based on what they know and as they experience the tools and information in a CWA. Developers must be able to "connect" their intended designs of constructs to what users already understand, at least partially to begin with, and to help users learn about the world.

The chapter begins by helping the reader understand the general nature of users based on their backgrounds and associated abilities, their information needs, and their expectations for a CWA. We continue the chapter by providing a bit more detail about how to conduct usability studies as a systematic approach for evaluating CWA user interfaces and capabilities accessible through those interfaces. We end the chapter by examining some of the pitfalls that developers might incur while developing user interfaces.

UNDERSTANDING USERS

When understanding use of GIS-based CWA applications, whether they are supported by single-user workstation or web-based technologies, it is crucial to take into account the nature of users. Users can be described in terms of characteristics of people as in their abilities, user needs for information, and expectations of users (Nyerges, 1993). Combining some of those characteristics helps us form user groups, whether we consider the whole of the group or individuals. When understanding user groups, some developers might emphasize an "audience" perspective while others might emphasize an individual user perspective. An audience is a user group whereas the user is an individual with certain qualities. One way to bridge the divide between the two perspectives is to articulate "prototypical users," as we can never fully understand all the details of single users, nor all the characteristics of an aggregated audience. We need to make some simplifying assumptions about who will use a CWA. Toward that end, the subsections that follow emphasize how we characterize user abilities, the information needs of users, and the expectations of users.

User Abilities

Users have different abilities. For example, some people are more technically-skilled than others, and thus can understand complex information displays as part of the user interface of tools, while other users are more challenged to understand such displays. Some users have more experience problem solving within a particular substantive area, while others have less experience with such problems. Such qualities are rather difficult to track, and thus difficult to generalize across, hindering our understanding of users. More general qualities that can give us a better understanding about user abilities include user background and perspective. Our understanding of background and perspective make it easier to develop user group categories.

USER BACKGROUND: A QUALITY INHERENT TO A USER (GROUP)

People have certain qualities gained through living their lives based on choices and constraints in social settings.

- Age – years of personal experience being exposed to various topics
- Education/experience – formal / informal training
    - Problem solving ability
- Number of years addressing a problem
    - Technical/computer ability

- Number of years working with computerized information systems
- Hours per week spent working with computerized information systems
- Culture – community context / worldview based on upbringing

User Perspective: A Quality Characterizing how People View Information

People have roles largely due to responsibility, authority, and/or interest in a topic, for example as technical specialists, executives, members of a public, and educators. A person might have one or more of those roles at any given time.

- Technical Specialists focus on problem articulation/elucidation
    - Scientists
    - Resource analysts/researchers
- Executives focus on responsibility for managing community well-being
    - Policy/Decision makers
    - Elected officials
- Public at large focus on valued concerns about their place/identity in the world
    - Stakeholder interest groups
    - Property owners
- Educators focus on what/how/why people learn
    - Primary and secondary school age
    - Higher education
    - Life-long (social) learning

An enumeration of the types of user groups targeted by the developers of a selected set of CWA is shown in Table 1 (O'Dea *et al.*, 2007). A variety of user groups have been targeted. Clearly, different organizations have different purposes in mind for respective CWA's.

Developing software systems for all intended user groups identified in Table 1 is not easy, particularly if developers want to gain empirical insight into the goals and motivations of groups. How we collect information about users and what they like, do not like, how they perform tasks and work with information are key concerns. However, in this chapter, we want to continue to explore more about the general issues associated with a usable atlas. To do that we recognize that the groups listed in Table 1 have different responsibilities to work with information. As such, the different groups have different information needs. We now turn to how to understand better user information needs.

User Needs

User needs for information differ depending upon the user group targeted. Thus, the first step is to know your audience as discussed in the previous section. Once the target audiences have been identified, then information needs can be elicited for each of the groups. Below is an example from experience with the Belgian Coastal Atlas.

- Scientist users need detailed and complex information (about diverse aspects of the coast) perhaps presented using graphs, maps, etc.
- Policy and decision maker users need selected information directed towards policy advice, with clear interpretations and/or indications. The interface must be user-friendly and less scientific, avoiding technical complexity with clear messages and comprehensible figures. The atlas could be integrated with policy supporting tools (e.g., sustainability indicators, scenario building, case studies, etc.).
- Public users need easily accessible and transparent capabilities that contain basic information of interest for a wide audience.

Table 1. User groups (target audience) of selected CWA's (based on targeted audiences identified in O'Dea et al., 2007)

User groups	Coastal Web Atlas[1]					
	A	B	C	D	E	F
General Public		√	√			√
Tourists		√	√			
Students		√	√			
Researchers/Scientists		√	√	√	√	√
NGOs		√	√	√	√	√
Government/Public Bodies	√	√	√	√	√	√
Commercial/Industry	√	√	√			
Consultancies	√	√	√	√	√	√
Coastal/Environmental Managers	√	√	√	√	√	√
Decision Makers	√	√	√	√	√	√
Other:				Outreach		

[1]Coastal Web Atlas column identifier
A. The UK Coastal and Marine Resource Atlas
B. DE Kustatlas Online, Belgium
C. The Marine Irish Digital Atlas
D. The Oregon Coastal Atlas
E. North Coast Explorer, Oregon
F. Mapping Tools for Coastal Management, Virginia

- Educational users need information prepared in ready to use education tools/packages.

Once a general idea of the content has been articulated, a next step is to define the geographic area(s) to be covered by the themes and overall by the atlas including both the seaward and landward boundary. The Belgian Coastal Atlas was unique in its approach, as it covered the land as well as the seaside of the coast from the very beginning. Many atlases still cover either the land or the seaside. From an integrated coastal zone management perspective, it is important to stress the need for integrated coastal and marine maps and information because of the intricate link between the two environments and the need for an integrated policy and management over the land-sea interface. This has been stressed in the European Maritime Policy and in the integrated coastal zone management communication (European Commission, 2000). Opportunities for stimulating integrated management and policies are missed if developers do not consider the landward side as well as the coastal seaward side.

Having defined the area, the more detailed themes/sectors that are to be included need to be considered. Will the content cover one theme/sector (biological atlas, social atlas) or several themes? One way to enumerate the themes is to aggregate the topics across a series of use cases.

The concept of use case has been developed as a way of characterizing complex tasks addressed through information technology capabilities. At its core, a use case contains a description of a series of invoked software actions to accomplish a task that are characterized from the point of view of a user. A use case scenario sets specific assumptions/parameters about a use case. The scenario helps to refine (putting constraints on) the use case actions to address the task goal.

As an example, a coastal erosion use case is being developed as part of the initial ICAN effort addressing data interoperability (Wright *et al.*, 2007). The initial user focus of ICAN is on regional planners/resource managers, property owners, emergency response teams, and local CWA system administrators (aka atlas administrators) that address coastal erosion. Hazard-related information and the boundaries of regulatory jurisdictions are routinely required for land and ocean planning, regulatory, and enforcement work. Eventually, the outcomes are meant to improve the ability of agency staff to quickly and efficiently analyze local geographic patterns of hazards, community development, and jurisdiction in a regulatory and/or planning context. The use case can be used to characterize and evaluate issues and impacts related to coastal erosion, but could also be used to inform and educate the public and coastal zone management community. Generalizing across the information needs of the information users listed above, a collection of key datasets for this use case includes the following:

- Coastal access and recreation
- Coastal armoring
- Cadastral datasets with assessor attribution
- Geology
- Land use and zoning
- Current shoreline position
- Historic shoreline positions
- Permit tracking systems and a dynamic link to cadastral data
- Aerial imagery
- Streams
- Beaches
- Bluff and dune fields
- Regulatory jurisdictions
- Community development
- Geomorphology profiles
- Erosion Risk study results – Risk Zones or Lines
- Topography
- Wave climate data
- Shallow water bathymetry
- Transportation networks
- Public utilities
- Public lands

Aggregating the data themes across applications provides a first pass summary of the information content of interest to users, or what users might expect to find in an application.

User Expectations

Software can have widely varying degrees of consistency. User expectation refers to the consistency that users expect from products. Interaction design deals with the organization of design elements, such as CWA constructs employed for a user interface, particularly when implemented as a sequence of actions. A good *design principle* to use in interaction design is to follow the "Principle of least astonishment". In the case of interactive software applications, for example, users form expectations based on their experience with similar kinds of software. Effective interaction design aims to conform to norms for user behavior about software interfaces and responsiveness. Many design features were identified at the 1st Coastal Mapping and Informatics Workshop that are relevant to the expectations of a user-centered interaction design (O'Dea *et al.*, 2007). These features were organized using a framework for strengths, weaknesses, opportunities, and threats (SWOT).

Strengths of CWA Design

- Intuitive structure of web sites and map pages;
- Inclusion of contextual information in order to better understand the data;
- Hierarchical data organization;

- Multiple user pathways to retrieve maps and layers of interest;
- Tools for data analysis and creating reports.

Weaknesses of CWA Design

- The cartography / design challenge of displaying many layers;
- Inadequate database management system for efficient management of information; metadata and data;
- Inadequate search functions for data and content;
- Failure to meet user needs where atlas developments are technology-driven;
- Lack of distributed systems to enable data owners to share and manage their own data.

Opportunities for Better CWA Design

- Improved cartographic display of large quantities of layers in coastal atlases;
- Potential for sharing data through distributed networks (e.g., utilizing Web Map Services and Web Feature Services);
- Potential to develop regional nodes that tie in with larger atlases (e.g., national or statewide).

Threats to CWA Design

- Keeping up with design expectations of users (e.g., Google Earth);
- User interpretation: misunderstanding of how to use atlases or their components.

Developers can address surprises with user expectations by detailing *use cases* to the extent that both users and developers can agree on the sequencing of CWA actions. Having users participate in the articulation of use cases with developers, called user consultation, fosters shared understanding among users and developers with regard to atlas capabilities, and will also make it easier to identify pitfalls and eventually the problems once the system is developed. Influences on use case action sequencing might consider each of the SWOT issues raised above. Developers and users should agree on which to address and which can be left for another time. More details about developing use cases are presented below.

USER-CENTERED DESIGN

User-centered design can be used to help guide the development of a CWA (Lazar, 2006). By the nature of its name, user-centered design is an approach to system design that makes users important participants in the design process. User-centered design is a design philosophy and a process in which the needs, wants, and limitations of the end user of an interface or document are given extensive attention at each stage of the design process. User-centered design can be characterized as a multi-stage problem solving process that not only requires designers to analyze and foresee how users are likely to use an interface, but also to test the validity of their assumptions with regards to user behavior in real world tests with actual users. Such testing is necessary as it is often very difficult for the designers of an interface to understand intuitively what a first-time user of their design experiences (Wikipedia, 2008). When one considers users to be an important part of the broader process of implementing, testing and evaluating systems, then one is engaging in a user-centered development process.

A productive approach for engaging users within design and development of a CWA is the use of a LOGIC Model. A LOGIC model is a structured process that integrates project design and evaluation (Mayeske & Lamber, 2001; McLaughlin & Jordan, 1999). Developers pose the design from input provided by users. Evaluation involves an assessment of how well the design features support the use case (user) actions.

Kramer (2008) describes one type of LOGIC model consisting of three stages for user-centered design within the context of user-centered development of the Canadian Atlas. The first stage is an examination of business requirements. The second stage is detailed user requirements. The third stage involves systems design, including product design. Others might see more steps, but seldom would we have fewer steps in the development process. An important point, no matter whose version of user-centered design one might take, the user is always placed "front and center" in the overall process.

The University of Wisconsin Sea Grant Institute and its partners are utilizing a LOGIC Model in developing the Wisconsin CWA. The process includes conducting an inventory of potential CWA partners to ascertain data capabilities, contact information, etc.; forming an advisory committee with broad representation of coastal constituencies; completing the LOGIC model for the project; documenting objectives, outputs, and outcomes; and utilizing collaborative technologies to share the LOGIC Model with members of the ICAN for review and critique. It is worth noting that the U.S. National Oceanic and Atmospheric Administration (NOAA) has adopted the LOGIC Model as a means to ensure that coastal management projects are well designed and that it is easy to measure the performance of projects as they are implemented and offers training courses on project design and evaluation that features the LOGIC Model (NOAA CSC, 2009).

In the user-centered design paradigm, some of the users become actual or *de facto* members of the design team. The term *user friendly* is often used as a synonym for *usable*, though it may also refer to accessibility of capabilities. *Usability* is a term that denotes the ease with which people can employ a particular tool, display, or other capability to achieve a particular goal. For several years Haklay (2009) has been researching the advantages of usability engineering for development of geographic information systems (GIS), including web GIS. That research has "…focused on the way in which 'common users' of GIS and geospatial technologies use these systems. The aim is to understand how the interfaces work and how to improve them so they will be effective, efficient and enjoyable to use" (Haklay, 2009). Usability can also refer to the methods of measuring usability and the study of the principles behind an object's perceived efficiency or elegance. To continue the theme of what makes a usable atlas, we focus on the issues concerned with the former rather than the latter.

The primary notion of usability is that an object designed with the users' psychology and physiology in mind is, for example:

- More efficient to use – it takes less time to accomplish a particular task;
- Easier to learn – operation can be learned by observing the object;
- More satisfying to use – a sense of productivity; and
- Understanding CWA usability develops from understanding both user-centric and technology-centric issues. User-centric issues address the abilities, needs, and expectations of users. Technology-centric issues address the capabilities presented to users for addressing their needs and expectations.

Usability evaluation starts with the development of a set of use cases (defined earlier in the user needs section) that are representative of the kinds of activities users will perform. Use cases can be drafted and adopted jointly by users and developers as "agreements" about the kinds of information linked to the kinds of capabilities to be offered within a CWA. A single use case as a narrative can contain many elements. However, to make a narrative somewhat systematic, some designers recommend using templates for developing use cases. Below is a template based upon the Wikipedia description that is quite thorough in comparison to several academically-oriented descriptions of a use case (Wikipedia, 2008). Developers can use this template as a quick guide for detailing use cases, adding or removing characteristics as appropriate to each case.

- *Use case name*: A unique descriptor, short and to the point.
- *Version*: Versions help users and developers track what changes, as use cases can be revised through iterative steps of creating and/or updating an atlas.
- *Goal*: With a user in mind, in its simplest sense the goal of a use case could take the form of a question that is posed. Questions beg answers. Since every (or at least almost any) statement can be transformed easily into a question, we use the questions because they motivate users to seek information.
- *Summary*: The brief description of the use case that can be consulted when someone wants to scan an overview.
- *Actors*: The user groups identified in section 17.1 form the basis for describing actors in a use case. As mentioned above various user groups have different needs for information, and thus the questions are likely to vary depending on user group actors.
- *Preconditions*: Preconditions are the basic inputs to a question, perhaps assumptions being made about the circumstances under which a question would be asked.
- *Triggers*: Triggers are what activate the questions, perhaps in the circumstances set up as part of preconditions.
- *Basic course of actions*: The basic course of actions in the use case is the sequence of capabilities to be invoked by a user group actor once the preconditions are in place. These events would be the button pushing, keystrokes or screen picks made to initiate computer activities within the atlas.
- *Alternative paths*: A work-around to the basic course of events, if it is possible.
- *Postconditions*: What results from the execution of the basic course of events?
- *Business rules*: Business rules are the general guidelines established by an organization for the ways of carrying out activities. Such rules might or might not apply to the use of certain information within certain conditions of seeking information, e.g., constrains on accessing information from certain sources.
- *Notes*: An ancillary information that could help interpret the use case, e.g., if there are special circumstances for its inclusion in development effort.
- *Author and date*: Sometimes, the template has variations, and it would be good to know whose creative input was used to establish that variation.

It is nearly impossible to specify all the capabilities and information to be designed, and thus a diverse collection of use cases, each with perhaps two or more scenarios, provide a sampling of what is to be created, and when tested, what was actually created in the software. Use cases can be prioritized to provide developers and users with a shared understanding about what is more important and less important.

WEB INTERFACE PITFALLS

All CWAs are websites, and thus need to consider and avoid common web interface mistakes that can materially detract from any user experience. In addition, due to the importance of maps and interactive map interfaces in CWA implementation, certain interface "caution areas" related to maps should also be considered.

In regards to common web interface mistakes, Nielsen & Loranger (2006) present a long list of web interface characteristics that can pose significant usability problems for users. They note that over the past decade, some old interface problems have become no longer relevant, while many others continue to be serious. If possible, designers should avoid or minimize use of features that those authors deem persistent "high-impact usability problems". Such serious issues include the following:

- Links that don't change color when visited.
- Breaking the back button.
- Opening new browser windows.
- Pop-up windows.
- Design elements that look like advertisements.
- Violating web-wide conventions (e.g., what can be clicked, and how).
- Non-existent content and empty claims.
- Dense content and unscannable text.

These items are provided as a checklist overview of typical web usability problem areas, and it is recommended that designers research the issues more thoroughly to understand how these problems might impact their design.

With regards to the particular hazards of incorporating maps and interactive map interfaces into a CWA, the issues are different depending whether static or interactive maps are being considered.

With static maps, many of the important issues to consider are those that are standard in conventional cartography. Legibility, ease of interpretation, and communication of essential information such as scale, map projection, metadata, etc. are all typical issues. These are complicated slightly on the web by issues of screen resolution (different screen resolutions can alter the scale of a displayed map if it has been rendered to display at a certain number of pixels per inch), available colors (some web image type palettes may be limited, causing maps to render poorly), and the inconsistencies of print vs. on-screen display renderings (typical rendering for print requires higher resolution than screen display, and this can alter the size of map elements such as labels and symbols). Coping with these cartographic challenges is inherent in any map-making exercise, while mastering the specific challenges of web maps will require close study of and experimentation with the specific map rendering software employed by each CWA.

When discussing interactive map interface hazards, all those items that pertain to static maps remain of concern, and in addition, the designers must consider the usability difficulties that might come with the addition of interactions such as panning, zooming, feature identification, feature search, and any other advanced map widget used in the mix. Harrower & Sheesley (2005) make the point that an interactive map that feels "natural" or "intuitive" to the user is not something based purely on design, but is a combination of the predisposition of the user (their level of need to accomplish a task, and their prior experience with similar tasks), as well as their level of exposure to any one specific design and the amount of repetition (or practice time) they have had with that design. As a result, testing of complex interactive map tools with real users asked to perform real tasks is the primary way that designers can obtain insight into how specific map interface features perform for the intended audience.

CONCLUSION

Knowing who the user groups are for coastal web atlas design, development, and evaluation is a necessary factor in successful user-centered design, but it is not sufficient for success. A systematic approach to user-centered design should use a logic model to frame the series of steps for engaging with users. Usability comes about by having users test software in multiple phases.

Inevitably, web interface pitfalls will arise. Sometimes this occurs because of "feature creep", that is, users ask for this and that feature as the system moves forward in development. Designers should stay vigilant for the "high impact" usability problems as part of the design; as such vigilance can reduce and/or eliminate problems before they enter the development phase.

REFERENCES

De Kustatlas Vlaanderen/België. (2005). Retrieved December 11, 2008 from Kustatlas web site: www.kustatlas.be.

European Commission. (2000). *Communication from the Commission to the Council and the European Parliament on Integrated Coastal Zone Management: a Strategy for Europe (COM/2000/547)*, adopted 27 September, 2000. Retrieved July 2, 2009, from the European Commission web site: http://ec.europa.eu/environment/iczm/ comm2000.htm.

Haklay, M. (2009). *Human-Computer Interaction and Usability of GIS*. Retrieved February 12, 2009 from UCL web site: http://homepages.ge.ucl.ac.uk/~mhaklay/usability.htm.

Harrower, M., & Sheesley, B. (2005). Designing better map interfaces: A framework for panning and zooming. *Transactions in GIS*, *9*(2), 77–89. doi:10.1111/j.1467-9671.2005.00207.x

Kramers, R. E. (2008). Interaction with maps on the Internet – A user centered design approach for the Atlas of Canada . *The Cartographic Journal*, *45*(2), 98–107. doi:10.1179/174327708X305094

Lazar, J. (2006). *Web usability: a user-centered design approach*. Boston: Pearson Addison Wesley.

Mayeske, G. W., & Lambur, M. T. (2001). *How to Design Better Programs: A Staff Centered Stakeholder Approach to Program Logic Modeling*. Crofton, MD: The Program Design Institute.

McLaughlin, J. A., & Jordan, G. B. (1999). Logic models: A tool for telling your program's performance story. *Evaluation and Program Planning*, *22*(1), 65–72. doi:10.1016/S0149-7189(98)00042-1

Nielsen, J., & Loranger, H. (2006). *Prioritizing Web Usability*. Berkeley, CA: New Riders.

NOAA Coastal Services Center. (2009). Project design and evaluation . *Coastal Connections*, *7*(2), 1–3.

Norman, D. (1990). *The Design of Everyday Things*. New York: Double Day.

Nyerges, T. (1993). How do people use geographical information systems? In Medyckyj-Scott, D., & Hearnshaw, H. (Eds.), *Human Factors in Geographical Information Systems* (pp. 37–50). London: Belhaven Press.

O'Dea, L., Cummins, V., Wright, D., Dwyer, N., & Ameztoy, I. (2007). *Report on Coastal Mapping and Informatics Trans-Atlantic Workshop 1: Potentials and Limitations of Coastal Web Atlases*. University College Cork, Coastal & Marine Resources Centre: Cork, Ireland. Retrieved May 7, 2009, from the ICAN web site: http://ican.science.oregonstate.edu/node/47.

Wikipedia, (2008). *Use Case*. Retrieved December 11, 2008 from the Wikipedia web site: http://en.wikipedia.org/wiki/Use_case.

Wright, D. J., Watson, S., Bermudez, L., Cummins, V., Dwyer, N., O'Dea, L., et al. (2007). *Report on Coastal Mapping and Informatics Trans-Atlantic Workshop 2: Coastal Atlas Interoperability*. Oregon State University: Corvallis, Oregon. Retrieved July 29, 2009 from the ICAN web site: http://ican.science.oregonstate.edu/node/46.

KEY TERMS AND DEFINITIONS

Logic Model: A structured process that integrates project design and evaluation as part of the overall steps in development of software systems.

Usability: The ease with which people can employ a particular tool, display, or other capability to achieve a particular goal.

Use Case: Contains a description of a series of invoked software actions to accomplish a task.

User: A person that makes use of software.

User Abilities: The level of skills that a particular user possesses, but can also be attributed to a user group.

User Background: A collection of characteristics relevant to a particular user group.

User-Centered Design: A perspective on software system design that places the user 'front and center' in the design process.

User Expectations: Refers to the consistency that users expect from products.

User Group: A collection of users with the same set of characteristics for which the system is designed and intended.

User Needs: A collection of information (composed of information products and/or outcomes) that is relevant to a particular user and/or a user group.

User Perspective: An outlook on information that derives from responsibility and/or role within an organization.

Web interface: The style and tools presented to users for the purpose of interacting with the atlas.

Chapter 18
Improving a Growing Atlas

Tanya C. Haddad
Oregon Coastal Management Program, USA

Declan Dunne
University College Cork, Ireland

ABSTRACT

As coastal web atlas (CWA) projects grow over time, designers will have to track and adapt to changes in site activity, user capabilities, and progress in available data and technology. The web server may log all user activity on a CWA and these logs may be analyzed and interpreted by the CWA designers to better understand how users are encountering the materials and data provided by the CWA. In addition, CWA users may be surveyed on their use of the web site either by direct observation studies or written surveys. Both the surveying of users and the analysis of server logs may expose trends or confirm patterns of use for a CWA, and these patterns may be valuable feedback for designers, and provide important reporting metrics for funding bodies and institutional sponsors.

INTRODUCTION TO ACTIVITY TRACKING

The world of online user activity tracking is a large and complex one, and, in an advanced setting, can incorporate very sophisticated techniques involving tracking cookies and other tools. Websites that involve chains of interactions that must be tied to specific users throughout a site visit (such as any site that involves shopping) are under a heavy burden to maintain knowledge ("state") of all the selections a user makes throughout a particular visit, and must carry that knowledge through to the completion of any final transaction (e.g., payment and checkout). Most such websites have an option for user login that makes it possible for a user to save certain selections or personal information for use in another visit, and some may employ knowledge of a user's past activity to tailor the views presented to the user upon subsequent visits.

Contrast this with the situation of most CWAs, which operate on a much more open and "stateless" basis with their user community. Since the primary objective of most CWAs is informational,

DOI: 10.4018/978-1-61520-815-9.ch018

Copyright © 2011, IGI Global. Copying or distributing in print or electronic forms without written permission of IGI Global is prohibited.

rather than commerce based, the necessity of closely and directly tracking individual users is lessened. Having said this, it is possible to track basic user interactions with any web site simply by logging all server activity and examining the patterns that emerge when these logs are studied. In the case of most CWAs, this sort of basic server logging is all that is needed to gather a detailed picture of how users are employing the resources of the CWA.

SERVER LOGS

Basically, a server log file is a file in which every request to a web server is recorded in text format. The information recorded with each "hit" or request to the server differs depending on the configuration of the server. However, common data recorded includes the date and time of the request, the Internet Protocol (IP) address of the computer making the request, and the name of the requested resource. The file is built chronologically, and, if a web site experiences heavy traffic, the files can grow quite large. Most web masters employ a regular archiving method to prevent any single log file from becoming too large. A site might begin a new log file once a day, once a week, or once a month, depending on the level of traffic it experiences. The two most popular web servers, The Apache Software Foundation's Apache HTTP Server and Microsoft's Internet Information Server (IIS) have different default settings for how log files are managed. In Apache, the default is to build a single "access.log" file that can grow quite large if not managed with a regular archive program, while the default in IIS is to archive the log file daily, which can lead to large numbers of smaller files named by date.

The two web servers also record different information by default, and CWA web masters who are hoping to gather detailed information from their server logs must pay attention to the server settings in order to ensure that the server is set to record all the types of useful information that are desired for any subsequent reporting. For example, by default, the Apache web server does not record the "referring site" or "user agent" in the log. By default both these pieces of information are recorded in IIS logs, although a web master would be wise to confirm this early in a project's deployment. A "referring site" is the site from which a user came to a CWA, such as a search engine or link in another web site. "User agent" is the technical term for the web browser application of the computer making the request. Knowing the referring site can help CWA administrators discover which external sites are driving substantial amounts of traffic to the CWA project site. Knowing the user agents can help inform on the computer platforms employed by the CWA audience (e.g., percentage of Windows versus Mac users) as well as provide a good indication of which browsers need to be tested for compatibility during design and construction. If this information is of interest, then web masters are advised to check the information being recorded in logs to confirm that they are getting what they require.

Once user activity has begun to be logged by a CWA server, it is possible to make summaries of the traffic that a website experiences over time, and to begin to observe patterns that might lead to insights on how the CWA is meeting or not meeting the needs of its users. Such summaries can be made manually for low traffic sites by sorting log file information in spreadsheets and performing simple counts, but most often the task is performed by scripts custom designed for that purpose. The script may be written (e.g., in PHP) to open all log files within a certain period and pull out specific pieces of information such as number of hits, number of unique visitors or number of returning visitors. Other types of information that might be useful to pull out are the name of specific files downloaded and the number of times each was requested. Due to the fact that every entry in a server log has a date and time recorded, it is also possible to calculate the duration of a user's visit and the proportion of time spent on any particular page. A list of the most commonly summarized log statistics might include the following:

- Number of hits per day/month/year
- Number of visits per day/month/year
- Number of unique visitors per day/month/year
- Number of returning visitors per day/month/year
- Bandwidth per day/month/year
- Time of day, or day(s) of week when the CWA is most frequently visited
- Hits or visits per country
- Hits or visits per host computer
- Duration of visits
- URLs most frequently visited
- Users' operating systems
- Users' web browsers
- Search engines used to find the CWA (e.g., Google, Yahoo)
- Search engine keywords used to find the CWA
- Direct links from related referring websites

Once a web log has become large, or the desired analysis too complex to accomplish with simple summaries or basic scripts, a CWA web master might look to more advanced analysis options such as can be provided by the various levels of web log analytic software packages. Some such packages can not only produce the summary statistics most often of interest to web masters, but can also produce well formatted reports on the fly, complete with charts and tables of data that tell the story of the web site over a particular period of time. Another advantage of using an analytics package is that they often simplify the task of summarizing where users are coming from. This information is inherent in the recorded IP address of the CWA user, but used alone, an IP address is meaningless to most people. Typically an IP address can be used, at least, to determine the country of origin of a user by using a method called reverse DNS lookup. This can give reasonable results; however, an even more reliable approach is to use a geo-location database to look up IP information. This can provide not only the country but often the state or sometimes the company of origin as well. Geo-location databases may be free (e.g., GeoLite Country http://www.maxmind.com/app/geolitecountry) or commercial (e.g., GeoIP Country http://www.maxmind.com/app/country) and can be used in conjunction with web analytics software to undertake robust analysis of both web metrics and user origins.

Web log analytic packages fall into two groups: those that can be accessed online and which can track a website's use remotely via a script (usually JavaScript) snippet inserted into the template of each page, and those that are deployed as stand alone desktop software packages that require access to the raw server logs and must be run each time a report summary is required. In the first category, a widely used example is the free Google Analytics service. To use this service, the CWA web master simply registers their site with the Google service and inserts a small piece of provided JavaScript tracking code into the pages that need to be tracked. The results of tracking are then available by logging into the Google Analytics account for the CWA, and are thus available remotely. In contrast, stand alone web analysis software must be run on a computer that has access to the raw server logs of a website and must be configured to report on whatever time period or metrics are desired. The range of standalone desktop analytics software is quite diverse and includes a variety of Open Source, e.g., AWStats (http://awstats.sourceforge.net), Analog (http://www.analog.cx), Webalizer (http://www.mrunix.net/webalizer/), and W3Perl (http://www.w3perl.com), as well as commercial, e.g., Webtrends Analyzer (http://www.webtrends.com) options. As mentioned, analytic packages are capable of generating well-formatted reports with graphics in a variety of output types (e.g., text files, Microsoft Word, or HTML). It should be noted that if log reports are created in a web accessible format, they should not be made available on the public website, but rather communicated only via intranet. This is because log reports could contain information that might be a potential breach of confidentiality for CWA users, and therefore any public log results should be made anonymous.

User Surveys

Another way to gauge CWA success is to survey users directly. User surveys as defined here can take many forms involving either controlled conditions (as in a focus group or observed user test setting) or less controlled written surveys (involving either structured multiple choice or unstructured, open-ended questions). In either case, there are two key elements to successful use of user surveys in atlas design:

- Take care in designing a useful survey (ask useful questions) and,
- Commit to, and follow through with, incorporating feedback into site revisions.

Without attention to these two elements, the goal of employing user surveys to improve CWA functionality will not be met. While the design of a useful survey is a topic far beyond the scope of this book, designers who may be interested in delving into the world of user surveys as it might apply to CWAs might benefit from browsing the range of questions contained in the ICAN user survey available at http://ican.science.oregonstate.edu/ican_assess.

LEARNING FROM ATLAS USE PATTERNS

Observation of website users and analysis of web server logs may reveal some interesting patterns in how a CWA is being used by visitors. Tidwell (2006) lists common behavior patterns observed among software users that designers could be on the lookout for. Examples of related user patterns in CWA use and lessons that might be learned from such patterns include:

- Users who hear about a site via word of mouth or advertisements may type in a URL simply to check out the look and feel of a site, and to satisfy some curiosity about content. Such users may only stay a short time and may only visit the home page. Featuring popular content in special feature areas of the home page may help draw such transient visitors deeper into the CWA content holdings and entice them to delve a few pages beyond the home page.
- Users referred to a CWA via a search engine may be more likely to explore for specific types of information or seek out specific data downloads. When this is the case, it is useful to track what search terms are leading users to a CWA in an effort to understand the kinds of questions that potential users are seeking to have answered. These sought-after areas of a site can then be reviewed with an eye to further improving the user experience they provide, and may also be considered for special featuring on the home page.
- When a CWA provides a tool or other structured interface to explore a specific topic or dataset, CWA designers typically have a roadmap of sequential pages that they expect a user to follow. Confirmation via server logs that these sequences are indeed being followed from beginning to end, and in an expected time frame, is valuable confirmation of a successful design. Alternately, disruptions, short-circuits, or dead-end user paths discovered via log analysis or user observation studies are likely to indicate a usability problem that needs attention or redesign in order to

achieve full success in meeting CWA user needs.

As with the use of user surveys results, the observation of common user patterns and disruptions in expected patterns is only the first step. A successful CWA designer will act upon the lessons that such patterns teach and incorporate the knowledge gained back into the CWA product in order to further improve performance for the benefit of users.

CHANGE OVER TIME

As time passes, the audience of a CWA as well as the capabilities of CWA users, their computers and connection speeds will evolve. Word of mouth and utility of CWA products will build a following of users who find the products and information provided by a CWA helpful. Over time the network connection speeds, and computer and browser capabilities of these core users should improve. This opens up the possibility that CWA designers may be able to re-visit early design decisions that may no longer be relevant and expand the contents and services of a CWA to meet the new capability of the audience. Nielsen and Loranger (2006) provide an excellent overview of changes in technology that may impact usability and design options over time. Examples of a few very basic changes in CWA audience technology capabilities to be on the lookout for include:

- Increased connection speeds as users gain access to broadband
- Improved screen resolution as users upgrade hardware
- Improved client side processing as users upgrade hardware
- Higher cross-browser compatibility as users upgrade to more standards-compliant browsers

Changes such as these allow CWA designers to evolve the offerings of their atlas to take advantage of new audience capabilities. Pages can have more images when connection speeds improve, and page layouts can be wider when screen sizes increase and support more resolutions. Interfaces can include more functions when cross-browser compatibility is high, as applications do not have to be re-written multiple times for multiple incompatible browsers.

Another aspect of CWA projects that may change over time is the breadth and availability of data. In the early days of a CWA, data holdings may be limited to small collections of data related to specific CWA projects, and attracting additional data from other contributors may be difficult. Providing attribution as well as feedback to external data custodians (e.g., number of downloads and other web statistics) encourages sharing of data and can aid data custodians in their search for more funding. Over time, as a CWA becomes a key gateway for information about the coast, external data stewards, rather than needing to be approached, may offer to allow a CWA to display their data or information. Similarly, as time passes and costs of acquisition fall for data sets such as orthophotography, bathymetry, or light detection and ranging (LIDAR) data that were formerly extremely expensive, a CWA may have the option to host datasets or offer analysis capabilities that were formerly only a distant dream.

One last aspect of change over time that can be a challenge for CWA designers is the issue of keeping up with the pace of change in the technology used to deploy the CWA. Once a technology path has been chosen for a project, future radical technological shifts may be difficult for a CWA host to implement for a variety of reasons. Since a CWA involves hardware, software, and personnel resources, a lack of flexibility or capability in any one of these areas can prevent adaptation to changes in the technological landscape.

For example, if a CWA uses proprietary software and invests a lot of time and money in customizing it, what happens if the company goes out of business or otherwise alters the product so that upgrading would force large changes upon the CWA design? Will there be time and money to perform the upgrade and / or redesign the site? This is not a rare issue, so at a minimum CWA designers should be aware and think about the probable impact if the CWA project is expected to be deployed for longer than 3 years. Users familiar with the world of desktop GIS cannot deny the often central presence that products from Environmental Systems Research Institute (ESRI) have in all sectors of GIS application. Some estimates are that ESRI products have one third of global market share and have been used by up to 80 percent of GIS professionals in the world (Wikipedia, 2009). Customers of ESRI web products are very familiar with the problem of changing technology affecting the consistent deployment of web projects over time. This type of potential vulnerability is something that aspiring CWA hosts should carefully evaluate at the outset of a project.

Advances in technology can also affect user expectations, which in turn can create pressure on a CWA host to alter a design or add new features that were not previously envisioned or planned for. The introduction and popularity of both Google Maps and Google Earth products are the classic example of such a revolutionary shift in user expectations. In fact, the debut of Google Maps in early 2005 effectively marked the end of the web map 1.0 era, and coupled with the Google Maps API caused a giant leap forward in the use and popularity of geographic information on the web. For the most part, enough time has now passed so that implementation of a web 2.0 framework in a new web map project is fairly simple in both the proprietary and open source software spheres, however, for existing projects the transition is more complicated and it may be inevitable that some prior investments will be lost. In addition, to some extent the Google Earth impact on the web mapping landscape is still playing out, with many implementation options and new features still evolving. This presents yet another set of challenges for CWA designers to consider.

Finally, it is already evident that both mobile devices and touch screen technology will be playing a large role in user expectations of what web maps should be able to do the near future. Further technology shifts are impossible to predict, but logic indicates that they are inevitable. As a result, designers should simply be comfortable with the notion that technology is never going to be static for any long duration, and plan for change by being open to periodic project revaluation and budgeting for technology transition.

DOCUMENTING PROGRESS

As discussed above, any CWA project that is deployed for a long period of time will experience change. Both server logs and user surveys can be used to document change if used in a consistent manner. Server logs alone are a simple tool for tracking basic metrics that can illustrate a project's progression. User surveys, if kept consistent over time, can also display trends in CWA use.

Common parlance often refers to web site "hits" as a measure of a website's popularity. A "hit" is any request to a web server for any resource. However, since one page on a website may contain many resources, a more accurate measure of how much material is being viewed on a website is "page views". Furthermore, CWA managers are most likely to want to know how many unique visitors they are serving, and how many of those unique visitors return to the CWA week after week or month after month. Using server log metrics to document that such visitors are growing over time is powerful evidence that a CWA project is serving its audience with a product of value to them and helps to make a case for continued support to funding bodies and institutional partners.

CONCLUSION

Once a CWA is established, there might be an inclination to consider the project a job well done, and move on to focus on other topics. However, because a CWA is a website and users have certain expectations of websites that must be met for success, a CWA project is seldom a static product. As time passes, the capabilities of CWA users, their computers, and connection speeds will change, and this opens up the possibility that CWA designers may be able to re-visit design decisions that are no longer relevant, and expand the contents and services of a CWA to meet the new capability of the audience. Metrics gleaned from server log analysis and user surveys can help document and confirm these changes over time and help build the case for design evolution, improvements in maps interfaces, or even suggestions for new CWA products. In addition, documentation of audience size and dependence on CWA resources, as well as the popularity of specific data sets and tools, is valuable for CWA hosts seeking to obtain or renew funding and continued institutional support.

REFERENCES

W3Perl. (2009). Retrieved October 15th, 2009 from the W3Perl - Free Logfile Analyzer web site: http://www.w3perl.com.

Analog. (2009). Retrieved October 15th, 2009 from Analog: The Most Popular Logfile Analyzer in the World web site: http://www.analog.cx

AWStats. (2009). Retrieved October 15th, 2009 from AWStats - Free log file analyzer for advanced statistics (GNU GPL) web site: http://awstats.sourceforge.net

Environmental Systems Research Institute. (2009). Retrieved October 15th, 2009 from the ESRI: GIS Software that Gives You The Geographic Advantage web site: http://www.esri.com

Geo, I. P. Country. (2009). Retrieved October 15th, 2009 from MaxMind - GeoIP Country Geolocation IP Address to Country web site: http://www.maxmind.com/app/country.

GeoLite Country. (2009). Retrieved October 15th, 2009 from MaxMind - GeoLite Country | Open Source IP Address to Country Database web site: http://www.maxmind.com/app/geolitecountry

Google Earth. (2009). Retrieved October 15th, 2009 from the Google Earth web site: http://earth.google.com/.

Google Maps. (2009). Retrieved October 15th, 2009 from Google Maps web site: http://maps.google.com/.

ICAN User Survey. (2009). Retrieved October 15th, 2009 from the ICAN Atlas Assessment Working Group web site: http://ican.science.oregonstate.edu/ican_assess.

Nielsen, J., & Loranger, H. (2006). *Prioritizing Web Usability*. Berkeley, CA: New Riders.

Tidwell, J. (2006). *Designing Interfaces*. Sebastopol, CA: O'Reilly Media.

Webalizer. (2009). Retrieved October 15th, 2009 from the Home of the Webalizer web site: http://www.mrunix.net/webalizer.

WebTrends Analyzer. (2009). Retrieved October 15th, 2009 from the W3Perl - Free Logfile Analyzer Web site: http://www.webtrends.com.

Wikipedia. (2009). *ESRI*. Retrieved October 15th, 2009 from Wikipedia: http://en.wikipedia.org/wiki/Esri.

KEY TERMS AND DEFINITIONS

Hit: An individual request to a web server for a specific resource such as an image or a downloadable document or an html page.

Page View: An individual request to a server for an HTML page, including all the components of a page, such as text, images, and embedded links.

Referring Site: The web site from which a web user was led to a destination web site (e.g., a search engine site, or a web site with a link to the destination web site).

Returning Visitors: Visitors who are known to have visited a site at a prior time, usually because their IP address has been previously logged.

State: The current status of a specific user's current transaction with a web server. A server that maintains knowledge of user state is aware of the context of any specific transaction within the transaction history of every user.

Stateless: Web transactions with a server that are treated as independent, unrelated requests. The server is not configured to be aware of what transactions it has with a user before or after any given transaction.

Unique Visitors: Web site visitors that are uniquely distinguished from each other, usually by IP address.

User Agent: The web browser application (e.g., Firefox 2.2 for Windows) of the computer making a web request to a web server.

Chapter 19
Supporting a Successful Atlas

Roger Longhorn
Info-Dynamics Research Associates Ltd., UK

Dawn J. Wright
Oregon State University, USA

Kathy Belpaeme
Coordination Centre on Integrated Coastal Zone Management, Belgium

ABSTRACT

The content and technical features of coastal web atlases (CWA) were presented in the introductory chapters of the book and the previous two chapters visited user-oriented issues involved in creating and improving a usable atlas, including user requirements, audience definition, learning from use of the atlas, etc. However, in the absence of confirmed long-term resources or a guaranteed income stream to fund the on-going development and maintenance of an atlas, how do you secure the long-term viability and success of a CWA where the intention is that the atlas is not just a one-off "display," but can develop organically? Several issues arise, including governance, institutional support, partnerships, funding and continued promotion. Ideally, these issues should be resolved prior to the web atlas being created, especially funding and institutional support (human and technical resources), if the atlas is to be successful in the long run, and not just another short-term 'project' that disappears from view when support dwindles. Also, over time, as atlas projects mature into on-going, demand-driven, sustainable information services, institutional and funding arrangements evolve as well. This chapter looks at some of the issues involved in attaining and maintaining long-term support for a successful online atlas.

PARTNERSHIPS AND INSTITUTIONAL SUPPORT

Establishing strong partnerships should be a precursor to creating the coastal web atlas as a service. Successful partnerships will span content providers, major user groups, the atlas developers, and those hosting the service. Identifying key partners is only the first step. Securing their commitment, especially in the long-term, is even more important.

The web atlas developer must determine who, among potential partners, can provide the strongest support, especially in a sustainable way.

DOI: 10.4018/978-1-61520-815-9.ch019

Copyright © 2011, IGI Global. Copying or distributing in print or electronic forms without written permission of IGI Global is prohibited.

Many organizations may come forward wishing to participate as partners, but not all will have the institutional capacity to be an effective long-term partner. Assessing the value of potential partners, i.e., what they can bring to the atlas, is a task for the developer or owner of the project.

One useful tool may be a simple "Statement of Intent" or Memorandum of Understanding (MoU) to which potential partners are requested to subscribe prior to joining the web atlas project. This approach was used in the UK, where the initial partners to a coastal/marine spatial data infrastructure (SDI) project – the Marine Data and Information Partnership (MDIP) – were required to sign a Statement of Intent setting out a range of basic principles to which all MDIP partner organizations were expected to accede. While MDIP has since been replaced by a new initiative, the Marine Environmental Data and Information Network (MEDIN), it is worth reviewing the terms established for MDIP in 2005, as these reflect sound principles to be followed in developing a Coastal Web Atlas that involves multiple contributors and supporters. Of course, an MoU or similar statement of principles is of less value unless enforcement measures are also included, and neither MDIP nor MEDIN have such requirements.

Participants in the Marine Data and Information Partnership (MDIP) were asked to commit to the following principles:

- To accept that there are benefits to be gained from improved harmonization of and access to marine data and information.
- To implement the agreed recommendations of the Expert Group.
- To accept that the specific purpose of the Partnership is to provide a mechanism for long-term stewardship and access to UK marine and coastal data.
- The Partnership will include UK Government holders of marine data at its core, but is open to all marine data providers on an equal basis.
- Long-term stewardship will be achieved by convergence to a set of agreed data management and dissemination standards, protocols and operating procedures.
- The management structure of the mechanism will provide overall governance, standards and procedures, data stewardship audits, etc.
- Contributors will work towards the mutually agreed standards and procedures of the Partnership.
- Contribution to the Partnership will not restrict commercial use and exploitation of contributors' data by themselves, but data or information contributed within the framework will be available for use for the public good.
- Some contributors will act as Data Archiving Centers (DACs), undertaking to receive, hold, conserve, undertake long-term stewardship / custodianship and make available, data submitted by data holders. Identification of DACs will be against criteria agreed by the Partnership.
- The Partnership will establish the standards and procedures and liaise with the leading key data providers in the harmonization of the overall approach.
- Key data providers will ensure selected data holdings conform to the standards and provide them to the MDIP to facilitate wider use.
- Once established and deployed, the Partnership will expand in coverage by enlarging the contributing base and the links to contributors.
- The establishment of the Partnership is in three parts: (a) the creation of the management structure, (b) the data related implementation by key Government data providers, (c) and the development of the proposed "Gateway."

- The Partnership management structure will drive improvements in web-based access to the data.
- In all aspects of its operation, the Partnership will strive to be an exemplar of best practice in distributed use of data.

The terms of such agreements can be as specific or as general as required by the atlas developer. They should set out clearly what is expected of partners, and there may be different classes of partnership with different levels of expectations, i.e., for content providers versus promoters, user groups, etc. One very important class of partner is the funding body.

Next consider the level of institutional support that is required and made available to the web atlas project - and the resulting product and service – in order that it is both successful and sustainable. Institutional support includes issues such as who pays for the atlas development, who hosts the resulting atlas during development and who helps in promotion initially and over time. Ideally, a web atlas should progress from being an interesting 'project' to becoming a viable and appreciated information service embedded in the work program and information infrastructure of those institutions who need, or can benefit from, access to the information provided by the atlas, whether government, academia, research, businesses or citizens. Long-term viability has implications for funding, either via an assured funding agreement, typically with a government body, or via an income stream arising from services provided to users.

What is the ideal level of partnership and institutional support required for a successful CWA? Firstly, key partners are the data owners or custodians who have content to be offered by the atlas. Without content there is no atlas, and without good quality content that is relevant to targeted users, the long-term success of the atlas will be threatened. An important partner in delivering and maintaining a successful web atlas service is the atlas "owner" or official custodian. Ownership is a significant motivation for that partner to maintain the initiative. In some cases, multiple organizations are responsible for maintaining an atlas, sometimes from different sectors, i.e., an academic or research institution may be the owner/operator of the atlas, new services can be developed by different sources (public or private), while a government agency or a private company provides hosting. Where there is no clear "ownership," the atlas may suffer. Such institutional arrangements should be formally agreed at the outset, then monitored for continued relevance and perhaps even changed as the atlas matures, so that all those involved know specifically what is expected of them.

Sustained institutional support will depend to some extent on how well the atlas meets the expectations of users, current or potential, which should have been identified in the early stage of atlas development. For example, an atlas that provides easy and transparent access to quality-controlled information of specific value to a government department or other institutions in fulfilling their mandate(s) is far more likely to succeed than one that simply presents an interesting collection of information to a wide audience. This is because the mandated requirement often underpins the request from the atlas owner or custodian for sustainable funding streams in order to support those requirements. Assessing use of the atlas via metrics acceptable to funding bodies is equally important in securing continued financial support.

Many nations around the globe are now involved in creating formal national Spatial Data Infrastructures (NSDIs), of which coastal and marine information should be an important element for all coastal states. If the CWA can be institutionally embedded in SDI initiatives, there is greater opportunity for securing long-term support and success while promoting the focus on coastal/marine information requirements within the NSDI. This raises issues relating to data standards, interoperability of web services, and data or information policy, some of which were visited in earlier chapters.

It should be noted that different authors, leading to confusion in interpretation, often use the terms "data policy" and "information policy" interchangeably. In this chapter, the term data policy is used in reference specifically to policy issues affecting data or datasets, e.g. policy on data and metadata standards, exchange formats, and other aspects of maintaining data collections. The term information policy is used to encompass both the underlying data policies and wider policy issues regarding provision of information to various communities of users, such as managing intellectual property rights (IPR), legal obligations, rights and responsibilities relating to information infrastructure (national or organizational), etc.

Where government information is provided in a web atlas, various requirements may come into play, for example relating to pre-existing metadata and systems interoperability standards. If the atlas is proposed as an official contribution to a national SDI, which could help secure funding and longer term, sustainable support, then such action is an important consideration in both atlas design and implementation, with implications for partnerships and other institutional arrangements. For example, government agencies might be required by statute to collect, store and/or present their data holdings according to a prescribed set of standards which do not apply to academic or research institutions or other potentially valuable data providers. Resolving such issues is not merely technical, but involves policy and related institutional issues. Longhorn (2004, 2005, 2009) reports on the status of coastal and marine SDI activities in key countries from which lessons could be learned relating to Coastal Web Atlas development as part of such initiatives.

INTELLECTUAL PROPERTY RIGHTS

Any information service, especially one present on the Web that will have wide access, needs a well-developed and transparently documented information policy. In such policy statements, a key issue is that of intellectual property rights (IPR) relating to the content. IPR exists to protect the value of the content for the original owners or content providers. Various levels of IPR exist, from 'public domain', determined under various national Freedom of Information acts, to strongly protected rights subject to internationally agreed conventions. Public domain is a legal definition meaning that the copyright holder, (i.e., the originator of the content), has surrendered all rights to the content, which can then be used freely by anyone for any purpose, usually without even requiring attribution.

A more formalized type of public domain involves "copy left" schemes, such as those promoted by the Creative Commons initiative (Creative Commons, 2009) for content and the Free Software Foundation's GNU General Public License (GPL) and adaptations of that license for software (Free Software Foundation, 2007). Using these licenses, both data (content) and the software used to access, process or present that data can be made widely available, at no cost to the user, as long as other terms are adhered to, such as attribution or making available to the general public any value added to a computer program.

IPR protection enacted in national legislation typically adheres to the articles of the Berne Convention for the Protection of Literary and Artistic Works administered by the World Intellectual Property Organization (WIPO, 1979) and/or to TRIPS, the Agreement on Trade-Related Aspects of Intellectual Property Rights administered by the World Trade Organization (WTO, 1994). While the Berne Convention protects the intellectual property rights for content generally, the TRIPS agreement extended such protection specifically to computer programs and compilations of data. In

Europe, further protection is offered specifically to electronic databases via the European Union's Directive on legal protection of databases of 11 March 1996 (European Commission, 1996), which took effect across the EU Member States in January 1998.

Where a global web atlas is concerned, which might contain content from multiple nations operating under multiple – and not always consistent – IPR regimes, complex issues may arise regarding access to, and use of, the data in the atlas, especially with regard to downloading data from the atlas. Even a coastal atlas developed within a single nation will typically have content providers from many sectors of society, including government, academia, research, business and citizens' groups. The information policy adopted for the atlas, relating to IPR for such diverse providers, needs to be carefully considered so that potentially valuable content is not excluded due to inadequate IPR protection, while access is not adversely restricted due to overly protective IPR terms. Thus, web atlas developers need to understand the information policies, both current and evolving, in their home countries and in the legal jurisdictions in which other important content providers may be located. Trends change, rules change, legislation changes – a successful CWA that survives for a number of years may need to adapt its original information policies accordingly.

Unfortunately, implementing a CWA information policy that adequately addresses IPR issues is only the first step. Much more difficult to achieve is intellectual property control, technically. Most on-line services today offer some degree of intellectual property management using only access control, via for example licenses with terms to which users have agreed prior to using the service, or perhaps by 'click use' on-line license regimes created by companies (such as Amazon.com or eBay.com) or government agencies (e.g., the UK's Office for Public Sector Information). Unless a data resource is fully in the public domain, i.e., can be used by anyone in any way that they wish, IPR rights apply to greater or lesser extent, as per the content owner's intention. These range from being quite openly available via a Creative Commons license, to strictly controlled for many resources, with every view, access or record downloaded incurring some charge.

In the geospatial realm, progress has been made in recent years relating to IPR management, via the work of the Open Geospatial Consortium, Inc. working group on Geospatial Digital Rights Management (GeoDRM), which met from 2004 to 2006, and published a GeoRM Reference Model (Vowles, 2006). This work was later taken up by the International Organization for Standardization (ISO) Technical Committee 211 (ISO/TC 211) which deals with geographic information standards at global level, who released a draft standard for "Geographic information - Rights expression language for geographic information - GeoREL," in August, 2008 (ISO, 2008). Various sectors of the geospatial community - GIS vendors, commercial spatial data providers and organizations offering their spatial content for a fee - are investigating how best to implement GeoREL.

ATLAS PUBLICITY

When developing a web atlas, the priority of many initiators is to think carefully about the content of the atlas, navigation ease and visual layout. However, having an interesting, good-looking atlas that is easy to navigate is not enough. People need to be aware of the existence of the web atlas, and promoting the atlas is just as important as development and presentation. Information that is unused because potential users do not know about it is information wasted.

In most of the examples described in this book, there is no printed version of the atlas available. The atlas is thus not a tangible product for people to lay their hands on, and the on-line version is the only product people can see - and they can see it only by knowing that it is there! Therefore, it is important to promote the web atlas, which can be done in several ways. For instance:

- officially launch the web site, possibly linked to a related event,
- contact the press and send out press releases;
- write articles for specialized and non-specialized magazines, e-newsletters, etc.;
- make presentations at relevant conferences, workshops and seminars;
- post announcements on relevant electronic mailing lists, including those that address potential users with an interest in the coastal environment, to libraries, etc. ;
- create gadgets or "give-aways" to promote the web site, such as post cards or high-quality prints of striking content from the atlas.

Some examples for promoting web atlases follow.

The Marine Irish Digital Atlas (MIDA) organized an official launch linked to an international workshop of the International Coastal Atlas Network (ICAN). The launch took place on 25 July 2006 at University College Cork. It celebrated four years of development and data collection and the beginning of MIDA's next phase of integrating new technology, tools and information. Prominent speakers at the launch were Prof. Gerard T. Wrixon, President, UCC; Joe Breen, Head of Aquatic Services, Environment and Heritage Service, Northern Ireland; and Valerie Cummins, Director, CMRC. A press release was sent out. The Irish Times, the main Irish quality newspaper, published the news on 26 July 2006 through their Marine Correspondent (Lorna Siggins) based on her experience with the atlas. Also, there have been articles published in other newspapers and magazine before and after the launch.

The MIDA team clearly saw a huge increase in visitors to the web site in July of that year and the impact of the launch public relations lasted for about three months as shown in Figure 1. Since then there has been no further major publicity event and no more obvious "bounces" in the number of users that might have resulted from such activity, indicating that publicity needs to be ongoing and continuous, not a one-off activity.

The launch of the Belgian Coastal Atlas (De Kustatlas) in 2005 was linked to the presentation of a photo atlas published by the consultancy involved with the development of the online atlas, Aquaterra. There was also a press release and a presentation. A set of postcards with aerial photographs of the coast was printed, mentioning the web address. These postcards were distributed to key players and sold in book shops on the coast.

Figure 1. Usage statistics for MIDA post-launch. (© 2009, University College Cork. Used with permission.)

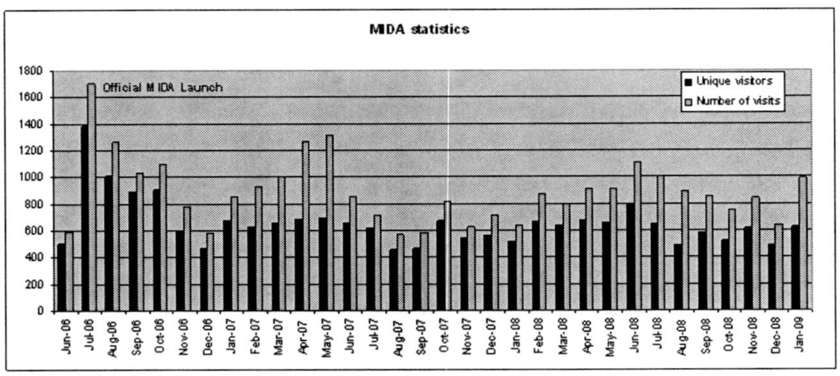

This promotion certainly helped to bring the Belgian web atlas to the attention of the public, but there was another and far more important factor, in the opinion of the developers - the atlas was also published as a book (Figure 2). The printed atlas was released the year before, in 2004. Tangible products like a book have proven very useful to draw the attention of a wide audience, scientists as well as politicians, managers, etc. Not only locally, but also at conferences and exhibitions, books or other printed material will draw the attention much more efficiently than yet another computer screen display. Visitors can touch the book and browse through it at their own pace.

The same experience can be found with the atlas for the Venice lagoon (http://atlante.silvenezia.it/en/index_ns.html), which was also published first as a printed atlas, and only in a second phase developed as a web application.

Figure 2. The Belgian Coastal Atlas – De Kust Atlas Vlanderen Belgie.(© 2004. Used with permission.)

FUNDING FOR COASTAL WEB ATLASES

Many agencies are responsible for the management and conservation of the coastal and marine environment, arising out of a commitment to meet international agreements under a wide range of conventions, as well as national or regional legislation, as in the European Union. High quality environmental information in an atlas can be crucial for sound management decisions, for monitoring the effects of these decisions, and reporting the results to constituents and stakeholders. Ideally an agency has a line item within its budget to support an atlas as an integral part of the agency mission, i.e., covering salaries, benefits, hardware and software upgrades, and appropriate fees for materials, supplies, and licensing. In some cases, atlases are funded long-term by an institution or network of partners, but the focus may be on innovation and technology developments rather than user requirements, data delivery and updates (O'Dea *et al.,* 2007). There is certainly a need to fund atlases on a long-term basis in order to guarantee their stability and ultimate effectiveness.

The reality is that funding is often extremely volatile, with a CWA being funded as a project, only for initial development and perhaps a short pilot operational period, thereby running the risk of going quickly out-of-date due to the lack of resources for site maintenance beyond the initial development stage. Ideally, a CWA should be seen as an important addition to information infrastructure, not simply as another ICT project. There are many negative impacts associated with unstable funding, as pointed out in O'Dea *et al.* (in prep.), as well as in Chapter 15, including: (1) staff turnover results in lost expertise; (2) inability to fund maintenance of projects (always needing to move to the next new funded activity); (3) short-term (often annual) budget cycles of many agencies who typically will not or cannot commit to long-term partnerships, making maintenance even more difficult. As a result, atlas managers and administrators are often forced into diverting their time and resources to seeking funding to maintain and grow atlases.

One recommendation made by O'Dea *et al.* (in prep.) is that different financial models should be examined to determine the best methods for continued CWA support. Consideration should be given to sponsorship by key organizations; obtaining multiple funding streams; providing "subscriber only" areas for advanced functionality (at a cost); or developing spin-off initiatives, such as the publication of a CWA in print media. Table 1 provides some suggestions as adapted from Dwyer & Wright (2008), where the discussion was primarily on the International Coastal Atlas Network (ICAN) as a whole. Oftentimes it is advisable to partner regionally with other atlases to seek regional funding instead of acting alone. In this way collaborators can investigate appropriate program announcements and different individuals can take the lead on writing and submitting grant proposals. However, there are implications to implementing this strategy that include both challenges, as well as opportunities, such as meeting the objectives of differing funding programs, and coordinating proposal development, submission and project execution, if the proposal is successful. These issues should not be underestimated and also apply to single atlas development proposals which seek funding from multiple sources.

In a proposal it is often best to highlight not only the driving factors for initial CWA development, i.e., why it *was* needed at the outset, but also the continuing benefits that are still to be derived from atlas availability, i.e., why it is *still* needed. Chapter 1 discussed some of the driving factors for initial CWA development, in most cases worldwide, e.g. speedy access to multiple sources of coastal data and information and more efficient use of time by avoiding individual contact with different data holders. Some factors that might be considered generically as future benefits in creating a CWA (after unpublished notes from the Institutional

Table 1. Some potential funding programs for coastal Web atlases

Networking and Workshops
National
• Irish Marine Institute, Sea Change Networking Initiative (http://www.marine.ie/home/funding/FundingCalls/closedcalls/NetworkingInitiative.htm)
• United States, National Science Foundation, Office of International Science and Engineering, (http://www.nsf.gov/od/oise/about.jsp)
International
• European Science Foundation, Research Networking Program (http://www.esf.org/activities/research-networking-programmes/rnp-call-for-proposals.html)
• European Union, Seventh Research Framework Program, Coordination and Support Activities http://cordis.europa.eu/fp7/
• European Union, Competitiveness and Innovation Framework Program 2007-2013 (CIP), Policy Support Program (PSP), Geographic Information (http://ec.europa.eu/ict_psp)
Technical Development
National
• United States, National Science Foundation, Office of Cyberinfrastructure (http://www.nsf.gov/dir/index.jsp?org=OCI)
International
• European Science Foundation – COST (http://www.cost.esf.org/index.php)
• European Union, Seventh Research Framework Program – Environment and ICT themes (http://cordis.europa.eu/fp7/dc/index.cfm)
• INTERREG, European Interregional development programs. (http://ec.europa.eu/regional_policy/index_en.htm)
Outreach and Training
National
• United States, The William and Flora Hewlett Foundation (http://www.hewlett.org/programs/environment-program)
International
• IOC's IODE Project[1] – Ocean Teacher Program (http://www.oceanteacher.org)
• International Geographical Union (IGU), Commission on Coastal Systems – (http://www.igu-ccs.org)
• Europe, EUCC – Coastal & Marine Union – CoastLearn Project (http://www.coastlearn.org)
• European Science Foundation - Marine Board (http://www.esf.org/research-areas/marine-board.html)

Capacity Breakout group of ICAN Workshop 1, O'Dea *et al.*, 2007) include:

- Demonstrate added value of atlases by exploiting opportunities to highlight best practices which can increase efficiencies and save money.
- Heightened visibility of certain partners, for example universities, can help market and build credibility for projects.
- Highlight emerging trends in e-government/digital government and knowledge-based economies, whereby CWA content can provide geospatial underpinning for many government activities affecting the coastal and marine environments.
- Articulate how a CWA will continue to help deliver government policy, e.g. implementing a national or regional ICZM mandate.
- Open data licenses can lead to new products, which increases economic development.
- Leverage data acquisition, e.g., added opportunities to secure resources to get more or better datasets.
- Reduce unnecessary duplication, e.g. sharing computer code, as well as data.

It will also be useful to consider focusing on one or two high-profile issues in order to attract funders/advocates (Dwyer &Wright, 2008). One such issue would be climate change *impact,* especially decision support for coastal communities vulnerable to climate change, island nations threatened by sea level rise, and the like. An atlas team can make a strong case by articulating the practical, applied products that it makes available. For example this can include available solutions to help an agency or organization do better what they are *already* mandated to do, provision of the knowledge gained as a result of working closely with partners on a common, proven approach, etc. It will also be helpful to ascertain real numbers in terms of who is using and benefiting from the atlas, i.e., how many users, who and where they are, how much money they are saving as a result of using our atlases, etc.

And finally, the inclusion of actual scenarios of prior success in the form of user profiles or stories are an added benefit to a proposal. A proposal might also show a sampling of the maps that have been created by users of the atlas. For example, with the Oregon Coastal Atlas (OCA), the administrative team has been motivated by the ways in which the Oregon coastal community has already used the atlas, sometimes beyond the intended audience of coastal resource managers and planners, and beyond the initial intention of hazards management, coastal watershed assessment, and ocean policy (to the edge of the Territorial Sea, 3 miles offshore). Below are four example stories explaining this effect (P. Klarin, pers. comm., 2001):

(1) Staff from other state and federal agencies report using the OCA when doing permit reviews and reports, primarily because: (a) they don't have an online system; (b) don't have or use GIS; (c) the data that they do have are limited to that produced by their only agency; but most of all (d) they find the OCA to be easier, faster and more reliable. Oftentimes several people can discuss a particular site together, with all of them looking at the same map data on the OCA. Our OCA team member from the Oregon Coastal Management Program (OCMP), coastal planner Paul Klarin, does this all the time with various state agency staff located at coastal field offices, whenever they discuss a specific site or proposal. Klarin conferred with a coastal field representative from Waldport, on the central Oregon coast, about a coastal erosion problem at a waste treatment plant in Seaside, on the north Oregon coast. They were able to see significant changes in the shoreline between

the two communities from a series of digital orthophotos available from the OCA.

(2) OCA team members have seen maps produced by the OCA being used by private individuals and their legal representatives in public hearings. The OCMP permit review staff has also seen OCA maps being used by individuals who are commenting on or producing permit applications.

(3) Coastal residents who are alarmed about the potential for a tsunami are using the OCA to access and print out the tsunami inundation maps and the tsunami evacuation route maps that were recently made available in the portal.

(4) One OCA team member has received phone calls from private individuals, some from out-of-state, who are seeking advice regarding coastal properties that they are interested in buying. While on the phone they have both used the OCA to look at the same site and talk about any special circumstances that may be evident, using the data available.

REGIONAL GOVERNANCE AND PARTNERSHIPS

Many Coastal Web Atlases are being developed in multinational environments, such as the African Marine Atlas and Caribbean Marine Atlas projects of the Intergovernmental Oceanographic Commission's IODE (International Oceanographic Data and Information Exchange) Program. Web portals will also be part of the developing European Marine Observation and Data Network (EMODNET) promoted by the European Union. Managing the institutional relationships between multiple organizations based in different countries typically requires more resources than are needed for a web atlas developed in a single country. Funding also becomes more complicated, as funds from multiple sources are typically required, including contributions from individual institutions, for which different financial terms and budgeting cycles apply, among other challenges.

In the case of the African and Caribbean Marine Atlas projects, the IOC's IODE has taken a leading management role, including securing funding and offering relevant training to the institutions involved in these projects. In the case of EMODNET development, the European Commission will be the lead organization, with funding most likely to be derived from funded projects enacted within multiple European Union programs, such as the Seventh Framework RTD Program, environmental programs and inter-regional development programs. Major projects that have a direct impact on coastal and marine information across national boundaries are already underway, such as SeaDataNet, which involves 49 partners from 35 countries and is funded for several million Euro by the EU's 6th RTD Framework Program.

The International Coastal Atlas Network (ICAN) initiative bridges the gap between many such projects, as representatives from many of these national and regional web atlas projects participate in ICAN meetings and on-going development of ICAN principles and tools. The European Environment Agency (EEA) has become a strong supporter for ICAN and Coastal Web Atlas development, having sponsored the 3rd ICAN Workshop at their headquarters in Copenhagen, Denmark, in July, 2008 and offering support for the 4th ICAN Workshop in Trieste, Italy, in November, 2009 (Wright & Dwyer, 2009). Recognizing the value of the ICAN initiative, the IOC-IODE has offered to host the 5th ICAN Workshop in Oostende, Belgium in 2010. With IOC projects now also involved in ICAN, this initiative has taken on a truly global aspect which should pay dividends in helping to create a network of federated CWAs, using common tools and standards, offering fully interoperable services.

CONCLUSION

This chapter looked at some of the issues relating to securing long-term support for a CWA, providing some guidance based on existing practice and experience with CWA developments at national and international levels. Creating effective partnerships is crucial for both initial development and securing long-term success of a CWA. Ideally, partners are committed to the project and have sufficient resources to participate fully both initially and over time. Sadly, securing sufficient resources for long-term operation of a CWA remains a problem for many initiatives. The preceding sections on funding and governance, which are closely related, hopefully provide some guidance on how to proceed with your own CWA project, especially in relation to other existing spatial information infrastructure initiatives within which the CWA may play an important role for coastal and marine communities.

Because most CWAs will comprise data sources and perhaps even information access services from multiple organizations, understanding the implications of managing intellectual property rights is an important consideration, and one which many information infrastructure projects have yet to fully address, let alone resolve. Progress is being made in this regard with new standards-based initiatives such as the Open Geospatial Consortium's (OGC) work in Geospatial Digital Rights Management (GeoDRM), now also taken up by the International Organization for Standardization (ISO) in their Geographic information - Rights Expression Language (GeoREL) draft standard (ISO 19149). However, implementations of this reference model and applications that can use GeoREL are still some years away.

Finally, we draw attention to the value and importance of publicizing the existence of a Coastal Web Atlas, as widely as possible, and as often as practicable. Potential users cannot use a resource unless they know that it exists! This obvious fact escapes many who create web services in numerous realms, not only those working with CWAs. Advertise, publicize, and gain 'readership' in order to help secure long-term funding and sustainability for your CWA.

REFERENCES

African Marine Atlas. (2009). Retrieved October 31, 2009, from the African Marine Atlas web site: http://www.africanmarineatlas.net.

Atlas of the Venice Lagoon. (2009). Retrieved October 31, 2009, from the Atlas of the Lagoon - Venice Between Land and Sea web site: http://atlante.silvenezia.it/en/index_ns.html.

Belgian Coastal Atlas. (2009). Retrieved October 31, 2009, from the Belgian Coastal Atlas web site: http://www.kustatlas.be.

Creative Commons. (2009). *Creative Commons Licenses – Attribution, Share Alike, Noncommercial, No Derivative Works*. Retrieved August 6, 2009, from Creative Commons web site: http://creativecommons.org/about/licenses/

Dwyer, N., & Wright, D. J. (2008). *Report of International Coastal Atlas Network Workshop 3 on Federated Coastal Atlases: Building on the Interoperable Approach*, European Environment Agency, Copenhagen, Denmark. Retrieved July 29, 2009, from ICAN web site: http://ican.science.oregonstate.edu/ican3_final_rpt

European Commission. (1996). *Directive 96/9/EC of the European Parliament and of the Council of 11 March 1996 on the Legal Protection of Databases*. Official Journal of the European Community, L 77, 27.3.1996, p. 20–28. Retrieved August 6, 2009, from the European Commission web site: http://bit.ly/40ertR.

European Marine Observation and Data Network (EMODNET). (2009). Retrieved October 31, 2009, from the European Commission web site: http://ec.europa.eu/maritimeaffairs/eu-marine-observation-data-network-mission_en.html.

Free Software Foundation. (2007). *GNU General Public License*. Retrieved August 6, 2009, from Free Software Foundation web site: http://www.fsf.org/licensing/licenses/gpl.html

IACMST. (2005, January). *Marine Data and Information Partnership (MDIP) Statement of Intent (Rev. 02)*. Presented to the Inter-Agency Committee for Marine Science and Technology, MDIP Steering Group meeting, London, January, 2005.

International Oceanographic Data and Information Exchange (IODE). (2009). Programme of the Intergovernmental Oceanographic Commission (IOC) of UNESCO. Retrieved October 31, 2009 from the IODE web site: http://www.iode.org.

ISO. (2008). *Geographic information—Rights expression language for geographic information—GeoREL, draft standard ISO/CD 19149*. Geneva: International Organization for Standardization.

Longhorn, R. (2004). *Integrated Coastal/Marine Spatial Data Infrastructure*. Paper presented at the ECO-IMAGINE Conference, 13-15 May 2004, Seville, Spain. Retrieved October 7, 2009, from the Gisig web site: http://www.gisig.it/eco-imagine.

Longhorn, R. (2005). Coastal Spatial Data Infrastructure. In Bartlett, D., & Smith, J. (Eds.), *GIS for Coastal Zone Management*. London: Taylor & Francis.

Longhorn, R. (2009in preparation). Coastal and Marine Spatial Data Infrastructure. In Green, D. (Ed.), *Coastal Zone Management*. London: Thomas Telford Books.

Marine Irish Digital Atlas. (2009). Retrieved October 31, 2009, from the Marine Irish Digital Atlas web site: http://mida.ucc.ie.

O'Dea, L., Cummins, V., Wright, D., Dwyer, N., & Ameztoy, I. (2007). *Report on Coastal Mapping and Informatics Trans-Atlantic Workshop 1: Potentials and Limitations of Coastal Web Atlases*. Cork, Ireland: University College Cork, Coastal & Marine Resources Centre. Retrieved May 7, 2009, from the ICAN web site: http://ican.science.oregonstate.edu/node/47.

Oregon Coastal Atlas. (2009). Retrieved October 31, 2009, from the Oregon Coastal Atlas web site: http://www.coastalatlas.net.

SeaDataNet. (2009). Retrieved October 31, 2009 from the British Oceanographic Data Centre web site: http://www.bodc.ac.uk/projects/european/seadatanet.

Vowles, G. (Ed.). (2006). *Geospatial Digital Rights Management Reference Model (GeoDRM RM)*. Wayland, MA: Open Geospatial Consortium, Inc.

WIPO. (1979). *Berne Convention for the Protection of Literary and Artistic Works*. World Intellectual Property Organization, Geneva. Retrieved August 6, 2009 from WIPO web site: http://www.wipo.int/export/sites/www/treaties/en/ip/berne/pdf/trtdocs_wo001.pdf

Wright, D. J., & Dwyer, E. (2009). *ICAN Workshop 4 and EEA National Reference Centers Meeting in Trieste, Italy, November 2009*. Retrieved October 31, 2009 from the International Coastal Atlas Network web site: http://ican.science.oregonstate.edu/ican4.

Wright, D. J., Watson, S., Bermudez, L., Cummins, V., Dwyer, N., O'Dea, L., et al. (2007). *Report on Coastal Mapping and Informatics Trans-Atlantic Workshop 2: Coastal Atlas Interoperability*. Oregon State University: Corvallis, Oregon. Retrieved July 29, 2009 from the ICAN web site: http://ican.science.oregonstate.edu/node/46.

WTO. (1994). *Agreement on Trade-Related Aspects of Intellectual Property Rights*. World Trade Organization, Geneva. Retrieved August 6, 2009, from WTO web site: http://www.wto.org/english/docs_e/legal_e/27-trips.pdf.

KEY TERMS AND DEFINITIONS

Copyright: Copyright is a legal term describing enforceable rights (commercial and moral) given to creators of literary and artistic works, such as novels, poems, plays, reference works, newspapers and computer programs; databases; films and musical compositions; artistic works such as paintings, drawings, photographs and sculpture; architecture; and advertisements, maps and technical drawings.

Creative Commons: Creative Commons is an that promotes "copyleft" principles to support the building of a richer public domain for content by providing an alternative to the "all rights reserved" principle of copyright with a less restrictive "some rights reserved" claim covered by various forms of Creative Commons license (see creativecommons.org).

Financial Models and Funding: Experience shows that various and sometimes multiple financial models are followed in creating and maintaining coastal/marine atlases, as with other types of information infrastructure, including public funding (by government), public-private finance, and some element of self-funding (charging users for access or value added services).

Geospatial Digital Rights Management (GeoDRM): Geospatial digital rights management (GeoDRM) involves the use of technology to manage access to digitally stored geospatial information, based on software tools, techniques, models and standards developed by the geospatial standards community (Open Geospatial Consortium, Inc. and ISO Technical Committee 211).

Governance: The act of governance includes decisions that define expectations, grant or restrict power, and specify and enforce performance of actions relating to some theme or area of responsibility, comprising management and leadership processes to ensure consistent management and implementation of cohesive policies and processes in an accepted and legitimized decision-making regime.

Information Infrastructure: Information infrastructure refers to the communications networks and associated information technology, standards, policies and regulations that support interaction among people and systems.

Institutional Support: Institutional support comprises the components that underpin implementation and on-going operation for an activity, including sustainable funding, providing human resources, al support, promotion, information system maintenance and governance,

Intellectual Property: Intellectual property (IP) refers to creations of the mind, i.e. inventions, literary and artistic works, symbols, names, images and designs used in commerce, and is divided into two categories: industrial property, which includes inventions (patents), trademarks, industrial designs, and geographic indications of source; and copyright (see definition above).

Intellectual Property Rights: Intellectual property rights are rights granted to creators and owners of intellectual property (see above), which protect the owner's moral and commercial interests in the works created, and include copyright, patents, trademarks, design rights, and protection of confidential information, plus special rights relating to computer software and electronic databases.

Compilation of References

Aditya, T., & Kraak, M. J. (2006). Geospatial Data Infrastructure Portals: Using National Atlases as a Metaphor. Cartographica, 41(2), 15–133.

African Marine Atlas. (2009). Retrieved October 31, 2009, from the African Marine Atlas web site: http://www.africanmarineatlas.net.

Alexander, P. (2009). Core Technologies for Ontologies. Retrieved April 09th, 2009 from The MMI Guides: Navigating the World of Marine Metadata website http://marinemetadata.org/guides/vocabs/ont/coretech

Alexander,C., &Tolson,J.P.(1990). COMPAS - NOAA's Coastal Ocean Management, Planning, and Assessment System National Ocean Service, National Oceanic and Atmospheric Administration. Unpublished. Galveston Bay Information Center, (pp. 213-214).

Analog. (2009). Retrieved October 15th, 2009 from Analog: The Most Popular Logfile Analyzer in the World web site: http://www.analog.cx

Arc Marine Data Model. (2009). Retrieved October 31, 2009 from the OSU Davey Jones' Locker web site: http://dusk.geo.orst.edu/djl/arcgis

Asch, K., Brodaric, B., Laxton, J. L., & Robida, F. (2004). An International Initiative for Data Harmonization in Geology. In 10th EC GI & GIS Workshop. Warsaw, Poland: ESDI State of the Art.

Askew, D., Evans, S., Matthews, R., & Swanton, P. (2005). MAGIC: a Geoportal for the English Countryside. Computers, Environment and Urban Systems, 29, 71–85.

Athanasis, N., Kalabokidis, K., Vaitis, M., & Soulakellis, N. (2008). Towards a semantics-based approach in the development of geographic portals. Computers & Geosciences. doi:.doi:10.1016/j.cageo.2008.01.014

Atlantic Coastal Zone Information Steering Committee & GeoConnections. (2009). Coastal and Ocean Information Network Atlantic. Retrieved May 7, 2009, from Coinatlantic web site: http://coinatlantic.ca/

Atlas of the Venice Lagoon. (2009). Retrieved October 31, 2009, from the Atlas of the Lagoon - Venice Between Land and Sea web site: http://atlante.silvenezia.it/en/index_ns.html.

AWStats. (2009). Retrieved October 15th, 2009 from AWStats - Free log file analyzer for advanced statistics (GNU GPL) web site: http://awstats.sourceforge.net

Barne, J., Davidson, N. C., Hill, T. O., & Jones, M. (1994). Coastal and Marine UKDMAP Datasets: A User Manual. JNCC Report No. 209. Joint Nature Conservation Committee, Coastal Conservation Branch.

Bartlett, D. J. (1999). Working on the Frontiers of Science: Applying GIS to the Coastal Zone . In Wright, D. J., & Bartlett, D. (Eds.), Marine and Coastal Geographical Systems (pp. 11–24). London: Taylor & Francis.

Bartlett, D., & Bruce, E. (2001, July). An Evaluation of Progress in Coastal Policies at the National Level. A Transatlantic and Euro-Mediterranean Perspective. Paper presented at Quality Control in Coastal GIS, NATO Advanced Research Workshop, Ljubljana, Slovenia.

Beckett, D. (2004). RDF/XML Syntax Specification, W3C Recommendation 10 February 2004. Retrieved July 25th, 2009 from W3C web site: http://www.w3.org/TR/rdf-syntax-grammar/

Copyright © 2011, IGI Global. Copying or distributing in print or electronic forms without written permission of IGI Global is prohibited.

Beckett, D., & Berners-Lee, T. (2008). Turtle – Terse RDF Triple Langauge, W3C Team Submission 14 January 2008. Retrieved July 24th, 2009 from W3C web site http://www.w3.org/TeamSubmission/turtle/

Belgian Coastal Atlas. (2009). Retrieved October 31, 2009, from the Belgian Coastal Atlas web site: http://www.kustatlas.be.

Belpaeme, K., & Konings, P. (Eds.). (2004). The Coastal Atlas Flanders-Belgium. The Co-ordination Center for Coastal Zone Management. Retrieved May 7, 2009, from Kustatlas website http://www.kustatlas.be

Belpaeme, K., & Konings, Ph. (2004). De kustatlas Vlaanderen-België. Oostende, Belgium: Coordination Centre on Integrated Coastal Zone Management.

Bermudez, L. E., Bogden, P., Bridger, E., Galvarino, C., Graybeal, J., Forest, D., et al. (2009). Web Feature Service (WFS) and Sensor Observation Service (SOS) comparison to publish time series data. International Symposium on Collaborative Technologies and Systems - Workshop on Sensor Web Enablement, Baltimore, MD.

Bianchini, D., De Antonellis, V., & Melchiori, M. (2003). Domain ontologies for virtual knowledge sharing and service composition in virtual district. In Proceeding of the International Workshop on Web Semantics (WebS2003), Prague, Czech Republic, September 1-5 2003.

Bishr, Y. (1998). Overcoming the semantic and other barriers to GIS interoperability. International Journal of Geographical Information Science, 12(4), 299–314. doi:10.1080/136588198241806

Brewer, C. A. (2005). Designing Better Maps: A Guide for GIS Users. Redlands, CA: ESRI Press.

Brickley, D., & Guha, R. V. (2004). RDF Vocabulary Description Language 1.0: RDF Schema, W3C Recommendation, 10 February 2004. Retrieved July 16th, 2009 from W3C website http://www.w3.org/TR/rdf-schema/

British Columbia Ministry of Sustainable Resource Management (BCMSRM). (2009). COINPacific Web Mapping Application. Retrieved June 29, 2009, from Land Information B.C. website http://aardvark.gov.bc.ca/apps/coin/

British Oceanographic Data Centre. (1991). Manual of Inventory of Moored Current Meter Data. Plymouth, UK: British Oceanographic Data Centre.

British Oceanographic Data Centre. (1992). United Kingdom Digital Marine Atlas. Version 2.0. User Guide. Plymouth, UK: British Oceanographic Data Centre.

British Ordnance Survey. (2009). Retrieved October 31, 2009 from the British Ordnance Survey web site: http://www.ordnancesurvey.gov.uk.

Brown, A., & Feringa, W. (2002). Colour Basics for GIS Users. New York: Prentice Hall.

Buitelaar, P. Eigner Thomas, & Declerck T. (2004). OntoSelect: A Dynamic Ontology Library with Support for Ontology Selection. In Proceedings of the Demo Session at the International Semantic Web Conference, Hiroshima, Japan.

Caldwell, K. (1988). Perspectives on ecosystem management for the Great Lakes: A reader. Albany, NY: State University of New York Press.

Canessa, R., Butler, M., Leblanc, C., Stewart, C., & Howes, D. (2007). Spatial Information Infrastructure for Integrated Coastal and Ocean Management in Canada. Coastal Management, 35(1), 105–142. doi:10.1080/08920750600970537

Caribbean Marine Atlas Team. (2009). Project Proposal for Submission to the Government of Flanders (FUST 2009-2013) Version 2.1. UNESCO/IOC.

Caribbean Marine Atlas. (2009). Retrieved October 31, 2009 from the Caribbean Marine Atlas web site: http://www.caribbeanmarineatlas.org.

Cartoweb. (2009). Retrieved October 15th, 2009 from Cartoweb - Advanced Geographical Information System for the Web web site: http://www.cartoweb.org/

CEN. (2009). European Committee for Standardisation. Retrieved Septembr 24th, 2009 from the website http://www.cen.eu

Center for Coastal Resources Management. (2009). Shoreline Manager's Assessment Kit. Virginia Institute of Marine Science. Retrieved on July 3, 2009, from VIMS website: http://ccrm.vims.edu/gis_data_maps/interactive_maps/disclaimer_smak.html

Centre for Marine and Coastal Zone Management. (2009). Retrieved October 31, 2009 from the University of Aberdeen web site: http://www.abdn.ac.uk/cmczm.

Cerami, E. (2002). Web Services Essentials - Distributed Applications with XML-RPC, SOAP, UDDI & WSDL. Sebastopol, CA: O'Reilly Media.

Chameleon. (2009). Retrieved October 31, 2009, from the Chameleon web site: http://chameleon.maptools.org.

Channel Coastal Observatory. (2009). Retrieved October 31, 2009 from the Channel Coastal Observatory web site: http://www.channelcoast.org

Chesapeake Bay Foundation. (2003). Chesapeake Bay Fact Sheet. Retrieved May 15, 2009 from the Chesapeake Bay Foundation web site: http://www.cbf.org/site/PageServer?pagename=resources_facts_general.

Coastal & Marine Resources Centre (CMRC). (2009). The Marine Irish Digital Atlas. University College Cork. Retrieved May 7, 2009, from MIDA web site: http://mida.ucc.ie

Coastal Geospatial and Education Mapping System. (2007). Virginia Coastal Zone Management Program, Virginia Department of Environmental Quality. Retrieved June 3, 2009 from the Virginia Department of Environmental Quality web site: http://www.deq.state.va.us/coastal/coastalgems.html.

Commission of the European Communities. (2007). Communication from the commission to the European parliament, the Council, the European economic and social committee and the committee of the regions- An Integrated Maritime Policy for the European Union, COM(2007) 575 final.

Commission of the European Communities. (2007a). SEC (2007) 1278 – Accompanying Document to the Communication from the Commission to the European Parliament, the Council, the European Economic and Social Committee and the Committee of the Regions, An Integrated Maritime Policy for the European Union. European Commission, Official Journal of the European Community, Luxembourg.

Commission of the European Communities. (2007b). An Integrated Maritime Policy for the European Union. Luxembourg: Office for Official Publications, Retrieved April 24, 2009 from the European Commission web site: http://ec.europa.eu/maritimeaffairs/policy_documents_en.html.

Commission of the European Communities. (2007c). Directive 2007/2/EC of the European Parliament and of the Council of 14 March 2007 Establishing an Infrastructure for Spatial Information in the European Community (INSPIRE). Official Journal of the European Community, 50(108). Retrieved August 6, 2009 from the European Commission web site: http://bit.ly/2lBuET

Commission of the European Communities. (2008). Communication from the Commission to the Council, the European Parliament, The European Economic and Social Committee and the Committee for the Regions - Towards a Shared Environmental Information System (SEIS), COM (2008) 46 final, Brussels, 1.2.2008. Office of Official Publications, Luxembourg. Retrieved April 4, 2009 from http://bit.ly/Q8hw0

Commission of the European Communities. (2009). SEC (2009) 499 final - Building a European marine knowledge infrastructure: Roadmap for a European Marine Observation and Data Network, European Commission, Official Journal of the European Community, Luxembourg. Retrieved August 6, 2009 from the European Commission web site http://ec.europa.eu/maritimeaffairs/pdf/roadmap_emodnet_en.pdf

Cornaert, M. H., & Moreira, J. M. (1992). La coordination de l'information sur l'environnement à la C.E.E. au travers de projets de systèmes d'informations géorepérées. L'experience CORINE-SinambA. Rev. Regions d'Europe. Special Environnement et Monde Rural, 5, Besançon, France.

Craglia, M., Goodchild, M. F., Annoni, A., Camara, G., Gould, M., & Kuhn, W. (2008). Next-generation Digital Earth: A position paper from the Vespucci Initiative for the Advancement of Geographic Information Science. International Journal of Spatial Data Infrastructures Research, 3, 146–167.

Creative Commons. (2009). Creative Commons Licenses – Attribution, Share Alike, Noncommercial, No Derivative Works. Retrieved August 6, 2009, from Creative Commons web site: http://creativecommons.org/about/licenses/

Crossett, K. M., Culliton, T. J., Wiley, P. C., & Goodspeed, T. R. (2004). Population Trends along the Coastal United States: 1980-2008. Coastal Trends Report Series. Silver Spring, MD: U.S. Department of Commerce, National Oceanic and Atmospheric Administration.

CSA Group/Circa Group/Swiftsure. (2005). An Assessment of the optimal use and application in the immediate to long term future of the Irish National Seabed Survey Deliverables to Date (Phase 1) and to contribute towards the development of phase 3. Retrieved June 29, 2009, from Infomar website: http://www.infomar.ie/documents/Assessment_of_the_INSS.pdf

CSML. (2009). Climate Science Modelling Language. Retrieved September 20th, 2009 from the website http://ndg.nerc.ac.uk/csml/

Current Meter Series Database. (2009). Retrieved October 31, 2009 from the British Oceanographic Data Centre web site: https://www.bodc.ac.uk/data/online_request/current_meters

Data Archive for Seabed Species and Habitats. (2009). RetrievedOctober31, 2009 from the Data Archive for Seabed Species and Habitats web site:http://www.dassh.ac.uk.

Data Integration System for Marine Pollution Water Quality. (2009). RetrievedOctober31, 2009 from the Nansen Environmental and Remote Sensing Center web site:http://www.nersc.no/dismar

Data.gov. (2009). Retrieved October 31, 2009 from the Data.gov web site: http://www.data.gov.

De Kustatlas Vlaanderen/België. (2005). Retrieved December 11, 2008 from Kustatlas web site: www.kustatlas.be.

De Kustatlas. (2009). Retrieved October 7, 2009, from De Kustatlas Online Web site: http://www.kustatlas.be/

de la Beaujardière, J. (2004). OGC Web Map Service Interface (Version 1.3.0), Open Geospatial Consortium Inc. Retrieved May 18th, 2009 from OGC portal http://portal.opengeospatial.org/files/?artifact_id=4756

Deliiska, B. (2007). Thesaurus and domain ontology of geoinformatics. Transactions in GIS, 11(4), 637–651. doi:10.1111/j.1467-9671.2007.01064.x

Dent, B. D. (2002). Cartography with ArcView GIS Software: Thematic Map Design: With ArcView GIS Software. New York: McGraw-Hill Higher Education.

Department of Environment, Food, and Rural Affairs. (2002). Delivering Integrated Marine Mapping for the UK: Report of DEFRA-funded workshop held at Church House, London, 11 September 2002, compiled by F.L. Franklin, DEFRA, Burnham Laboratory.

Digital Coast. (2009) Retrieved October 31, 2009, from NOAA Coastal Services Center web site: http://www.csc.noaa.gov/digitalcoast.

Directory of Coastal Web Atlases Worldwide. (2009). RetrievedOctober31, 2009 from the International Coastal Atlas Network website:http://ican.science.oregonstate.edu/atlases.

DM Solutions Group. (2009). Retrieved October 15, 2009, from DM Solutions Group web site: http://www.dmsolutions.ca/.

Drupal. (2009). Retrieved October 15th, 2009 from Drupal.org – Community Plumbing web site: http://www.drupal.org.

Dwyer, N., & Wright, D. J. (2008). Report of International Coastal Atlas Network Workshop 3 on Federated Coastal Atlases: Building on the Interoperable Approach, European Environment Agency, Copenhagen, Denmark. Retrieved July 29, 2009, from ICAN web site: http://ican.science.oregonstate.edu/ican3_final_rpt

Dwyer, N., O'Dea, L., & Cummins, V. (2003). The Marine Irish Digital Atlas: a Web Portal to Coastal and Marine Data in Ireland. Paper presented at CoastGIS

2003, Genova, Italy. Retrieved June 24, 2009, from MIDA website: http://mida.ucc.ie/assets/documents/MIDA_CoastGIS2003.pdf

Ecotrust. (2009). Retrieved October 15, 2009, from Ecotrust web site: http://www.ecotrust.org/.

Egenhofer, M. (2002). Toward the Semantic Geospatial Web, Tenth ACM International Symposium on Advances in Geographic Information Systems. New York: ACM Press.

Environmental Climatology Subsystem (CLIMA). (2009). Retrieved October 31, 2009 from the CLIMA web site: http://www.juntadeandalucia.es/medioambiente/clima (in Spanish).

Environmental Information Network of Andalusia (REDIAM) Biodiversity Subsystem. (2009). Retrieved October 31, 2009 from the REDIAM web site: http://bit.ly/4exovu (in Spanish).

Environmental Information Network of Andalusia (REDIAM) Water Subsystem. (2009). Retrieved October 31, 2009 from the REDIAM web site: http://bit.ly/3T6iEQ (in Spanish).

Environmental Systems Research Institute. (2009). Retrieved October 15th, 2009 from the ESRI: GIS Software that Gives You The Geographic Advantage web site: http://www.esri.com

Erdas Imagine. (2009). RetrievedOctober31, 2009 from the Erdas, Inc. web site:http://www.erdas.com.

Erle, S., Gibson, R., & Walsh, J. (2005). Mapping Hacks Tips & Tools for Electronic Cartography. O'Reilly.

ESRI. (2009). Retrieved October 15th, 2009 from Wikipedia: http://en.wikipedia.org/wiki/Esri.

European Commission. (1996). Directive 96/9/EC of the European Parliament and of the Council of 11 March 1996 on the Legal Protection of Databases. Official Journal of the European Community, L 77, 27.3.1996, p. 20–28. Retrieved August 6, 2009, from the European Commission web site: http://bit.ly/40ertR.

European Commission. (2000). Communication from the Commission to the Council and the European Parliament on Integrated Coastal Zone Management: a Strategy for Europe (COM/2000/547), adopted 27 September, 2000. Retrieved July 2, 2009, from the European Commission web site: http://ec.europa.eu/environment/iczm/comm2000.htm.

European Commission. (2002). 413. Recommendation of the European Parliament and of the Council of 20 May 2002 concerning the implementation of Integrated Coastal Zone Management in Belgium, 2002/413/EC.

European Commission. (2007). 456 FINAL. COMMUNICATION FROM THE COMMISSION TO THE EUROPEAN PARLIAMENT pursuant to the second subparagraph of Article 251 (2) of the EC Treaty concerning the Common position of the Council on the adoption of a European Parliament and Council Directive establishing a Framework for Community Action in the field of Marine Environmental Policy (Marine Strategy Framework Directive)-2005/0211 (COD).

European Commission. (2007). An Integrated Maritime Policy for the European Union. Commission of the European Communities. Brussels, 10 October 2007, COM (2007) 575 final. Retrieved July 29, 2009 from the European Commssion website http://eur-lex.europa.eu/LexUriServ/site/en/com/2007/com2007_0575en01.pdf

European Commission. (2007). Communication from the Commission to the European Parliament, the Council, the European Economic and Social Committee and the Committee of the Regions: An Integrated Maritime Policy for the European Union. Commission of the European Communities. Brussels, 10 October 2007, COM (2007) 575 final. Retrieved July 29, 2009 from the European Commission web site: http://eur-lex.europa.eu/LexUriServ/site/en/com/2007/com2007_0575en01.pdf

European Commission. (2009). European Commission Joint Research Center African Marine Information System (AMIS). Retrieved 22 July 2009, from the AMIS website http://amis.jrc.ec.europa.eu/index.php

European Environment Agency. (2009). Water Watch: Eye On Earth. Retrieved July 3, 2009, from EEA web site: http://eoe.eea.europa.eu/

European Environment Information and Observation Network. (2009). Retrieved October 31, 2009 from the European Environment Agency web site: http://www.eionet.europa.eu/partners.

European Marine Observation and Data Network (EMODNET). (2009). Retrieved October 31, 2009, from the European Commission web site: http://ec.europa.eu/maritimeaffairs/eu-marine-observation-data-network-mission_en.html.

European Union. (2003). Directive 2003/98/EC of the European Parliament and of the Council on the re-use of public sector information. Official Journal of the European Union. L&C, L345(90), 46.

European Union. (2007). Directive 2007/2/EC of the European Parliament and of the Council establishing an Infrastructure for Spatial Information in the European Community (INSPIRE). Official Journal of the European Union. L&C, L108(1), 50.

Federal Geographic Data Committee. (2009). FGDC Content Standard for Digital Geospatial Metadata. Retrieved October 31, 2009 from the Federal Geographic Data Committee website http://www.fgdc.gov/metadata/csdgm.

Federal Geographic Data Committee. (2009). Retrieved October 15, 2009, from Federal Geographic Data Committee web site: http://www.fgdc.gov/.

FGDC Marine and Coastal Spatial Data Subcommittee. (2009). Retrieved October 31, 2009, from NOAA Coastal Services Center web site: http://www.csc.noaa.gov/mcsd

Flanders Marine Institute. (2009). Retrieved October 7, 2009, from the VLIZ Web site: http://www.vliz.be

Free Software Foundation. (2007). GNU General Public License. Retrieved August 6, 2009, from Free Software Foundation web site: http://www.fsf.org/licensing/licenses/gpl.html

Garcia-Molina, H., Papakonstantinou, Y., Quass, D., Rajaman, A., Sagir, Y., & Ullman, J. (1997). The TSIMMIS approach to mediation: Data models and Languages. Journal of Intelligent Information Systems, 8, 117–132. doi:10.1023/A:1008683107812

GDAL. (2009). Retrieved October 15, 2009, from GDAL - Geospatial Data Abstraction Library web site: http://www.gdal.org.

General Bathymetric Chart of the Oceans GEBCO Digital Atlas. (2009). RetrievedOctober31, 2009 from the Global Bathymetric Chart of the Oceans websitehttp://www.gebco.net/data_and_products/gebco_digital_atlas

Geo, I. P. Country. (2009). Retrieved October 15th, 2009 from MaxMind - GeoIP Country Geolocation IP Address to Country web site: http://www.maxmind.com/app/country.

GeoBC Spatial Analysis Branch. (2009). British Columbia Coastal Resource Information Management System. Integrated Land Management Bureau, Ministry of Agriculture and Lands. Retrieved June 29, 2009, from Spatial Analysis Branch B.C. website http://www.ilmb.gov.bc.ca/cis/coastal/others/crimsindex.htm

Geo-Information Decision Support Processing and Dissemination System. (2009). RetrievedOctober31, 2009 from the University of Aberdeen web site:http://www.abdn.ac.uk/~geo402/index.htm

GeoLite Country. (2009). Retrieved October 15th, 2009 from MaxMind - GeoLite Country | Open Source IP Address to Country Database web site: http://www.maxmind.com/app/geolitecountry

GeoNetwork Opensource. (2009). Retrieved October 15, 2009, from the GeoNetwork Opensource web site: http://geonetwork-opensource.org/.

GeoSciML. (2009). Retrieved September 20th, 2009 from the web site http://www.geosciml.org/

GeoServer. (2009). Retrieved October 31, 2009, from the GeoServer web site: http://GeoServer.org

Geospatial One Stop. (2009). Retrieved October 31, 2009, from the Geospatial One Stop web site http://geodata.gov

Global Bathymetric Chart of the Oceans. (2009). RetrievedOctober31, 2009 from the Global Bathymetric Chart of the Oceans web site:http://www.gebco.net

Goasdoue, F., Lattes, V., & Rousset, M.-C. (2000). The Use of CARIN Language and Algorithms for Information Integration: The PICSEL System. International Journal of Cooperative Information Systems, 9(4), 383–401. doi:10.1142/S0218843000000181

Godiva. (2009). RetrievedOctober31, 2009 from the Natural Environment Research Council of the UK website http://www.nerc-essc.ac.uk/godiva

González Del Campo, A. (2009). Current Practice and Potential on the Application of Geographic Information Systems as a Support Tool in Strategic Environmental Assessment of Irish Land Use Plans. GISEA Manual – Consultation Draft (1st April 2009). Ireland: Environmental Protection Agency.

Goodchild, M. (2007). Citizens as sensors: The world of volunteered geography. GeoJournal, 69(4), 211–221. doi:10.1007/s10708-007-9111-y

Google Earth. (2009). Retrieved October 15th, 2009 from Google Earth web site: http://earth.google.com.

Google Maps Application Programming Interface. (2009). Retrieved October 31, 2009, from the Google Maps Application Programming Interface web site: http://code.google.com/apis/maps.

Google Maps Legal Notices. (2009). Retrieved October 15, 2009, from Google Maps Legal Notices web site: http://www.google.com/intl/en_ALL/help/legalnotices_maps.html.

Google Maps. (2009). Retrieved October 15th, 2009 from Google Maps web site: http://maps.google.com/

Graybeal, J. (2009). Achieving Semantic Interoperability. Retrieved July 29, 2009 from The Marine Metadata Interoperability web site, Guides: Navigating the World of Marine Metadata section: http://marinemetadata.org/guides/vocabs/cvchooseimplement/cvsemint.

Great Lakes Information Network. (2009). Retrieved October 31, 2009, from the Great Lakes Information Network web site: http://glin.net.

Green, D. R. (1994a). The United Kingdom Digital Marine Atlas: A review. GeoCal, (June), 280-30. Green, D.R. (1994b). Geographic and marine information systems: Some recent developments and progress in the United Kingdom. In Proceedings Hydro '94, The Ninth Biennial International Symposium of the Hydrographic Society, UK Branch (pp. 13-15). Aberdeen Exhibition and Conference Centre. Special Publication No. 33.

Green, D. R. (1994c). Using GIS to construct and update an estuary information system. In Proceedings of Management Techniques for the Coastal Zone. Centre for Coastal Zone Management, University of Portsmouth. October 24-25, (pp. 129-162).

Green, D. R. (1995). Internet, the WWW and Browsers: The Basis for a Network-Based Geographic Information System (GIS) for Coastal Zone Management. In Proceedings of AGI'95 Conference. Expanding Your World. 21-23 November 1995. (pp. 5.1.1- 5.1.12).

Green, D. R., & King, S. D. (1998). The UK Coastal Map Creator: ArcIMS provides a solution to coastal map and data delivery for the United Kingdom's coastal community. In Proceedings of the ESRI User Conference, San Diego, CA.

Gruber, T. (1993). A Translation Approach to Portable Ontology Specification. Knowledge Acquisition, 5(2), 199–220. doi:10.1006/knac.1993.1008

Gulf of Maine Mapping Portal. (2009). Retrieved October 31, 2009, from the Gulf of Maine Ocean Observing System web site: http://www.gommap.org.

Haddad, T. C., Wright, D. J., Dailey, M., Klarin, P., Marra, J., Dana, R., & Revell, D. (2006). The tools of the Oregon Coastal Atlas . In Wright, D. J., & Scholz, A. J. (Eds.), Place Matters: Geospatial Tools for Marine Science (pp. 133–149). Corvallis, OR: Oregon State University Press.

Haklay, M. (2009). Human-Computer Interaction and Usability of GIS. Retrieved February 12, 2009 from UCL web site: http://homepages.ge.ucl.ac.uk/~mhaklay/usability.htm.

Hamre, T., Krasemann, H., Groom, S., Dunne, D., Breitbach, G., & Hackett, B. (2009). Interoperable web GIS services for marine pollution monitoring and forecasting. Journal of Coastal Conservation, 13, 1–13. doi:10.1007/s11852-009-0046-y

HarborView Web Mapping Application. (2009). Retrieved October 31, 2009, from the HarborView web site: http://glos.us/hview.

Harrower, M., & Sheesley, B. (2005). Designing better map interfaces: A framework for panning and zooming. Transactions in GIS, 9(2), 77–89. doi:10.1111/j.1467-9671.2005.00207.x

Hart, D. (2000). Building a Horizontally and Vertically Integrated Coastal GIS using Local Government Spatial Data: The Case of Coastal Erosion Hazards on the Lake Michigan Coast of Wisconsin. Doctoral dissertation. Madison, WI: University of Wisconsin-Madison.

Heydari, N., Mansourian, A., Taleai, M., & Fallahi, G. R. (2009). Ontology-based GIS web service for increasing semantic interoperability among organizations involving drilling in city of Tehran. In GSDI 11 World Conference, Rotterdam, The Netherlands.

Hills, J. M., Alcock, D., Higham, T., Kirkman, H., Le Tissier, M., & Pagdilao, C. (2006). Capacity building for integrated coastal management in Asia-Pacific: The case for case studies. Coastal Management, 34(3), 323–337. doi:10.1080/08920750600686703

IACMST. (2005, January). Marine Data and Information Partnership (MDIP) Statement of Intent (Rev. 02). Presented to the Inter-Agency Committee for Marine Science and Technology, MDIP Steering Group meeting, London, January, 2005.

ICAN User Survey. (2009). Retrieved October 15th, 2009 from the ICAN Atlas Assessment Working Group web site: http://ican.science.oregonstate.edu/ican_assess.

ICAN. (2009). International Coastal Atlas Networks. Retrieved October 21th, 2009 from the website http://ican.ucc.ie/

IMIS, Integrated Marine Information System. (2009). Retrieved October 7, 2009, from the VLIZ Web site: http://www.vliz.be/imis/imis.php?module=institute.

Infoterra. (2009). Retrieved October 31, 2009 from the Infoterra website http://www.infoterra.co.uk/.

INSPIRE. (2009). Retrieved September 24th, 2009 from the website http://inspire.jrc.ec.europa.eu

Institute for Natural Resources. (2005). North Coast Basin Prototype Web site Project Completion Report. Oregon State University. Retrieved May 7, 2009, from OSU website http://inr.oregonstate.edu/download/OWEB_project_completion_report.pdf

interactive Spatial Explorer and Administrator. (2009). RetrievedOctober31, 2009 from the Center for Environment, Fisheries and Aquaculture Science, UK websitehttp://www.cefas.co.uk/isea

Intergovernmental Oceanographic Commission. (2003). ODINAFRICA III: An Integrated Ocean Observation and Service Network for Africa. Project Proposal. Intergovernmental Oceanographic Commission (of UNESCO), Paris, France. Retrieved 29 July 2009, from the ODINAFRICA web site: http://www.odinafrica.org/

Intergovernmental Oceanographic Commission. (2007). IODE/ICAM Workshop on the Development of the Caribbean Marine Atlas (CMA), IOC Workshop Report No. 205. Oostende, Belgium: UNESCO IOC.

International Coastal Atlas Network. (2009). About ICAN. Retrieved October 31, 2009 from the International Coastal Atlas Network web site: http://icoastalatlas.net or http://ican.science.oregonstate.edu.

International Coastal Atlas Network. (2009). Retrieved October 15, 2009, from the International Coastal Atlas Network web site: http://ican.science.oregonstate.edu.

International Council for the Exploration of the Sea. Denmark ICES Fishmap. (n.d.). Retrieved fromhttp://www.ices.dk/marineworld/ices-fishmap.asp.

International Hydrographic Bureau. (2008). IHO Geospatial Standard for Hydrographic Data, Version 0.0.0 – January 2008. International Hydrographic Bureau, Monaco, 2008, Special Publication No. 100, Retrieved April 24, 2009 from the International Hydrographic Organization web site http://www.iho.int/COMMITTEES/CHRIS/TSMAD/S-100-Feb08.zip

International Hydrographic Organization. (2008). Marine Spatial Data Infrastructure Working Group (MSDIWG) Terms of Reference. International Hydrographic Organization, IHB, Monaco. Retrieved April 24, 2009 from the International Hydrographic Organization web site: http://www.iho.shom.fr/COMMITTEES/CHRIS/MSDIWG/MSDIWG-TOR.pdf

International Oceanographic Data and Information Exchange (IODE). (2009). Programme of the Intergovernmental Oceanographic Commission (IOC) of UNESCO. Retrieved October 31, 2009 from the IODE web site: http://www.iode.org.

International Organization for Standards (ISO). (2003). International Standard ISO 19115, geographic Information. Metadata.

Internet Mapping Framework. (2009). Retrieved October 15th, 2009 from Moxi Media Inc. - Internet Mapping Framework web site: http://www.moximedia.com/imf.html

InterRisk. (2009). RetrievedOctober31, 2009 from the Nansen Environmental and Remote Sensing Center websitehttp://interrisk.nersc.no

Irish Marine Institute. (2009). Marine Data Online. Retrieved May 7, 2009, from Marine Data Online website http://www.marinedataonline.ie/

Isenor, A. W., & Spears, T. W. (2007). A conceptual model for service-oriented discovery of marine metadata descriptions. CMOS Bulletin, 35(3), 85–91.

Isenor, A., & Neiswender, C. (2009). Categories of Controlled Vocabularies. In The MMI Guides: Navigating the World of Marine Metadata. Retrieved September 15th, 2009 from Marine Metadata Interoperability website http://marinemetadata.org/guides/vocabs/voctypes/voccat

ISO. (2008). Geographic information — Rights expression language for geographic information — GeoREL, draft standard ISO/CD 19149. Geneva: International Organization for Standardization.

ISO/IEC. (1993). ISO 2382-1: Information technology – Vocabulary – Part 1: Fundamental terms. International Organization for Standardization, 3.0 edition.

ISO/TC 46/SC 4. (2003). ISO 15836:2003: Information and documentation – The Dublin Core metadata element set. Amsterdam: International Organization for Standardization.

ISO/TC211 Geographic information/Geomatics. (2009). Retrieved September 24th, 2009 from the website http://www.isotc211.org

ISO/TC211. (2003). [– Geographic Information – Metadata. Amsterdam: International Organization for Standardization.]. ISO, 19115, 2003.

ISO/TC211. (2006). ISO/PRF TS 19139 – Geographic information – Metadata – XML schema implementation. Amsterdam: International Organization for Standardization.

ISO/TC211. (2007). [Geographic information – Geography Markup Language][GML][. Amsterdam: International Organization for Standardization.]. ISO, 19136, 2007.

IWEA. (2008). Best Practice Guidelines for the Irish Wind Energy Industry. The Irish Wind Energy Association and Sustainable Energy Ireland. Retrieved from http://www.iwea.ie

Jena. (2009). Jena 2 Inference support. Retrieved September 14th, 2009 from the website http://jena.sourceforge.net/inference/

Jones, M. T., Tabor, A. R., & Weatherall, P. (1994). Supporting Volume to the GEBCO Digital Atlas. British Oceanographic Data Centre (BODC). Swindon, UK: Natural Environment Research Council.

Joomla! Content Management System. (2009). Retrieved October 15, 2009, from Joomla! - the dynamic portal engine and content management system web site: http://www.joomla.org.

Juda, L. (2005). The report of the U.S. Commission on Ocean Policy: State perspectives. Coastal Management, 34(1), 1–16. .doi:10.1080/08920750500364930

Junta de Andalucía – Consejería de Medio Ambiente. (2009). Compendio de Cartografía y Estadísticas Ambientales de Andalucía. 2 Edición 2008 (de Información Ambiental de Andalucía, R., Ed.). Sevilla: REDIAM.

Kahane, L., de Laat, W., Myncke, R., & Staelens, P. (2009). Second sojourn at the coast: applied research on the littoral profile, the effects and toolkit for sound coastal policy (original title: Tweede verblijven aan de Kust: onderzoek naar het profiel, de effecten en een toolkit voor het beleid). Brugge, Belgium: Idea Consult.

ka-Map! (2009). Retrieved October 15, 2009, from Ka-MapMapTools.org website http://ka-map.maptools.org/

KATU. (2009). Retrieved October 15, 2009, from KATU.com - News, Weather and Sports - Portland, Oregon web site: http://www.katu.com/.

Kemp, Z., Tan, L., & Whalley, J. (2007). Interoperability for Geospatial Analysis: a Semantics and Ontology-based Approach. In J. Bailey & A. Fekete, (Eds.), Proc. Eighteenth Australasian Database Conference (ADC 2007), Ballarat, Australia, (CRPIT, Vol. 63, pp. 83-92). ACS.

Kirk, T., Levy, A. Y., Sagiv, Y., & Srivastava, D. (1995). The Information Manifold . In Knoblock, C., & Levy, A. (Eds.), Information Gathering from Heterogeneous, Distributed Environments. Stanford, CA: Stanford University.

Kraak, M. J. (2001). Setting and needs for web cartography . In Kraak, M. J., & Brown, A. (Eds.), Web Cartography: Developments and Prospectus (pp. 1–7). London: Taylor & Francis.

Kramers, R. E. (2008). Interaction with maps on the Internet – A user centered design approach for the Atlas of Canada . The Cartographic Journal, 45(2), 98–107. doi:10.1179/174327708X305094

Krygier, J., & Wood, D. (2005). Making Maps: A Visual Guide to Map Design for GIS. New York: The Guildford Press.

Lacy, L. W. (2005). OWL: Representing Information Using the Web Ontology Language. Victoria, Canada: Trafford Publishers.

Lake Superior Coastal Mapping Portal. (2009). Retrieved October 31, 2009, from the University of Wisconsin web site: http://maps.aqua.wisc.edu/lscmp

Lake, R., Burggraf, D. S., Trnini'c, M., & Rae, L. (2004). Geography Mark-Up Language (GML) – Foundation for the Geo-Web. New York: John Wiley.

Lassoued, Y., Pham, T., Bermudez, L., Stocks, K., O'Grady, E., Isenor, A., & Alexander, P. (2009this volume). Coastal atlas interoperability . In Wright, D., Dwyer, E., & Cummins, V. (Eds.), Handbook of Coastal Informatics: Web Atlas Design and Implementation. Hershey, PA: IGI-Global.

Lassoued, Y., Wright, D. J., Bermudez, L., & Boucelma, O. (2008). Ontology-based mediation of OGC Catalogue Service for the Web requests: A virtual solution for integrating coastal web atlases. In Proceedings of ICSOFT 2008, the 3rd International Conference on Data and Software Engineering, Berlin: Springer.

Lazar, J. (2006). Web usability: a user-centered design approach. Boston: Pearson Addison Wesley.

Library (Computing). (2009). Retrieved October 15, 2009 from Wikipedia web site: http://en.wikipedia.org/wiki/Library_(computing).

Longhorn, R. (2002). Global Spatial Data Sharing Frameworks: the case of the Intergovernmental Oceanographic Commission (IOC). In Proceedings of GSDI-6, Budapest, Hungary, 16–19 September 2002. Retrieved April 21, 2009 from the Global Spatial Data Infrastructure web site: http://bit.ly/yqhZM.

Longhorn, R. (2004). Integrated Coastal/Marine Spatial Data Infrastructure. Paper presented at the ECO-IMAGINE Conference, 13-15 May 2004, Seville, Spain. Retrieved October 7, 2009, from the Gisig web site: http://www.gisig.it/eco-imagine.

Longhorn, R. (2005). Coastal Spatial Data Infrastructure . In Bartlett, D., & Smith, J. (Eds.), GIS for Coastal Zone Management. London: Taylor & Francis.

Longhorn, R. (2009). Coastal and marine data infrastructure . In Green, D. R. (Ed.), Coastal Zone Management (pp. 206–225). London: Thomas Telford Limited.

Longhorn, R. (2009in preparation). Coastal and Marine Spatial Data Infrastructure . In Green, D. (Ed.), Coastal Zone Management. London: Thomas Telford Books.

Lowry, R. K., Cramer, R. N., & Rickards, L. J. (1992). North Sea Project CD-ROM User's Guide. British Oceanographic Data Centre. Swindon, UK: Natural Environment Research Council.

Lowry, R., & Williams, G. (in press). Putting meaning in SeaDataNet. Information Sciences (special issue on the 2008 International Conference on Marine Data and Information Systems).

MacKenzie, S. (1996). Integrated Resource Planning and Management: The Ecosystem Approach in the Great Lakes. Washington, DC: Island Press.

Madej, E. (2000). Cartographic Design Using Arcview GIS. Albany, NY: OnWord Press.

Maedche, A., Motik, B., Stojanovic, L., Studer, R., & Volz, R. (2003). Ontologies for enterprise knowledge management. IEEE Intelligent Systems, 18(2), 26–33. doi:10.1109/MIS.2003.1193654

Maelfait, H., & Belpaeme, K. (Eds.). (2007). The Coastal Compass, Key Indicators as Guidelines for ICZM. Oostende, Belgium: Coordination Centre for Integrated Coastal Zone Management.

Mahony, O. C., Sutton, G., McMahon, T., O'Cinneide, M. & Nixon, E. (2008). Issues and Recommendations for the Development and Regulation of Marine Aggregate Extraction in the Irish Sea. Galway, Ireland: Marine Institute.

Maillard, C., Lowry, R., Maudire, G., Schaap, D. & the SeaDataNet Consortium (2007). SeaDataNet: Development of a Pan-European Infrastructure for Ocean and Marine Data Management. Contribution to EurOcean07, theme 5.0 subtheme 5.1 - Paper 070131-026.

Manninen, C. (1999, December 5-9). The Great Lakes Information Network: Trials and Triumphs of an Integrated Approach to Web Design. Paper presented at the Symposium on Water Resources and the World Wide Web, American Water Resources Association, Seattle, WA. Retrieved on October 25, 2009 from http://www.awra.org/proceedings/www99/w10/index.htm

Manolescu, I., Florescu, D., & Kossmann, D. (2001). Answering XML queries over heterogeneous data cources. In Proceedings of VLDB, (pp. 241–250).

MapBender. (2009). Retrieved October 15th, 2009 from MapbenderWiki web site: http://www.mapbender.org.

Mapping Science Committee. (2001). National Spatial Data Infrastructure Partnership Programs: Rethinking the Focus. Washington, DC: National Academy Press.

MapServer Optimization. (2009) Retrieved October 15, 2009, from MapServer Optimization web site: http://mapserver.org/optimization.

MapWindow. (2009). RetrievedOctober31, 2009 from the MapWindow websitehttp://www.mapwindow.org

Marine and Coastal Environmental Information Services. (2009). RetrievedOctober31, 2009 from the Marine and Coastal Environmental Services website http://gmes-marcoast.com

Marine Digimap. (2009). RetrievedOctober31, 2009 from the Marine Digimap websitehttp://edina.ac.uk/digimap/description/marine_overview.shtml

Marine Environment Data and Information Network. (2009). RetrievedOctober31, 2009 from the Marine Environmental Data and Information Network websitehttp://www.oceannet.org

Marine Irish Digital Atlas. (2009). Retrieved October 31, 2009, from Coastal and Marine Resources Centre, University College Cork web site: http://mida.ucc.ie.

Marine Irish Digital Atlas. (2009). Retrieved October 31, 2009, from the Marine Irish Digital Atlas web site: http://mida.ucc.ie

MarineMap Consortium. (2009). MarineMap: Decision Support for Marine Spatial Planning. Retrieved July 3, 2009, from marinemap web site: http://marinemap.org/

Maritime & Coastguard Agency & DEFRA. (2006). Coastal and Marine Resource Atlas - Final Report.

Retrieved May 7, 2009, from MAGIC web site: http://www.magic.gov.uk/Me1302_4030_FRP.pdf

Maritime & Coastguard Agency & DEFRA. (2009). UK Coastal and Marine Resource Atlas. Retrieved May 7, 2009, from MAGIC web site: http://www.magic.gov.uk/camra.html

Maryland Chesapeake and Coastal Program. (2005). Maryland Shorelines Online, Maryland Department of Natural Resources. Retrieved June 15, 2009 from the Maryland Department of Natural Resources web site: http://shorelines.dnr.state.md.us.

Masser, I. (2007). Building European Spatial Data Infrastructures. Redlands, CA: ESRI Press.

Mayeske, G. W., & Lambur, M. T. (2001). How to Design Better Programs: A Staff Centered Stakeholder Approach to Program Logic Modeling. Crofton, MD: The Program Design Institute.

McCormack, B. (2003). Irish Spatial Data Infrastructure, paper presented at Irish Organization for Geographic Information (IRLOGI) 2003 Conference, Dublin, Ireland. Retrieved June 24, 2009, from IRLOGI website: http://www.irlogi.ie/pdf/ISDI_IRLOGI_2003.pdf

McGuinness, D. L., & van Harmelen, F. (2004). OWL Web Ontology Language – Overview, W3C Recommendation 10 February 2004. Retrieved January 20th, 2009 from W3C website http://www.w3.org/TR/owl-features/

McLaughlin, J. A., & Jordan, G. B. (1999). Logic models: A tool for telling your program's performance story. Evaluation and Program Planning, 22(1), 65–72. doi:10.1016/S0149-7189(98)00042-1

Mees, J. (2008). Het regionaal, nationaal en internationaal belang van de Universiteit Gent in de mariene wetenschappen. Oostende, Belgium: Vlaams Instituut voor de Zee (VLIZ).

Memorial University. (1998). Guide to Map Design and Production Using CorelDRAW 7 and CorelDRAW 8. St. Johns, Newfoundland, Canada: Memorial University of Newfoundland.

Microsoft Virtual Earth. (2009). RetrievedOctober31, 2009 from the Microsoft Virtual Earth web site:http://www.microsoft.com/maps/.

Miles, A., & Bechhofer, S. (2009). SKOS Simple Knowledge Organization System Reference, W3C Recommendation 18 August 2009. Retrieved September 10th, 2009 from W3C website http://www.w3.org/TR/2009/REC-skos-reference-20090818/

Miller, S. P., Bermudez, L., Montgomery, E., Isenor, A., & Neiswender, C. (2009). Vocabularies: Dictionaries, Ontologies, and More. The MMI Guides: Navigating the World of Marine Metadata. Retrieved September 17th, 2009 from The Marine Metadata Interoperability website http://marinemetadata.org/guides/vocabs

Minnesota MapServer. (2009). Retrieved October 31, 2009, from the Minnesota MapServer web site: http://www.mapserver.org

Moreira, J. M. (1991). SinambA: an information system for the environment of Andalusia. Evaluation of the results of the CORINE Programme 1985-1990- Lessons for the European Environment Agency. Workshop organized by the Commission of the European Communities, Luxembourg, 28-30 May. 1991.

Moreira, J. M., Fernández-Palacios, A., & Rodriguez, M. (1997). Land use and land capability. Pressure and state indicators of the evolution of ecosystems. European Topic Centre on Land Cover. Workshop on Land Cover Applications. Needs and Use. Copenhagen, 12-13 May, 1997.

Mr. Sid. (2009). RetrievedOctober31, 2009 from the Lizard Tech websitehttp://www.lizardtech.com

Multi-Agency Geographic Information for the Countryside. (2009). Retrieved October 31, 2009 from the Multi-Agency Geographic Information for the Countryside web site: http://www.magic.gov.uk.

Multipurpose Marine Cadastre. (2009) Retrieved October 31, 2009, from NOAA Coastal Services Center web site: http://www.csc.noaa.gov/mbwg/htm/multipurpose.html.

MyOcean Service. 2009-2012. (2009). RetrievedOctober31, 2009 from the My Ocean web site:http://www.myocean.eu.org.

National Aeronautics and Space Administration World Wind. (2009). Retrieved October 15th, 2009 from World Wind JAVA SDK web site: http://worldwind.arc.nasa.gov.

National Oceanic and Atmospheric Administration Coastal Services Center. (2009). Retrieved October 15, 2009, from NOAA CSC web site: http://www.csc.noaa.gov/.

National Oceanic and Atmospheric Administration Coastal Zone Management Program. (2009). Retrieved October 15, 2009, from Office of Ocean and Coastal Resource Management web site: http://coastalmanagement.noaa.gov/.

National Oceanic and Atmospheric Administration, Coastal Services Center. (2003). Opening a gateway to Oregon's coastal information. Coastal Services, 6(6), 2–3.

National Research Council. (2008). Increasing Capacity for Stewardship of Oceans and Coasts: A Priority for the 21st Century, Committee on International Capacity-Building for the Protection and Sustainable Use of Oceans and Coasts (Feeley, M. H., & Pantoja, S. C., Eds.). Washington, D.C.: The National Academies Press.

National Spatial Data Infrastructure. (2009). Retrieved October 31, 2009, from the Federal Geographic Data Committee web site: http://www.fgdc.gov/nsdi/nsdi.html

Nebert, D., & Whiteside, A. (2005). OGC Catalogue Services Specification. Open Geospatial Consortium Inc. Retrieved May 6th, 2009 from OGC website http://portal.opengeospatial.org/files/?artifact_id=5929

Neiswender, C. (2009). What is a Controlled Vocabulary? The MMI Guides: Navigating the World of Marine Metadata. Retrieved September 17th, 2009 from Marine Metadata Interoperability website http://marinemetadata.org/guides/vocabs/vocdef

NERC Earth Observation Data Acquisition and Analysis Service. (2009). RetrievedOctober31, 2009 from the Natural Environment Research Council web site:http://www.npm.ac.uk.

Netcraft. (2009). Retrieved October 15, 2009, from Netcraft – Internet Research web site: http://news.netcraft.com/

New York Department of State. (2009a). Coastal Resources Online. Division of Coastal Resources. Retrieved June 29, 2009, from NYS DOS web site: http://nyswaterfronts.com/index.asp.

New York Department of State. (2009b). New York Coastal Atlas. Division of Coastal Resources. Retrieved May 7, 2009, from NYS DOS website http://nyswaterfronts.com/maps_relief.asp

New York Ocean and Great Lakes Ecosystem Conservation Council (NYOGLECC). (2009). New York Ocean and Great Lakes Atlas. Retrieved July 3, 2009, from NYOGATLAS website http://nyoglatlas.org

Nielsen, J. (2000). Designing Web Usability. Berkeley, CA: New Riders.

Nielsen, J., & Loranger, H. (2006). Prioritizing Web Usability. Berkeley, CA: New Riders.

NOAA Coastal Services Center. (2006). 2006 Coastal Resource Management Customer Survey. U.S. Department of Commerce (DOC), National Oceanic and Atmospheric Administration (NOAA), National Ocean Service (NOS), Coastal Services Center. CSC.

NOAA Coastal Services Center. (2009). nowCOAST: Web Mapping Portal to Real-Time Coastal Observations and NOAA Forecasts. Retrieved May 29, 2009, from NOAA website http://nowcoast.noaa.gov/

NOAA Coastal Services Center. (2009). Project design and evaluation . Coastal Connections, 7(2), 1–3.

NOAA Office of Ocean & Coastal Resource Management. (2009). The Many Faces of Coastal Issues. Retrieved May 7, 2009, from NOAA website http://coastalmanagement.noaa.gov/issues/welcome.html

NOAA Office of Oceanic and Atmospheric Research. (2007). Integrated Ocean Mapping. Retrieved October 31, 2009 from the NOAA Office of Oceanic and Atmospheric Research web site: http://www.oar.noaa.gov/oceans/t_oceanmapping.html.

Norman, D. (1990). The Design of Everyday Things. New York: Double Day.

North Sea Interdepartmental Consultation Directorate. (2009). North Sea Atlas. Retrieved June 29, 2009, from NordZeeLoket website http://www.noordzeeatlas.nl/en/index.html

Nyerges, T. (1993). How do people use geographical information systems? In Medyckyj-Scott, D., & Hearnshaw, H. (Eds.), Human Factors in Geographical Information Systems (pp. 37–50). London: Belhaven Press.

O'Dea, L., Cummins, V., & Dwyer, N. (2004). Developing an Informational Web Portal for Coastal Data in Ireland: Data Issues in the Marine Irish Digital Atlas. Paper presented at Oceanology International London 2004, 18 March, London, UK. Retrieved June 24, 2009, from MIDA website: http://mida.ucc.ie/assets/documents/oceanology2004_mida.pdf

O'Dea, L., Cummins, V., Wright, D., Dwyer, N., & Ameztoy, I. (2007). Report on Coastal Mapping and Informatics Trans-Atlantic Workshop 1: Potentials and Limitations of Coastal Web Atlases. Cork, Ireland: Coastal and Marine Resources Centre, University College Cork.

O'Dea, L., Dwyer, N., Cummins, V., & Wright, D. (2007). Potentials and limitations of coastal web atlases: Outcomes of a transatlantic coastal mapping workshop. In Proceedings of CoastGIS 2007, Santander, Spain.

O'Dea, L., Dwyer, N., Cummins, V., Dunne, D., & Tuama, Ó. É. (2004, May). Improving Access to Coastal Information: Metadata in the Marine Irish Digital Atlas. Paper presented at ECO-IMAGINE Conference, Seville, Spain.

Ocean and Coastal Rsource Management. NOAA. (2009). States and Territories Working With NOAA on Ocean and Coastal Management. Retrieved October 31, 2009 from the Ocean and Coastal Resource Management web site: http://coastalmanagement.noaa.gov/mystate/welcome.html

Ocean Data and Information Network for Africa (ODIN-AFRICA). (2009). African Marine Atlas. Retrieved May 29, 2009, from African Marine Atlas website http://www.africanmarineatlas.net/

Ocean Teacher. (2009). Retrieved October 31, 2009 from the Ocean Teacher web site: http://www.oceanteacher.org.

Office of Management and Budget, Executive Office of the President of the United States. (2009). Circular No. A16 Revised. Retrieved October 31, 2009 from the Office of Management and Budget web site: http://www.whitehouse.gov/omb/rewrite/circulars/a016/a016_rev.html.

OGC. (2009). Retrieved September 24th, 2009 from the website http://www.opengeospatial.org

OgleEarth. (2009). RetrievedOctober31, 2009 from the OgleEarth web site:http://www.ogleearth.com.

OGR. (2009). Retrieved October 15, 2009, from OGR Simple Feature Library web site: http://www.gdal.org/ogr/

Ojeda-Zújar, J., Cabrera-Tordera, A., Fraile-Jurado, P., & Loder, J. (2007) Design and functionality of a 2D/3D viewer for a Coastal management oriented GIS. In Proceedings of CoastGIS 2007, Cantabria, Spain.

Ominiverdi. (2009). Retrieved October 15, 2009, from Ominiverdi web site: http://www.ominiverdi.org/.

Onsrud, H. (Ed.). (2007). Research and Theory in Advanced Spatial Data Infrastructure Concepts. Redlands, CA: ESRI Press.

Open Geospatial Consortium. (2009). Retrieved October 15, 2009, from Open Geospatial Consortium web site: http://www.opengeospatial.org/.

Open Source Content Management Systems. (2009). Retrieved October 15, 2009, from Open Source Content Management Systems web site: http://www.opensourcecms.com

OpenLayers. (2009). Retrieved October 15, 2009, from OpenLayers: Free Maps for the Web web site: http://openlayers.org.

Oregon Coastal Atlas. (2009). Retrieved October 15, 2009, from the Oregon Coastal Atlas web site: http://www.coastalatlas.net/.

Oregon Coastal Management Program (OCMP), Oregon State University, & EcoTrust. (2009). Oregon Coastal Atlas. Retrieved May 7, 2009, from OCA website http://www.coastalatlas.net

Oregon Coastal Management Program. (2009). Retrieved October 15, 2009, from Oregon Coastal Management Program web site: http://www.oregon.gov/lcd/ocmp/

Oregon Department of Geology and Mineral Industries. (2009). Retrieved October 15, 2009, from Oregon Department of Geology and Mineral Industries web site: http://www.oregongeology.org/.

Oregon Department of Land Conservation and Development. (2009). Protecting and managing Oregon's coastal and ocean resources, 2007-2009 biennial report, Oregon Coastal Management Program. Salem, OR: Oregon Coastal Management Program.

Oregon State University Department of Geosciences. (2009). Retrieved October 15, 2009, from Oregon State University Department of Geosciences web site: http://www.geo.oregonstate.edu/.

Oregon State University. (2009). North Coast Explorer. Retrieved May 7, 2009, from North Coast Explorer website http://www.northcoastexplorer.info/

ORR. (2009). Pntology Registry and Repository. Retrieved September 10th, 2009 from the website http://mmisw.org/or

Osborne, M., & Pepper, J. (2008). Marine SDI and the International Hydrographic Community. IHO/CHRIS Marine Spatial Data Infrastructure Working Group (MSDIWG) Meeting, IHB, Monaco, 4-5 February 2008. Retrieved April 24, 2009 from the International Hydrographic Organization web site: http://bit.ly/4mx6Yk.

Peterson, G. N. (2009). GIS Cartography: A Guide to Effective Map Design. Boca Raton, FL: CRC. doi:10.1201/9781420082142

Pew Oceans Commission. (2003). America's Living Oceans: Charting a Course for Sea Change. A Report to the Nation. Arlington, VA: Pew Oceans Commission.

phpBB. (2009). Retrieved October 15th, 2009 from phpBB - Creating Communities Worldwide web site: http://www.phpbb.com

Plymouth Marine Laboratory. (2009). RetrievedOctober31, 2009 from the Plymouth Marine Laboratory web site:http://www.pml.ac.uk.

pmapper. (2009). Retrieved October 15th, 2009 from p.mapper - a MapServer PHP/MapScript Framework web site: http://www.pmapper.net

POL. (2009). Protege Ontology Library. Retrieved from the website http://protegewiki.stanford.edu/index.php/Protege_Ontology_Library

PostGIS. (2009). Retrieved October 15th, 2009 from PostGIS - Home web site: http://postgis.refractions.net

Postgre, S. Q. L. (2009). Retrieved October 15th, 2009 from PostgreSQL: The world's most advanced open source database web site: http://www.postgresql.org

Powers, S. (2003). Practical RDF. Sebastopol, CA: O'Reilly Media.

Protégé. (2009). Retrieved October 15, 2009, from The Protégé Ontology Editor and Knowledge Acquisition System web site: http://protege.stanford.edu/.

Proudman Oceanographic Laboratory Observatory. (2009). RetrievedOctober31, 2009 from the Proudman Oceanographic Laboratory Obervatory web site:http://cobs.pol.ac.uk/gmapper/

Provincie West-Vlaanderen. (2009). West-Flanders: Social and Economic Facts and Figures 2009 (original title: West-Vlaanderen sociaal economisch-feiten en cijfers editie 2009). Brugge, Belgium: Provinciebestuur West-Vlaanderen.

Purce, D., Wood, M., Kueck, J., Taylor, K., O'Dea, L., Wright, D., & Haddad, T. (2009). Proceedings of the West Coast Coastal Atlas Workshop, Seattle, WA. Retrieved Jul 1, 2009, from ICAN website http://ican.science.oregonstate.edu/westcoast/

Ray, E. T. (2003). Learning XML (2nd ed.). Sebastopol, CA: O'Reilly Media.

RDF. (2009). Resource Description Framework (RDF). Retrieved September 14th, 2009 from the website http://www.w3.org/RDF/

Regional Ministry for the Environment of Andalusia. (2009). Retrieved October 31, 2009 from the Regional Ministry for the Environment of Andalusia web site: http://www.juntadeandalucia.es/medioambiente (in Spanish).

Rink, M., Hart, D., & Miller, A. (1998). Customized geographic information systems training for coastal and marine resource managers. Paper presented at the 1998 ESRI User Conference, San Diego, CA. Retrieved on October 25, 2009 from the ESRI User Conference website http://proceedings.esri.com/library/userconf/proc98/PROCEED/TO650/PAP604/P604.HTM

Sahoo, S. S., Bodenreider, O., Rutter, J. L., Skinner, K. J., & Sheth, A. P. (2008). An ontology-driven semantic mashup of gene and biological pathway information: Application to the domain of nicotine dependence. Journal of Biomedical Informatics, 41(5), 752–765. doi:10.1016/j.jbi.2008.02.006

Scott, L., & Brown, M. (2009). African Marine Atlas. In Odido, M & Mazzilli, S. (Eds.), African Oceans and Coasts. Nairobi, Kenya: United Nations Educational, Scientific and Cultural Organization (UNESCO).

Scott, L.E.P, Ababio, S., Armah, A.K., Brown, M., Diallo, A., Dovlo, E.K., et al. (2009). The African Marine Atlas.

Scottish Marine Bill. (2009). RetrievedOctober31, 2009 from the Scottish government web site:http://www.scotland.gov.uk/Publications/2009/09/28115722/0.

SeaData Net. (2009). RetrievedOctober31, 2009 from the British Oceanographic Data Centre web site:http://www.bodc.ac.uk/projects/european/seadatanet/.

SeaZone. (2009). RetrievedOctober31, 2009 from the SeaZone Solutions web site:http://www.seazone.com.

Shape2Earth. (2009). RetrievedOctober31, 2009 from the Shape2Earth web site: http://shape2earth.com.

Shen, J., Gong, W., & Wang, H. (2005). Simulation of Hurricane Isabel using the Advanced Circulation Model (ADCIRC). In K.G. Sellner (Ed.), Hurricane Isabel in Perspective: Proceedings of a Conference (pp. 107-116). Baltimore, MD: Chesapeake Bay Consortium.

Shi, H. & Singh, A. (2003). Status and interconnections of selected environmental issues in the global coastal zones. AMBIO: A Journal of the Human Environment, 32(2), 145–152.

Smits, P., Düren, U., Østensen, O., Murre, L., Gould, M., & Sandgren, U. (2002). INSPIRE Architecture and Standards Position Paper. Ispra, Italy: JRC-Institute for Environment and Sustainability.

Snell, J., Tidwell, D., & Kulchenko, P. (2001). Programming Web Services with SOAP. Sebastopol, CA: O'Reilly Media.

Solutions, D. M., & Go, M. O. O. S. (2009). Gulf of Maine Mapping Portal. Retrieved May 7, 2009, from GOMMAP web site: http://www.gommap.org/.

Sowa, J. F. (1999). Knowledge Representation: Logical, Philosophical, and Computational Foundations. Pacific Grove, CA: Brooks Cole Publishing Co.

Sproule-Jones, M. (2002). Restoration of the Great Lakes: promises, practices, performances. Vancouver, BC: University of British Columbia Press.

Stone, J., Harrower, M., Hart, D., & Koch, T. (in press). Coastal erosion on the Great Lakes: Using remote sensing and geovisualization for public education . In Madden, M. (Ed.), Landscape Analysis Using Geospatial Tools. New York: Springer-Verlag.

SWSE. (2009). Semantic Web Search Engines. Retrieved September 14th, 2009 from the website http://esw.w3.org/topic/TaskForces/CommunityProjects/LinkingOpenData/SemanticWebSearchEngines

The Oregonian. (2009). Retrieved October 15, 2009, from The Oregonian newspater web site: http://www.oregonlive.com/oregonian/.

Tidwell, J. (2006). Designing Interfaces. Sebastopol, CA: O'Reilly Media.

Tikunov, V. S., Ormeling, F., & Konecny, M. (2008). Atlas information systems and geographical names information systems as contributants to spatial data infrastructure. International Journal of Digital Earth, 1(3), 279–290. doi:10.1080/17538940802291817

Tufte, E. R. (1983). The Visual Display of Quantitative Information. Cheshire, CT: Graphics Press.

UK Digital Marine Atlas. (2009). RetrievedOctober31, 2009 from the UK Digital Marine Atlas web site:http://www.bodc.ac.uk/products/bodc_products/ukdmap.

UK Marine Bill. (2009). RetrievedOctober31, 2009 from the Department of Environment, Food and Rural Affairs web site:http://www.defra.gov.uk.

UKSeaMap. (2009). RetrievedOctober31, 2009 from the Joint Nature Conservation Committee web site:http://www.jncc.gov.uk/page-3663.

UNEP. (2008). Africa: Atlas of Our Changing Environment. Nairobi, Kenya. Division of Early Warning and Assessment (DEWA). United Nations Environment Program (UNEP).

UNESCO. (1989). Bilko Software. RetrievedOctober13, 2009, from the UNESCO-Bilko web site:http://www.noc.soton.ac.uk/bilko.

United Nations Environment Program. (2009). United Nations Environment Program GEO Data Portal. Retrieved 22 July 2009, from the UNEP web site: http://geodata.grid.unep.ch

United Nations. (1992). Agenda 21: The United Nations Programme of Action from Rio. New York: United Nations.

US Congress. (2009). Public Law 111-11: Omnibus Public Land Management Act of 2009, Section 12201: Ocean and Coastal Mapping Integration Act. Retrieved August 6, 2009 from the US Government Publications Office web site http://bit.ly/1Smb23.

Vaez, S. S., Rajabifard, A., & Williamson, I. (2009). Seamless SDI model -- Bridging the gap between land and marine environments . In van Loenen, B., Besemer, J. W. J., & Zevenbergen, J. A. (Eds.), SDI Convergence: Research, Emerging Trends, and Critical Assessment, 48 (pp. 239–252). Delft, the Netherlands: Netherlands Geodetic Commission and Global Spatial Data Infrastructure Association.

van der Vlist, E. (2002). XML Schema - The W3C's Object-Oriented Description for XML. Sebastopol, CA: O'Reilly Media.

Van Elzakker, C. P. J. M. (2001). Use of maps on the Web . In Kraak, M. J., & Brown, A. (Eds.), Web Cartography: Developments and Prospectus. London: Taylor & Francis.

van Loenen, B., Besemer, J. W. J., & Zevenbergen, J. A. (2009). SDI Convergence: Research, Emerging Trends, and Critical Assessment (Vol. 48). Delft, the Netherlands: Netherlands Geodetic Commission and Global Spatial Data Infrastructure Association.

Vivero, J. S., & Mateos, J. R. C. (2007). The Atlas of European Seas and Oceans - Marine Jurisdictions, Sea Uses and Governance. Ediciones del Serbal. Retrieved July 29, 2009 from the EuroOcean web site: http://www.eurocean.org/np4/file/18/Atlas_of_the_European_Seas.pdf

Von Meyer, N., Foote, K. E., & Huebner, D. J. (1999). In Information Quality Considerations for Coastal Data. In D.J. Wright & D. Bartlett (Ed.), Marine and Coastal Geographical Systems (pp. 295-308). London: Taylor & Francis.

Vowles, G. (Ed.). (2006). Geospatial Digital Rights Management Reference Model (GeoDRM RM). Wayland, MA: Open Geospatial Consortium, Inc.

Vretanos, P. A. (2002). Web Feature Service Implementation Specification. Open Geospatial Consortium Inc. Retrieved May 6th, 2009 from OGC website http://portal.opengeospatial.org/files/?artifact_id=7176

W3Perl. (2009). Retrieved October 15th, 2009 from the W3Perl - Free Logfile Analyzer web site: http://www.w3perl.com.

Wang, H. V., Cho, J., Shen, J., & Wang, Y. P. (2005). What has been learned about storm surge dynamics from Hurricane Isabel model simulations? In K.G. Sellner, (Ed.), Hurricane Isabel in Perspective: Proceedings of a Conference (pp. 117-112). Baltimore, MD: Chesapeake Bay Consortium.

Washington State Department of Ecology. (2009). The Washington Coastal Atlas. Retrieved May 7, 2009, from Washington State Department of Ecology website http://www.ecy.wa.gov/programs/sea/sma/atlas_home.html

Wave Data Series. (2009). RetrievedOctober31, 2009 from the British Oceanographic Data Centre web site:https://www.bodc.ac.uk/data/online_request/waves/.

WaveNet. (2009). Retrieved October 31, 2009 from the Center for Environment, Fisheries and Aquaculture Science, UK web site: http://www.cefas.co.uk/wavenet-mapping.

Webalizer. (2009). Retrieved October 15th, 2009 from the Home of the Webalizer web site: http://www.mrunix.net/webalizer.

WebTrends Analyzer. (2009). Retrieved October 15th, 2009 from the W3Perl - Free Logfile Analyzer Web site: http://www.webtrends.com.

West Coast Governors' Agreement on Ocean Health. (2009). Retrieved October 31, 2009, from West Coast Governors' Agreement on Ocean Health web site: http://westcoastoceans.gov.

Western Channel Observatory. (2009). RetrievedOctober31, 2009 from the Western Channel Observatory web site:http://www.westernchannelobservatory.org.uk.

Wiederhold, G. (1992). Mediators in the Architecture of Future Information Systems. IEEE Computer, 25(3), 38–49.

Wikipedia, (2008). Use Case. Retrieved December 11, 2008 from the Wikipedia web site: http://en.wikipedia.org/wiki/Use_case.

Wikipedia. (2009). ESRI. Retrieved October 15th, 2009 from Wikipedia: http://en.wikipedia.org/wiki/Esri.

WIPO. (1979). Berne Convention for the Protection of Literary and Artistic Works. World Intellectual Property Organization, Geneva. Retrieved August 6, 2009 from WIPO web site: http://www.wipo.int/export/sites/www/treaties/en/ip/berne/pdf/trtdocs_wo001.pdf

Wisconsin Coastal Data Catalog. (2009). Retrieved October 31, 2009, from the Wisconsin Coastal Data Catalog web site: http://speedy.ersc.wisc.edu:8080/geonetwork.

Wisconsin Coastal Guide. (2009). Retrieved October 31, 2009, from the Wisconsin Coastal Guide web site: http://www.wisconsincoastalguide.org

Wordle. (2009). Retrieved October 15, 2009, from Wordle - Beautiful Word Clouds web site: http://www.wordle.net/.

WordPress. (2009). Retrieved October 15th, 2009 from WordPress - Blog Tool and Publishing Platform web site: http://www.wordpress.org

World Resources Institute. (2001). World Resources 2000–2001. Washington, DC: World Resources Institute.

World Wind, N. A. S. A. (2009). Retrieved October 31, 2009, from the NASA World Wind web site: http://worldwind.arc.nasa.gov

Wright, D. J. (2009). Spatial data infrastructures for coastal environments. In X. Yang (Ed.), Remote Sensing and Geospatial Technologies for Coastal Ecosystem Assessment and Management, (LNGC, pp. 91-112). Berlin: Springer-Verlag.

Wright, D. J., & Dwyer, E. (2009). ICAN Workshop 4 and EEA National Reference Centers Meeting in Trieste, Italy, November 2009. Retrieved October 31, 2009 from the International Coastal Atlas Network web site: http://ican.science.oregonstate.edu/ican4.

Wright, D. J., Cummins, V., & Dwyer, E. (2009a). About ICAN. Retrieved October 31, 2009 from the International Coastal Atlas Network web site: http://ican.science.oregonstate.edu/about_ican.

Wright, D. J., Watson, S., Bermudez, L., Cummins, V., & Dwyer, N. O'Dea, L., Nyerges, T., Benoit, G., Berman, M., Helly J., & Uhel, R. (2007). Report on Coastal Mapping and Informatics Trans-Atlantic Workshop 2: Coastal Atlas Interoperability. Oregon State University, Corvallis, Oregon, USA. Retrieved July 29, 2009 from http://workshop1.science.oregonstate.edu/final_rpt2.

Wright, D., & Scholz, A. (Eds.). (2005). Place Matters: Geospatial Tools for Marine Science Conservation and Management in the Pacific Northwest. Corvallis, OR: Oregon State University Press.

Wright, D., Blongewicz, M. J., Halpin, P. N., & Breman, J. (2007). Arc Marine: GIS for a Blue Planet. Redlands, CA: ESRI Press.

Wright, D., Dwyer, N., Haddad, T., O'Dea, L., Hart, D., LaVoi, T., et al. (2009b). Introducing the International Coastal Atlas Network. In Proceedings of Coastal Zone '09, Boston. Retrieved July 29, 2009 from the ICAN web site: http://ican.science.oregonstate.edu/node/40.

Wright, D., Longhorn, R., Dwyer, N., Lassoued, Y., & Boucelma, O. (2009). The International Coastal Atlas Network: An emerging spatial data infrastructure initiative. In Proceedings of the Third INSPIRE Conference, in conjunction with the Eleventh International Conference of the GSDI (GSDI 11), Rotterdam, The Netherlands. Retrieved October 31, 2009 from the International Coastal Atlas Network web site http://ican.science.oregonstate.edu/node/75

WTO. (1994). Agreement on Trade-Related Aspects of Intellectual Property Rights. World Trade Organization, Geneva. Retrieved August 6, 2009, from WTO web site: http://www.wto.org/english/docs_e/legal_e/27-trips.pdf.

Yétongnon, K., Suwanmanee, S., Benslimane, D., & Champin, P.-A. (2006). A Web-Centric Semantic Mediation Approach for Spatial Information Systems. Journal of Visual Languages and Computing, 17(1), 1–24. doi:10.1016/j.jvlc.2005.04.002

About the Contributors

Dawn J. Wright is a professor of geography and oceanography at Oregon State University, and the director of the Davey Jones' Locker Seafloor Mapping/Marine GIS Laboratory. Her research interests include geographic information science, coastal web atlases, benthic terrain and habitat characterization, tectonics of mid-ocean ridges, and the processing and interpretation of high-resolution bathymetry and underwater videography/photography. She serves on the editorial boards of the International Journal of Geographical Information Science, Transactions in GIS, Journal of Coastal Conservation, The Professional Geographer, and Geography Compass, as well as on the US National Academy of Sciences' Ocean Studies Board, Committee on Strategic Directions in the Geographical Sciences for the Next Decade, Committee on an Ocean Infrastructure Strategy for US Ocean Research in 2030, and the Committee on Geophysical and Environmental Data. She serves on the Technical Advisory Board of the Marine Metadata Interoperability project. Dawn's other books include Arc Marine: GIS for a Blue Planet (with M. Blongewicz, P. Halpin, and J. Breman, ESRI Press, 2007), Place Matters: Geospatial Tools for Marine Science, Conservation, and Management in the Pacific Northwest (with A. Scholz, Oregon State University Press, 2005), Undersea with GIS (ESRI Press, 2002), and Marine and Coastal Geographical Information Systems (with D. Bartlett, Taylor & Francis, 2000). Dawn holds a PhD in Physical Geography and Marine Geology from the University of California at Santa Barbara. She is a fellow of the American Association for the Advancement of Science.

Valerie Cummins is the director of the Coastal and Marine Resources Centre, University College Cork, which has a staff of over 30 researchers. Her research interests cover a range of coastal governance issues including public participation, intuitional aspects of geomatics, capacity building for coastal management, and the science and policy interface and ecosystems frameworks. She is currently reading for a PhD on organizational tools for sustainability science in coastal zone management. She was instrumental in securing funding for the development of the Marine Irish Digital Atlas and has overseen the project since its inception. This activity led to her being a co-founder of the International Coastal Atlas Network. She contributes to the editorial panel of the international Marine Policy journal published by Elsevier and is a member of the Marine Geography Commission of the International Geographic Union.

Edward (Ned) Dwyer has a MSc and a PhD in remote sensing and image processing technology. He has worked in the area of terrestrial and marine remote sensing applications for many years using data from both optical and synthetic aperture radar sensors. Since joining the CMRC in 2002 he has been the project manager of the Marine Irish Digital Atlas, with particular responsibility for metadata specification and data acquisition. He is also researching climate observation systems in Ireland in regard to the

Copyright © 2011, IGI Global. Copying or distributing in print or electronic forms without written permission of IGI Global is prohibited.

Global Climate Observing System requirements. He is currently the co-chair of the International Coastal Atlas Network and an occasional lecturer on remote sensing and GIS topics at University College Cork.

***

Paul R. Alexander studied Anthropology and received his BA from California State University, Monterey Bay and his MA from the University of Chicago. Using his background in Digital Anthropology and research methodologies, he now develops software in conjunction with researchers and scientists. Alexander was previously employed by the Marine Metadata Interoperability project, which is supported by an NSF grant to "To promote the exchange, integration and use of marine data through enhanced data publishing, discovery, documentation and accessibility." While with the project he worked extensively on the community website, both developing functionality and writing guidance for the community. Currently, Alexander is a web and user interface developer with the Stanford Center for Biomedical Informatics Research, working on BioPortal, a project housed in the National Center for Biomedical Ontology that stores, indexes, and build relationships between publically accessible ontologies.

Robert Bailey is Manager of the Oregon Coastal Management Program in the Department of Land Conservation and Development. From 1982 to 2003 he was Ocean Program Administrator working on a range of state and national ocean policy issues under five governors. From 1995- 2000 he led the Pacific Northwest Ecosystems Regional Study a $6 million multidisciplinary assessment of coastal ecosystem process funded by NOAA's Coastal Ocean Program. In 2004-2005 he was a member of the National Academy of Sciences Ocean Studies Board Sea Grant Review Panel. He has served on numerous regional and nation committees regarding marine and coastal science, policy and governance. Mr. Bailey was elected as a City Commissioner in Oregon City, Oregon, 2003-2006. He was born and raised on the Oregon coast and is a 1968 graduate of Portland State University in Earth Science. He is married with three grown children.

Kathy Belpaeme studied biology at the Free University of Brussels, Belgium. She worked as a PhD researcher for 3 years, focussing on mutagenetic effect of marine pollution. In 1999 she began working in the field of Integrated Coastal Zone Management. In 2001 she was instrumental in setting up the Coordination Center on Coastal Zone Management for Belgium, where she currently works. The Coordination Center's main objective is to stimulate and promote integrated management of the coastal zone in Belgium. The Center is collaboration between different authorities at local, regional and national level.

Marcia Berman came to the Virginia Institute of Marine Science in 1989 to direct the Comprehensive Coastal Inventory, a GIS and remote sensing program. The Coastal Inventory is charged with mapping conditions along the 16,100 km of tidal shoreline in Virginia. This effort has expanded to include the state of Maryland and parts of North Carolina and Delaware. Marcia's research interests in applied coastal science include the development of GIS based decision support tools to enhance coastal management at the local and regional planning levels within the Chesapeake Bay Watershed. Her work has spanned subjects such as shallow water use conflict, ecosystem risk assessment, shoreline erosion control, and assessment of climate change impacts to coastal landscapes. Marcia is a coastal geologist by training with a B.S. in Geology from Northeastern University, Boston, MA, and a M.S. degree in Oceanography from the School of Oceanography at Old Dominion University in Norfolk, Virginia.

Luis E. Bermudez received his BS in Industrial Engineering from Universidad de los Andes, Colombia an MS and PhD. in Hydroinformatics from Drexel University, Philadelphia, US. He leads technical initiatives related to coastal research at the Southeastern University Research Association (SURA). The coastal research program advances the understanding of coastal phenomena by deploying cutting edge information technology (IT) for data integration in support of ocean observing efforts and advancing distributed modeling test-bed efforts. Previous to SURA he was a software engineer at the Monterey Bay Aquarium Research Institute (MBARI) and leading the technical work at the Marine Metadata Interoperability project. Before MBARI, he was involved with moving forward standardization efforts for data interchange and semantics with the Consortium of Universities for the Advancement of Hydrologic Science (CUAHSI). Luis participates in various standardization-working groups at the Open Geospatial Consortium (OGC) and World Wide Web Consortium (W3C). He leads the OGC Ocean Science Interoperability Experiment, participates in OGC Sensor Observation Service (SOS) working group, and is member of the Semantic Sensor Network working group.

Declan Dunne has a BSc in Computer Systems from the University of Limerick and a MSc in Applied Science (Modeling and Numerical Computing) from University College Cork. Since joining the Coastal and Marine Resources Centre in 2002 he has worked on numerous Irish and European Union (EU) research projects involving computer science and geomatics expertise. Research interests include geospatial data integration, delivery and visualization using OGC standards. He developed the web-mapping GIS engine and database for the Marine Irish Digital Atlas. He was responsible for development of geospatial data and metadata web services for the EU Framework Program-6 InterRisk project using OGC standards, and lead architect and developer of a pan-European coastal observational data management system for the FP-6 ECOOP project. He is currently team leader and developer for the Geological Survey of Ireland funded Griffith Geomatics for Geosciences project which will advance standards-based data management solutions for geological data.

David R. Green wears many hats at the University of Aberdeen, Scotland: Director of the Centre for Marine and Coastal Zone Management (CMCZM), Director of the Applied Geospatial Technology (AGT) and Geographic Information Systems (GIS) M.Sc. Programmes, and Director of the Marine and Coastal Resource Management Degree Programme. He is currently the Editor-in-Chief of the Journal of Coastal Conservation: Planning and Management (JCCPM) and Vice-Chair of the European Centre for Nature Conservation (ECNC) Scientific Advisory Board, as well as past Editor-in-Chief of the British Cartographic Society's Cartographic Journal, past Chairman of the UK Association for Geographic Information (AGI), and past President of the EUCC-The Marine and Coastal Union. David is also External Examiner for Kingston University's GIS programme, the UNIGIS programme at Manchester Metropolitan University (MMU), and Falmouth Marine School's (FMS) marine programme. He has edited a number of international books on ICZM and GIS, on landscape ecology and GIS, and on GIS in education.

Tanya Haddad is a geospatial data and web application developer at the Oregon Coastal Management Program. She is a former NOAA Coastal Management Fellow, and has a BS in Biology and Environmental Science from Tufts University, and a M.E.M. degree in Coastal Environmental Management from Duke University. Since joining the Oregon Coastal Management Program in 2000 she has worked on numerous geospatial data and web information projects relating to coastal zone management and marine spatial planning. She is the chief information architect of the Oregon Coastal Atlas project and is involved in all facets of its maintenance and growth. In her spare time she is an avid sailor and loves to travel. Research interests include all aspects of coastal/marine geospatial data integration, delivery and visualization, in particular via the Internet using open-source tools.

David Hart is the Geographic Information Systems Specialist at the University of Wisconsin Sea Grant Institute. Through this position, he conducts research and outreach that supports sustainable coastal development along the Great Lakes. David's current research occurs at the intersection of geographic information science, urban planning, and environmental management. He has developed methods for discovery, inventory, acquisition, integration, and analysis of local geospatial data for use in decision-making about regional-scale issues. Prior to moving to Wisconsin in 1993, David served six years at the New Orleans City Planning Commission as project manager for the City's Multipurpose Land Information System. He earned a PhD in Land Resources from the University of Wisconsin-Madison, a Master degree in Urban and Regional Planning from the University of New Orleans, and a BS in Natural Resources Policy and Management from the University of Michigan.

Alejandro Iglesias-Campos is a GIS analyst at the University of Seville involved in the development of different projects for the implementation of the Water Framework Directive in coastal zones, the generation of environmental indicators for ICZM and the implementation of the Coastal and Marine Information System of Andalusia. Since 2005, he has worked at the European Environment Agency's Topic Centre on Land Use and Spatial Information, working in GIS support and coastal and marine management, and planning activities for different European and international projects. Additionally, Alejandro is the technical representative for the Ministry of Environment of the Junta de Andalucía at the ETC-LUSI and the EEA, and is also responsible for liaison with the Spanish National Ministry of Environment, Rural and the Marine, and other national authorities. He holds a MSc in Physical Geography, Coastal Management and Water Information Systems, with an Advanced Study Diploma for Doctoral Studies from the University of Seville, Spain.

Anthony W. Isenor currently works for Defence Research and Development Canada, an agency operating under the Canadian Department of National Defence (DND). He has extensive experience in geospatial data sets, metadata standards, data vocabularies, and data structures, and is currently working on applied research examining how technologies can influence user trust, within the context of maritime situational awareness. Nationally, Anthony is involved in efforts to implement information management solutions within the DND meteorological and oceanographic office and parent organization. Internationally, he is a member of North Atlantic Treaty Organization (NATO) research group investigating techniques to achieve semantic interoperability (group designation IST-075) among NATO coalition forces, and a second group investigating enabling technologies for NATO maritime situational awareness (group designation SCI-211). Anthony received his BSc in physics and MSc in physical oceanography from Dalhousie University in Nova Scotia, Canada.

Kathrin Kopke graduated with honors from the MSc in Ecosystem Conservation and Landscape Management at University College Cork in 2005 and then joined the CMRC. She has carried out research in the area of Integrated Coastal Zone Management (ICZM) including an investigation of a coastal Brownfield site, a recreational carrying capacity study in Cork Harbor as well as being involved in the facilitation of the Cork Harbor Forum, a stakeholder group for the Harbor. Kathrin also contributes to projects within the coastal governance group, particularly in regard to adaptation to climate change in coastal communities. In 2007, Kathrin joined the Marine Irish Digital Atlas (MIDA) team as the Atlas data manager. Her work involves working with GIS technologies and Open Source web-based mapping techniques and the supervision of GIS internships with CMRC. Kathrin is involved in the outreach and education work of the International Coastal Atlas Network (ICAN).

Yassine Lassoued is a computer science and GIS researcher interested in several fields, notably: geographic data integration, ontologies, metadata and data quality, and the use of Artificial Intelligence (AI) for data and metadata integration. Yassine graduated from the National Civil Aviation Faculty (Toulouse - France) with an engineering degree in Computer Science and Air Traffic, as well as a M.Sc. in computer science and Artificial Intelligence (AI), in 2000. In 2005, he graduated from University Aix-Marseille 1, with a PhD in Computer Science and GIS. During his PhD, he worked as a temporary researcher and teacher in Universities Aix Marseille 1 & 2, and the Institute of Advanced Internet Applications. After graduation he joined the Coastal and Marine Resources Centre in 2006 as a researcher in the GIS team. He has extensive background in data integration, metadata and data quality, OGC and ISO standards, as well as the Arc Marine Data model.

Tony LaVoi is the Chief of the Integrated Information Services (IIS) Division at the National Oceanic and Atmospheric Administration (NOAA) Coastal Services Center in Charleston, South Carolina. The Coastal Services Center was established in 1995 with a mission to support the environmental, social, and economic well being of the coast by linking people, information, and technology. The Center assists its primary customers, the US coastal resource managers, by providing access to information, technology, and training. The IIS division focuses its efforts on geospatial standards and interoperability, software application and database development, programming and visualization, and network and desktop information technology support. Tony serves as the NOAA representative to the Federal Geographic Data Committee (FGDC) and the Geospatial Line of Business. He chairs both the NOAA GIS Committee and the Marine and Coastal Spatial Data Subcommittee of the FGDC. Tony has a BS in Engineering from the University of Wisconsin.

Roger Longhorn is the director of Info-Dynamics Research Associates Ltd. (headquartered in West Flanders, Belgium) and the chief editor of GEO:connexion International magazine. He has been involved in the information and communications technology industry since 1976. He developed marine information systems globally (1976 to 1986), then worked as an expert in information services for the European Commission (EC) until 1999, and has remained involved in several EC research and development programs since then. Roger assisted in developing the European Spatial Data Infrastructure strategy (now called INSPIRE) since 1995 to the present. He authored a practical guide to spatial data legal issues (2002) and co-authored a book on value, pricing and consumption of geographic information (2008), plus chapters in books dealing with SDI and marine/coastal information (1999, 2003, 2005, 2009). He is chair of the Global Spatial Data Infrastructure Association's Legal and Socioeconomic

Committee, and Information Policy Advisor to the Coastal and Marine Union (EUCC). Roger holds BS and MS degrees in Ocean Engineering and Shipping Management from the Massachusetts Institute of Technology (MIT).

Hannelore Maelfait earned a MSc in Biology at the University of Leuven, Belgium. She came in contact with a variety of marine related research areas during her second international Master degree in Marine Science at the University of Ghent, Belgium. Since 2005 she has been working for the Co-ordination Center on Integrated Coastal zone Management for Belgium, as an adjunct adviser. Within the Center she is responsible for the development of local actions within an ICZM approach to stimulate the network between scientists, policy makers and stakeholders. Furthermore she is maintaining a set of sustainability indicators for the Belgian Coast and developing communication devices to stimulate the implementation of indicators and the use of data in a local government.

Gonzalo Malvárez-García was awarded a PhD in Physical Geography in 1997 by the University of Ulster, Northern Ireland, and worked at the university as a researcher, reader and lecturer. He then took the position of principal professor at the University Pablo Olavide in Seville, Spain. Since 2007 he has been Dean of the Faculty of Human Sciences at the University. Gonzalo has led and participated in many international projects on coastal and marine management and planning, and has many scientific papers published in international journals. He has also been author of nearly 30 chapters in various books on the coastal and marine environment. Among other projects at a regional level, Gonzalo participates in the implementation of the SIGLA (Coastal and Marine Information System of Andalusia).

Catherine McCall is a planner with the Maryland Department of Natural Resources' (MD DNR) Chesapeake & Coastal Program in Annapolis, Maryland. Since joining DNR, she has led a number of coastal habitat mapping and targeting initiatives including the development of a blue infrastructure assessment; marine spatial planning activities; and coastal habitat sea level rise adaptation planning. She has also been involved with a number of other programs at MD DNR related to marine protected areas, living shorelines, and coastal hazards. Catherine holds a BS from Pennsylvania State University and an MS from Georgetown University, both in biology with a focus on conservation ecology.

Andrus Meiner is the Project Manager of Regional Assessments and Geospatial Data in Natural Systems and Vulnerability, Land (NSV3) at the European Environment Agency. He is currently working on integrated spatial assessment and indicator development for the coastal zones. Specific responsibility of that task involves analysis of land cover changes and main sectoral policies with regard of their impact of state of coastal zones on European level. Andrus is also involved in development of EEA spatial data infrastructure and coordinates land use/cover information system. His professional experience includes numerous international assignments in the fields of modeling, environmental information, geospatial database development and integrated environmental assessments, he is co-author of an assessment report on state of coasts in Europe and a monograph on simulation modeling of "river catchment-coastal sea" system. Andrus is a graduate from Tartu University in Estonia and holds academic degree in environmental geography (1990). He has attended various international courses, including research fellowships in the US and Norway.

José Manuel Moreira-Madueño has, since 1988, combined his scientific research with other responsibilities at the Regional Ministry of Environment of Andalusia. He is currently Deputy Director for Sustainable Development and Environmental Information, and previously served as Head of Department for Remote Sensing and Environmental Information Systems. José Manuel has many scientific publications and articles on GIS, as well as technologies for decision support and environmental management in international journals. José Manuel is responsible for the Environmental Information Network of Andalusia (REDIAM), and also teaches courses in cartography and GIS as applied to environmental management at three universities in Andalusia, Spain. He holds a Ph.D. in Physical Geography from the University of Seville, Spain.

Joshua Murphy is a Management and Program Analyst with the National Oceanic and Atmospheric Administration (NOAA) Coastal Services Center in Charleston, South Carolina. Over the past eight years, he has supported the application of geographic information systems (GIS) to coastal issues through numerous capacity-building activities. As a trainer, Josh has taught introductory and intermediate-level GIS, global positioning systems (GPS) and metadata courses to hundreds of participants within the coastal zone. As a project manager, Josh has led the development of a variety of geospatial tools, such as the Benthic Terrain Modeler and the Chesapeake Bay Oyster Larvae Tracker (CBOLT). More recently, Josh's interests lie in marine spatial planning and strategic conservation planning using a green infrastructure approach. He currently serves as the project manager for the Digital Coast initiative, a community resource for coastal geospatial data, tools, and training. Josh holds a BS in Geography from Penn State University.

Elizabeth (Liz) O'Dea is a geographer with a taste for environmental mapping and making spatial information widely accessible. She has a MSc in Geography from Oregon State University and has worked with terrestrial, coastal and marine environments in Oregon, Washington, and Ireland. Whilst at the Coastal and Marine Resources Centre in Cork, Ireland, she worked as a designer and developer of the Marine Irish Digital Atlas. She dealt with all aspects of what it takes to create and manage a coastal atlas, including web design, data management, usability, and technology. Liz has been involved in ICAN since its inception and was the inspiration behind the original workshop series. She is currently a GIS Analyst with the Washington State Department of Ecology in Olympia, Washington, where she works with the Washington Coastal Atlas team. In her spare time she geocaches, travels, and sea kayaks whenever she has the opportunity.

Timothy Nyerges is Professor of Geography at the University of Washington where he specializes in teaching and research related to public participation geographic information systems (GIS) and coastal GIS, focusing on land use, transportation, and water resource related issues. He received his PhD from the Ohio State University in 1980 specializing in database management languages for GIS. For the past fifteen years he has had a stream of nationally-funded research projects to explore development and evaluation of networked GIS, particularly as supported by cyberinfrastructure technology, for enabling stakeholder participation in environmental decision support. He is currently president (2009-2011) of the University Consortium for Geographic Information Science (UCGIS).

Eoin O'Grady currently leads the Information Services and Development Section in the Marine Institute (Ireland) and is responsible for overseeing the Institute's IT and data management systems.

These includes systems to manage a wide variety of physical, biological and chemical marine data from environmental monitoring, seabed surveys, predictive modeling and fisheries survey programmes, and participation in projects such as SeaDataNet. Eoin also leads the Knowledge and Information Management programme of SeaChange the Irish national Marine Research, Knowledge and Innovation Strategy. This includes the development of information systems such as the Irish Spatial Data Exchange to support marine research and innovation. Eoin currently participates in the Irish Spatial Data Infrastructure working group. Previous to the Marine Institute Eoin has worked in a variety of software development and architecture roles for telecoms, manufacturing and financial services companies. Eoin received his B. Eng in Computer Engineering from the University of Limerick in 1995.

José Ojeda-Zújar is the principal professor of coastal and marine geomorphology, management and planning, GIS and remote sensing at the University Pablo de Olavide in Seville, Spain. He is also the director of the Coastal Planning and Territorial Information Technologies research group there. José has more than 100 scientific publications in international journals, and has also had a number of scientific placements at a number of international institutions that lead research activities in remote sensing and GIS (Nottingham and Aberdeen, UK), coastal and marine policy (University of Delaware, USA) and the National Center of Scientific Research or CNRS (France), among others. José was responsible for the implementation of the SIGLA (Coastal and Marine Information System of Andalusia), which is part of the Environmental Information Network (REDIAM). He holds a PhD in Physical Geography from the University of Seville, Spain.

Sean Padmanabhan is a Research Officer in the field of Geographic Information Systems (GIS) at the Institute of Marine Affairs, Trinidad & Tobago. With 9+ years experience in GIS, he assisted in inter-departmental and inter-organizational research; created and implemented new strategies for improving the generation, flow and dissemination of spatial data and products across both intra- and inter-organizational levels; and, developed new GIS and GIS-related marine research projects at the Institute and around the wider Caribbean Region. Sean has also been significantly involved in several regional initiatives in the Caribbean such as IODE/ODINCARSA, under which the Caribbean Marine Atlas Project falls, and the UNEP/CAR-RCU Programme responsible for the Protocol and National Programme of Action for Land-Based Sources of Pollution. He holds an MA in Marine Affairs and Policy/GIS, as well as an MS in Marine Science/Biology from the Rosenstiel School of Marine and Atmospheric Sciences, University of Miami.

Dr. Trung Pham is a computer science and GIS researcher. His research interests include geographic information integration, data quality and 3D GIS, and the application of GIS in environmental science, geology and urban management. His recent research activity has been focused on Web delivery and visualization of GIS data. Trung graduated from University Aix-Marseille I, with a PhD in computer science and GIS, in 2005. After graduation, he joined the CNRS (French National Centre for Scientific Research) as a post-doctoral researcher, Nantes, 2005-2007. He then worked as a researcher and teacher at University Paris Est Marne-La-Vallée, 2007-2008. In addition to his academic background, Trung has two years of experience in industry in Paris, as an engineer at IBM. He also worked in ALSY Paris as a development, maintenance and evolution engineer for the L'Oreal e-commerce websites with new Microsoft technologies (ASP.NET, C#, Biztalk, Shapepoint, SQL Server, etc.).

Greg Reed is the Executive Officer of the Australian Ocean Data Centre Joint Facility, a national multi-agency distributed data management system, and Deputy Director of the Ocean Data Services Group of the RAN Hydrography and Metoc Branch. Mr. Reed is an internationally recognized marine data expert and Co-chair of the IOC's International Oceanographic Data and Information Exchange (IODE) committee. He is also involved in the development of international standards for oceanographic data management and exchange. Mr. Reed is a lead author for Ocean Teacher, an online encyclopedia aimed at training students and mid-level professionals on marine data and information management. He has participated in a number of international capacity building activities as course coordinator and lecturer and in the development of marine atlas projects. He is a contributed to the African Marine Atlas.

Gabe Sataloff is a GIS Analyst and Data Manager at the NOAA Coastal Services Center in Charleston, SC. Gabe joined the Coastal Services Center in 2008 after finishing his Masters of Environmental Studies at the College of Charleston. He got involved on the coastal atlas assessment team of the International Coastal Atlas Network in 2008.

Lucy Scott coordinates national and regional data and information management activities for the United Nations Development Program (UNDP) Global Environment Facility (GEF) Agulhas and Somali Currents Large Marine Ecosystem (ASCLME) Project. Over the past ten years, she has worked in South Africa, Mozambique, Tanzania, Malawi, Comoros, Kenya, Seychelles, Mauritius, Maldives and Madagascar on a variety of projects and programs. These ranged from marine research expeditions to data management projects and the development of Geographic Information Systems, particularly for coastal resource mapping, aquaculture and conservation planning. Lucy has participated in GIS atlas projects at several scales and is currently one of five editors of the African Marine Atlas, a project involving participants from 16 African, European and North American countries. She has published in several fields, serves on the board of trustees of the Sustainable Seas Trust and is a member of the Global Oceanographic Observing System (GOOS)-AFRICA remote sensing working group.

Karen Stocks is a research scientist at the San Diego Supercomputer Center. Her interests include biodiversity informatics, data integration, and developing information systems for marine biogeography, seamount ecology, and oceanographic research vessel data. She is part of the Marine Metadata Interoperability project, and leads the development of the MMI Guides, a set of online resources introducing metadata topics and providing guidance on best practices for data documentation (http://marinemetadata.org/guides). Karen completed a Bachelors of Science degree in Wildlife and Fisheries Biology at the University of Massachusetts, and a doctorate in Biological Oceanography at Rutgers University, and has worked at the San Diego Supercomputer Center since 2000.

Ronan Uhel is the Head of Natural Systems and Vulnerability, at the European Environment Agency. He has 20 years in environmental and sustainable development information and analysis at the European and international level. He specializes in bridging between science (knowledge) and policy (actions), and in assessing the state-of-the-environment and policy effectiveness. Ronan has had coordination and editorial responsibility for many studies, reports and publications on these topics, with broad coverage from economic sectors to technologies to education. He has participated in many committees and working groups at the European and international levels on environmental governance, and has been a speaker/discussant at conferences and workshops covering all aspects of environment and development issues.

Ronan's academic background is in geography, physical planning, and oceanography, with additional training in EU environmental legislation and regional policies.

Kuuipo Walsh is in her first year as Associate Director of the GIScience Certificate Program at Oregon State University (OSU). Kuuipo brings to her role a rich and varied background in computer science, marine and coastal geographic information science, and natural resource management. Kuuipo is not new to OSU as she works on making spatial information accessible through the Oregon Explorer, a web-based, digital library co-managed by OSU Libraries and the Institute for Natural Resources (INR). As a project manager for INR, she recently launched the Oregon Hazards Explorer, which provides online maps and spatial information about many natural hazards relevant to Oregon including coastal erosion, floods and tsunamis. Before joining INR, Kuuipo was the Data/Metadata Librarian in Dawn Wright's Seafloor Mapping Marine and Coastal GIS Laboratory. She received her MSc in Marine Resource Management from Oregon State University and her BSc in Computer Science from California Polytechnic State University.

Index

Copyright © 2011, IGI Global. Copying or distributing in print or electronic forms without written permission of IGI Global is prohibited.

F

G

H

I

J

K

L

M

N

O

LaVergne, TN USA
01 August 2010
191549LV00004B/1/P